# Lecture Notes in Artificial Intelligence    7326

Subseries of Lecture Notes in Computer Science

Mohamed Kamel   Fakhri Karray
Hani Hagras (Eds.)

# Autonomous and Intelligent Systems

Third International Conference, AIS 2012
Aveiro, Portugal, June 25–27, 2012
Proceedings

 Springer

Series Editors

Randy Goebel, University of Alberta, Edmonton, Canada
Jörg Siekmann, University of Saarland, Saarbrücken, Germany
Wolfgang Wahlster, DFKI and University of Saarland, Saarbrücken, Germany

Volume Editors

Mohamed Kamel
Fakhri Karray
University of Waterloo
Department of Electrical and Computer Engineering
Waterloo, ON, Canada N2L 3G1
E-mail: {mkamel, karray}@uwaterloo.ca

Hani Hagras
University of Essex
Computation Intelligence Centre
Wivenhoe Park, Colchester CO4 3SQ, UK
E-mail: hani@essex.ac.uk

ISSN 0302-9743     e-ISSN 1611-3349
ISBN 978-3-642-31367-7     e-ISBN 978-3-642-31368-4
DOI 10.1007/978-3-642-31368-4
Springer Heidelberg Dordrecht London New York

Library of Congress Control Number: 2012940335

CR Subject Classification (1998): I.2.6, I.2.9, I.2, C.2, H.3, I.4, H.4.2, H.2.8

LNCS Sublibrary: SL 7 – Artificial Intelligence

*Typesetting:* Camera-ready by author, data conversion by Scientific Publishing Services, Chennai, India

Printed on acid-free paper

Springer is part of Springer Science+Business Media (www.springer.com)

# Preface

AIS 2012, the International Conference on Autonomous and Intelligent Systems, held in Aviero, Portugal, June 25–27, 2012, was the third edition in the AIS series of annual conferences alternating between Europe and North America. The main goals of these conferences are to foster collaboration and exchange between researchers and scientists in the broad fields of autonomous design and intelligent systems, and to address recent advances in theory, methodology and applications. AIS 2012 was organized at the same time and place as ICIAR 2012, the International Conference on Image Analysis and Recognition. Both conferences are organized by AIMI – Association for Image and Machine Intelligence—a not-for-profit organization registered in Ontario, Canada, and are cosponsored by the Center for Pattern Analysis and Machine Intelligence of the University of Waterloo, the Kitchener Waterloo Chapters of the IEEE Computational Intelligence Society, and the IEEE Systems, Man and Cybernetics Society.

For AIS 2012, we received a total of 48 full papers from 15 countries. The review process was carried out by members of the Program Committee of the conference. Each paper was reviewed by at least three reviewers and checked by the Conference Chairs. A total of 31 papers were finally accepted and appear in one volume of the proceedings. The high quality of the papers is attributed first to the authors, and second to the quality of the reviews provided by the experts. We would like to sincerely thank the authors for responding to our call, and to thank the reviewers for their careful evaluation and feedback provided to the authors. It is this collective effort that resulted in the strong conference program and high-quality proceedings.

We are very pleased to include in the conference program keynote talks by well-known experts including Wiro Nissen, Delft University of Technology, The Netherlands; Rudolf Kruse, University of Magdeburg, Germany, Germany; Bob Fisher, University of Edinburgh, UK; Bioucas Dias, Instituto Superior Técnico, Portugal. We would like to express our sincere gratitude to the keynote speakers for accepting our invitation to share their vision and recent advances in their respective areas of specialty. Special thanks are also due to the local Organizing Committee, and to the members of the committee for their advice and help. We are also grateful to Springer's editorial staff, for supporting this publication in the *Lecture Notes in Artificial Intelligence* (LNAI) series.

We would like to thank Khaled Hammouda, the webmaster of the conference, for maintaining the Web pages, interacting with the authors and preparing the proceedings. We would also like to acknowledge the professional service of Viagens Abreu in taking care of the registration process and the special events of the conference.

For those who were not able to attend AIS 2012, we hope this publication provides a good view of the research presented at the conference, and we look forward to meeting you at the next AIS conference.

Mohamed Kamel
Fakhri Karray
Hani Hagras

# Organization

## AIS 2012 – International Conference on Autonomous and Intelligent Systems

### General Co-chairs

Mohamed Kamel
University of Waterloo, Canada
mkamel@uwaterloo.ca

Fakhri Karray
University of Waterloo, Canada
karray@uwaterloo.ca

### Program Co-chairs

Hani Hagras
Essex University, UK
hani@essex.ac.uk

Aníbal de Matos
University of Porto, Portugal
anibal@fe.up.pt

### Special Session Chairs

#### Autonomous Sensors and Sensor Systems

Stoyan Nihtianov

Delft University of Technology,
The Netherlands

#### Autonomous Systems and Intelligent Control with Applications

Mohammad Biglarbegian
William Malek

University of Guelph, Canada
University of Waterloo, Canada

#### Intelligent Knowledge Management

Richard Khoury
Kacem Abida

Lakehead University, Canada
University of Waterloo, Canada

## Local Organizing Committee

Ana Maria Mendonça
University of Porto
Portugal
amendon@fe.up.pt

Pedro Quelhas
Biomedical Engineering Institute
Portugal
pedro.quelhas@gmail.com

Jorge Alves Silva
University of Porto
Portugal
jsilva@fe.up.pt

Gabriela Afonso
Biomedical Engineering Institute
Portugal
iciar10@fe.up.pt

António Pimenta Monteiro
University of Porto
Portugal
apm@fe.up.pt

## Publicity Chair

Shahab Ardalan
Gennum Corp., Canada

## Conference Secretariat

Viagens Abreu SA
Porto, Portugal
congresses.porto@viagensabreu.pt

## Webmaster

Khaled Hammouda
Waterloo, Canada
khaledh.aimi@gmail.com

# Co-Sponsorship

 AIMI – Association for Image and Machine Intelligence

 Department of Electrical and Computer Engineering
Faculty of Engineering
University of Porto
Portugal

 INEB – Instituto de Engenharia Biomédica
Portugal

 CPAMI – Centre for Pattern Analysis and Machine Intelligence
University of Waterloo
Canada

 Department of Electrical and Computer Engineering
University of Waterloo
Canada

 Computational Intelligence Society Chapter, IEEE Kitchener Waterloo Section

 Systems, Man, & Cybernetics (SMC) Chapter, IEEE Kitchener Waterloo Section

 Control Systems Chapter, IEEE Kitchener Waterloo Section

 Signal Processing Chapter, IEEE Kitchener Waterloo Section

## Advisory Committee

Pierre Borne              Ecole Centrale de Lille, France
Toshio Fukuda             Nagoya University, Japan
Elmer Dadios              De La Salle University, Philippines
Clarence de Silva         University of British Columbia, Canada
Mo Jamshidi               University of Texas, USA
Jong-Hwan Kim             Korea Advanced Institute for Science and
                            Technology, South Korea
T.H. Lee                  National University of Singapore, Singapore
Oussama Khatib            Stanford University, USA
Kauru Hirota              Tokyo Institute of Technology, Japan
Witold Perdrycz           University of Alberta, Canada

## Technical Program Committee

| | |
|---|---|
| Mohamed Abderrahim | Spain |
| Waleed Abdulla | New Zealand |
| Giovanni Acampora | Italy |
| Mohsen Afsharchi | Iran |
| Mohammed AlHaddad | Saudi Arabia |
| Mohammad Biglarbegian | Canada |
| Juan Antonio Botía Blaya | Spain |
| John Cabibihan | Singapore |
| Rui Camacho | Portugal |
| Walid Chainbi | Tunisia |
| Antonina Dattolo | Italy |
| Kerstin Dautenhahn | UK |
| Aníbal Castilho Coimbra de Matos | Portugal |
| Faiyaz Doctor | UK |
| Ruggero Donida | Italy |
| James Dooley | UK |
| Mehmet Önder Efe | Turkey |
| Hadi Firouzi | Canada |
| Giancarlo Fortinio | Italy |
| Maki Habib | Egypt |
| Hani Hagras | UK |
| Sousa João | Portugal |
| Panagiotis Karampelas | USA |
| Mehmet Kaya | Turkey |
| Richard Khoury | Canada |
| Keivan Kianmehr | Canada |
| Dana Kulic | Canada |
| Chang-Shing Lee | Taiwan |
| Xuelong Li | China |

| | |
|---|---|
| Honghai Liu | UK |
| Ahmed Lotfi | UK |
| Areej Malibari | Saudi Arabia |
| William Melek | Canada |
| Narges Noori | USA |
| Tansel Ozyer | Turkey |
| P.B. Sujit | Portugal |
| Fernando Lobo Pereira | Portugal |
| Rabie A. Ramadan | Egypt |
| Luis Paulo Reis | Portugal |
| John Ringwood | Canada |
| Agos Rosa | Portugal |
| Miguel Angel Salichs | Spain |
| Farook Sattar | Canada |
| Sabrina Senatore | Italy |
| Mae L. Seto | Canada |
| Insop Song | USA |
| Jiping Sun | Canada |
| Hooman Tahayori | Canada |
| Mehmet Tan | Turkey |
| Peter Won | Canada |
| Dongrui Wu | USA |

## Reviewers

| | |
|---|---|
| Kacem Abida | Canada |
| Jamil Abousaleh | Canada |
| Masoud Alimardani | Canada |
| Mohammad Azam Javed | Canada |
| Howard Li | Canada |
| Miao Yun Qian | Canada |

# Table of Contents

## Autonomous Sensors and Sensor Systems

## Autonomous Systems and Intelligent Control with Applications

## Intelligent Fuzzy Systems

## Intelligent Robotics

## Intelligent Knowledge Management

## Swarm and Evolutionary Methods

## Applications

# Spatially Correlated Multi-modal Wireless Sensor Networks: A Coalitional Game Theoretic Approach

Artemis Voulkidis[1], Spiros Livieratos[2], and Panayotis Cottis[1]

[1] National Technical University of Athens, Athens, Greece
avoulk@mail.ntua.gr, pcottis@central.ntua.gr
[2] School of Pedagogical and Technological Education (ASPETE), Athens, Greece
slivieratos@aspete.gr

**Abstract.** A coalition formation game theoretic method is proposed for efficient multi-service clustering and power control in QoS constrained Wireless Sensor Networks. The method is initiated by the wireless sensor nodes in a distributed way and, next, is successively optimized by a set of powerful nodes called *representatives*. The proposed method manages the inherent trade-off between energy efficiency and data accuracy to increase WSN lifetime at the cost of controllable loss of accuracy. Simulation results show that the proposed coalition formation method significantly increases WSN lifetime without necessitating significant communications overhead.

**Keywords:** Wireless Sensor Networks, Autonomous Operation, Coalition Formation Games, Constrained Optimization, Multi-modal WSNs.

## 1 Introduction

Recent advances in embedded systems and Micro-Electro-Mechanical Systems (MEMS) make feasible the development of wireless sensors that monitor different types of physical phenomena. These sensors are usually deployed in vast numbers to form large scale *Wireless Sensor Networks* (WSNs). The need for cost efficiency imposes severe limitations as to the computing and communications capabilities of the WSN nodes severely affecting the design and implementation of WSNs. The main constraint restricting WSN operation is imposed by the nodes battery lifetime. As energy consumption is mainly due to node transmissions, a primary WSN design objective is the elimination of unnecessary node transmissions without degrading the Quality of Service (QoS) of WSN operation [1].

A critical factor determining WSN operation is the spatiotemporal behavior of the target physical phenomenon [2]. Spatial correlation implies that data sensed by closely located nodes are expected to be correlated, whereas temporal correlation implies that subsequent measurements of the same node are very close. The present work proposes a coalition formation game theoretic WSN clustering scheme that exploits the spatial correlation characteristics of the sensed phenomenon to enhance WSN lifetime at the cost of controllable loss of reportings accuracy.

M. Kamel, F. Karray, and H. Hagras (Eds.): AIS 2012, LNCS 7326, pp. 1–9, 2012.
© Springer-Verlag Berlin Heidelberg 2012

## 2    Related Work

Although the trade-off between data accuracy and energy efficiency has been studied in the literature, the exploitation of the spatiotemporal correlation of sensed phenomena has not been seriously considered. *ELink*, presented in [3], is a lossy distributive clustering protocol that benefits from spatial correlation to create maximal clusters that marginally satisfy the QoS specifying WSN operation. Although the resulting clustering architecture is iteratively improved, optimality is not guaranteed. The authors in [4] present *CAG*, a lossy clustering protocol achieving a cluster formation that guarantees that the data reportings accuracy is kept above an acceptable error tolerance level. A dynamic clustering protocol exploiting spatiotemporal correlation is presented in [5]. Stating that the complexity of analytically finding the optimal WSN clustering is NP-Complete, the authors provide a greedy heuristic framework for near optimal cluster formation. Both the spatial and the temporal correlation of the sensed phenomenon are examined, without making any a priori assumption about the specific spatial correlation model followed each time. In [6] the authors argue that WSN lifetime can be increased via aggregation based on clustering to exploit spatiotemporal correlation of the sensed by the WSN phenomenon. The optimal WSN cluster size is analytically obtained; however, the mechanism of the cluster formation is not discussed. Data fusion exploiting the spatial characteristics of the sensed phenomenon to apply advanced data compression methods during message forwarding is examined in [7] whereas in [8] two combined schemes for routing and clustering in correlated WSNs with multiple sinks are proposed. Clustering is performed using data correlation estimators based on the differential entropy of the measurements of different nodes and targets at minimizing the overall WSN energy consumption. In [9], a hierarchical virtual clustering scheme is analyzed offering QoS constrained WSN lifetime optimization. Clustering is accomplished at low computational cost also resulting in energy consumption well balanced among the WSN nodes.

The present work proposes a coalition formation game theoretic approach on multi-modal WSN clustering aiming at increasing WSN lifetime. A cyclic, three phase coalition formation framework is proposed to exploit the spatial correlation characterizing the services supported to achieve the optimal coalition structure. To the knowledge of the authors it is the first time that coalition formation is considered to model multi-modal WSN operation.

## 3    Model Analysis

*Coalitional Game Theory* (CGT) can model a wide range of applications where cooperation among WSN nodes can improve network performance. Packet forwarding policies, aggregation techniques, data routing optimization and node transmission power control are some of the networking applications where node cooperation proves to be beneficial to network efficiency. Usually, CGT assumes that the coalition including all the players, referred to as the *Grand Coalition*

(GC), will be formed as soon as a set of value-related assumptions are satisfied. However, these assumptions are not valid in most practical situations; coalition formation games refer to games where structure and cooperative costs are not negligible and should, therefore, be considered in the analysis. Since WSNs are basically ad-hoc constrained networks, crucial assumptions such as *superadditivity* may not hold. Furthermore, the number of node transmissions, constituting the main source of energy consumption in WSNs, is crucial, while full network knowledge is usually unavailable. Taking into consideration the above restrictions, cooperation in WSNs is modeled as a special case of coalition formation games.

Consider a heterogeneous WSN consisting of two types of nodes. The first type of nodes encompasses simple wireless sensors, hereafter called nodes. The second category includes the *representatives*, i.e. nodes characterized by less stringent limitations as to energy consumption and computing power. The representatives may act both as simple nodes and as node coordinators. They also collect information from the nodes and forward it to the sink(s). Each node reports its measurements to the representative it is attached to. The sets of nodes and representatives are denoted by $\mathcal{N} = \{n_1, .., n_N\}$ and $\mathcal{R} = \{r_1, .., r_R\}$, respectively.

Suppose that the nodes constituting the WSN aim at measuring $M$ different physical phenomena characterized by distinct spatial correlation characteristics. For example, a node may report on both the temperature, the relative humidity, and the wind speed. For the rest of the analysis, the behaviors of the sensed physical phenomena are supposed to be independent from each other, i.e. no correlation exists between the sensed values of different sensing tasks.

## 3.1   Single Service Coalition Formation

The objective value of a node in a typical WSN derives from the number of transmissions it can offer within the context either of the overall WSN operation, i.e. by contributing to packet forwarding, or of its standard measurement reporting activity. The clustering scheme proposed in this paper attempts to extend WSN lifetime by drastically reducing the number of transmissions involved in the standard data reporting activity of the WSN nodes. The coalition formation process is accomplished in three phases, namely the *initialization*, the *optimization* and the *steady state* phase, and follows the notions presented in [10] to guarantee *stable* coalition formation.

**Initialization Phase.** During the initialization phase, the representatives broadcast expansion messages to the nodes. Upon reception of such messages, the nodes acknowledge their closest representative. Next, they send negotiation messages to their 1-hop neighbors, hereafter called neighbors. These messages constitute cooperation invitations within the coverage area of a common representative. The node, say $n_c$, initiating the procedure leading to the formation of coalition $S(k, r_j)$, is referred to as the *starter* node of $S(k, r_j)$, where $k$ denotes the id of the coalition. A node, say $n_i$, accepts the invitation of a node $n_j$ already participating in coalition $S(k, r_j)$ only if this action is to its own benefit and, at the same time, the resulting coalition formation satisfies the QoS constraint specifying proper WSN operation. Next, $n_i$ transmits negotiation messages to its neighbors inviting them to also join coalition $S(k, r_j)$, simultaneously notifying them about the measurement of $n_c$. Thus, coalition expansion continues until no node has the incentive to change its coalitional state.

To evaluate the benefit from its possible participation in a coalition, node $n_i$ calculates its utility acquired from its participation in the coalition. Let $\hat{u}(n_i, k, r_j)$ be the utility function indicating the expected cooperative value of node $n_i$ reporting to representative $r_j$ coordinating coalition $S(k, r_j)$. $\hat{u}(n_i, k, r_j)$ should be appropriately modeled in order to i. manage the accuracy-efficiency trade-off, ii. provide *fair* value allocation over the members of a coalition, iii. prohibit the formation of coalitions that could lead to reportings of unacceptable accuracy, and iv. ensure that the coalition expansion continues until no further coalition expansion is achievable.

Suppose that proper WSN operation is determined by a data precision error tolerance reflected by a QoS metric, say $q \in [0, 100\%]$. The proposed scheme defines that a coalition is valid only if the maximum data dissimilarity between the measurements of the cluster nodes and the starter node is below $1 - q$. This rule will be hereafter referred to as the $q$-*rule*. Since the measurements reported by the members of coalition $S(k, r_j)$ conform to the $q$-rule, the reporting of a random coalition member is sufficient each time to represent the relevant measurements of the cluster members. Therefore, instead of transmitting $|S(k, r_j)|$ messages every $T$ seconds, the measurements of only a single coalition member needs to be transmitted. Imposing probabilistic transmission to the members of coalition $S(k, r_j)$ at a frequency $f_t = f_o \cdot p_t$, where $f_o = 1/T$ is the original node transmission frequency and $p_t = 1/|S(k, r_j)|$ is the transmission probability, the lifetime of every node belonging to the coalition is increased by $|S(k, r_j)| - 1$ times.

This WSN lifetime increase comes at the cost of reduced reportings accuracy. Since spatial correlation is usually a decreasing function of distance [11], coalition expansion decreases the respective reportings accuracy. If $X_i$ and $X_c$ are the measurements of node $n_i \in S(k, r_j)$ and of the starter node, the measurements dissimilarity is quantified via $d(n_i, k) = |(X_i - X_c)/X_c|$.

**Definition of the Utility Function.** Taking into consideration the previous analysis, the utility function providing the expected coalitional value of node $n_i$ participating in coalition $S(k, r_j)$ is given by

$$u(\widehat{n_i, k}, r_j) = (|\widehat{S(k, r_j)}|_i - 1) \cdot f(d(n_i, q), k) \tag{1}$$

where $|\widehat{S(k, r_j)}|_i$ is the estimation of the size of $S(k, r_j)$ made by $n_i$ and $f(d(n_i, q), k)$ is a function quantifying the dependence of the node value on the required data accuracy, hereafter referred to as *accuracy function*. The first factor of (1) determines the effect of the coalition size on the node lifetime whereas through the accuracy function the effect on the measurements accuracy is taken into account.

Assuming that $n_i$, located $l$ hops away from $n_c$, is the latest node to enter $S(k, r_j)$, the current coalition size can be estimated as $|\widehat{S(k, r_j)}|_i = deg(n_i) \cdot l^2$, where $deg(n_i)$ is the number of 1-hop neighbors (*degree*) of node $n_i$ [12].

Since the nodes have the incentive to always join larger coalitions, the accuracy function should restrain unacceptable coalition expansion reflecting the effect of the quality loss to the coalition credibility. In this framework,

$$f(d(n_i, k), q) = \begin{cases} 1 - d(n_i, k) & d(n_i, k) \leq 1 - q \\ -1 & d(n_i, k) > 1 - q \end{cases} \tag{2}$$

is proposed as the accuracy function. Note that the negative value of $f(d(n_i, k), q)$ when $d(n_i, k) > 1 - q$ is proposed to assure that, in this case, node $n_i$ does not join $S(k, r_j)$.

**Optimization Phase.** When the nodes cease negotiating their admission to new coalitions, they transmit their coalitional status to the respective representatives, implicitly initiating the optimization phase. During this phase the representatives optimize the coalition formation in an attempt to further increase WSN lifetime. The coalitions are no longer viewed as collections of nodes, but are viewed as singleton entities. As stated earlier, coalition $S(k, r_j)$ can prolong its reporting activity by $|S(k, r_j)| - 1$ times whereas, as the reportings of each coalition satisfy the $q$-rule, every measurement is considered to be acceptably accurate. Therefore, the aggregate coalitional value of coalition $S(k, r_j)$ during the optimization phase is written as

$$V(k, r_j) = \sum_{n_i \in S(k, r_j)} (|S(k, r_j)| - 1) = |S(k, r_j)| \cdot (|S(k, r_j)| - 1) \tag{3}$$

As the representatives have *complete* knowledge of the finalized coalition formation, the above metric is deterministic.

Since the complexity of finding the optimal partitioning of a set of entities is known to be NP-complete [13], a standard optimization approach i.e. applying Mixed Integer Programming methods, is not feasible. Furthermore, since the $q$-rule limits the number of valid coalition structures (CSs), determining all possible CSs is not optimal and is therefore optimized following the algorithm proposed in [13].

**Steady State Phase.** Upon completion of the optimization phase, the representatives inform the nodes about the new CS and each node adjusts its transmission probability accordingly to match the coalition specifications they belong to after the optimization. Then, the nodes begin to perform their sensing and transmission tasks accordingly, bringing the network operation to the steady state phase. During this phase, the representatives post-process the received measurements in the attempt to verify that the current CS is both optimal and valid. If the reportings received from a specific coalition indicate that the $q$-rule is violated, a generic coalition split message is sent to the relevant coalition members, forcing coalition splitting. This process is followed by a new three-phase cycle of the proposed method performed only by the nodes of the split coalition.

## 3.2   Generalization to Multi-service WSN Operation

Having presented the single service analysis, the extension to multiservice WSN operation is straightforward.

Following the multi-service WSN operation paradigm, each node is able to sense $M$ physical phenomena characterized by different spatial correlation characteristics. Therefore, to exploit the spatial correlation of each sensed phenomenon, a node should be able to participate in $M$ coalitions, each one corresponding to a single service.

Let $\boldsymbol{M} = \{m_l | l \in [1, M]\}$ be the set of the WSN supported services and $\widetilde{\boldsymbol{S}}(n_i, r_j) = \{\boldsymbol{S}(k, m_l, r_j) | n_i \in \boldsymbol{S}(k, m_l, r_j) \forall l \in [1, M]\}$ be the set of all the coalitions that comprise $n_i$, where $\boldsymbol{S}(k, m_l, r_j)$ indicates a coalition with id $k$, adjusted to the spatial correlation characteristics of service $m_l$ reporting to representative $r_j$. Extending the single service analysis, the probabilistic reporting frequency of node $n_i$ for service $m_l$ will be $f_t(n_i, m_l) = f_o / |\boldsymbol{S}(k, m_l, r_j)|$. Evidently, it is to the nodes benefit to join the coalitions that maximize their expected utility for each service. The expected utility of node $n_i$ in the framework of its participation in coalition $\boldsymbol{S}(k, m_l, r_j)$ serving service $m_l$ is given by

$$\widehat{u}(n_i, k, m_l, r_j) = (|\widehat{\boldsymbol{S}(k, m_l, r_j)}|_i - 1) \cdot f(d(n_i, q), k, m_l) \tag{4}$$

As the operation of a WSN node is not restricted to providing only a single service, the marginal contribution of node $n_i$ to a coalition $\boldsymbol{S}(k, m_l, r_j)$ also depends on the coalitional behavior of node $n_i$ with regard to the rest of supported services $\boldsymbol{M} \setminus m_l$; the coalition values are no longer independent. Having defined the per service transmission frequency of node $n_i$, the average transmission frequency of the node is defined as $f_t(n_i, \boldsymbol{M}) = \sum_{l=1}^{M} f_t(n_i, m_l)$. Consequently, instead of sending $M$ messages per transmission round, $n_i$ will probabilistically transmit $f_t(n_i, \boldsymbol{M})$ messages only, extending its measurement reporting activity by $M / f_t(n_i, \boldsymbol{M}) - 1$ times.

Neglecting accuracy considerations, the optimization is limited to maximizing the aggregate coalitional value of all the coalitions, namely

$$\operatorname{argmax} \sum_{i=1}^{N} \left\{ M / f_t(n_i, \boldsymbol{M}) - 1 \right\} \tag{5}$$

under the constraint that the $q$-rule is not violated. Since the optimization problem described by (5) is also NP-Complete, the optimal CS identification should follow the analysis of the single service case presented in [13]. If $C(m_l, r_j)$ is the set of coalitions adjusted to the spatial correlation characteristics of service $m_l$ that report to $r_j$, then, assuming that the sensed phenomena are not correlated, the optimization process can be iteratively accomplished by optimizing the CS defined over a single service $m_l$, i.e. optimize $C(m_l, r_j), \forall m_l \in \boldsymbol{M}$.

# 4   Simulation Results

The simulation results evaluating the performance of the proposed coalition formation scheme have been obtained assuming a hypothetical WSN consisting of $N$ nodes and $R$ representatives of fixed transmission range $T_r$ deployed over an $H \times H$ square area. To achieve almost full network connectivity, the simulation parameters have been adjusted to yield a mean node degree $\overline{deg(n_i)} \approx 6$, by setting $N = 100$, $R = 4$, $T_r = 4m$ and $H = 30m$. In the simulations, the sensed physical phenomena are assumed to follow a normalized Gaussian distribution with regard to internodal distance $x$, namely

$$C(x) = e^{-x^2/2\sigma^2} \tag{6}$$

where $\sigma$ is the standard deviation. Large $\sigma$ values indicate highly correlated physical phenomena that may be exploited to increase WSN lifetime and vice versa. Five distinct services ($M = 5$) have been considered characterized by sigma values: $\sigma(m_1) = H/30$, $\sigma(m_2) = H/6$, $\sigma(m_3) = H/2$, $\sigma(m_4) = H$ and $\sigma(m_5) = 2H$. The modeling of the spatial correlation as a Gaussian function has been also used in [11]. However, any realistic pdf can be used without affecting the validity of the previous analysis or the relevant results. For each distinct WSN operation setup, 100 different simulations implementing random WSN topologies where carried out and the average WSN performance was taken into account.

In Fig.1 the performance of the proposed method is examined with regard to the number of services supported by the WSN. The aggregate WSN lifetime increase (WLI) compared to that of the non-optimized WSN operation and the number of overhead messages sent per node during the coalition formation process have been examined. Three scenarios were studied, where the WSN provides one service ($\mathcal{M} = \{m_2\}$), three services ($\mathcal{M} = \{m_2, m_3, m_4\}$) and five services ($\mathcal{M} = \{m_1, m_2, m_3, m_4, m_5\}$). Fig.1(a) indicates that the WSN operation is significantly affected by the accuracy constraint $q$; relatively low $q$ values relax the QoS constraint specifying proper WSN operation allowing for the formation of coalition structures consisting of fewer and larger coalitions; this leads to increased WLI. On the contrary, when $q \to 100\%$, the nodes tend to behave less cooperatively, limiting the benefit from coalition formation. The WLI increase with the number of services is attributed to the fact that high $\sigma$ values (corresponding to services $m_4$ and $m_5$) result in fewer, large coalitions and, consequently, in low average transmission frequencies.

Fig.1(b) shows that the number of overhead messages, exchanged by the WSN nodes to accomplish coalition formation, increases with the number of supported services. This is expected since the enhanced coalition expansion for services characterized by high correlation increases the number of messages sent per node. As confirmed by the plots on Figs 1(a) and 1(b), both the aggregate WLI and the number of overhead messages are decisively affected by the less correlated service.

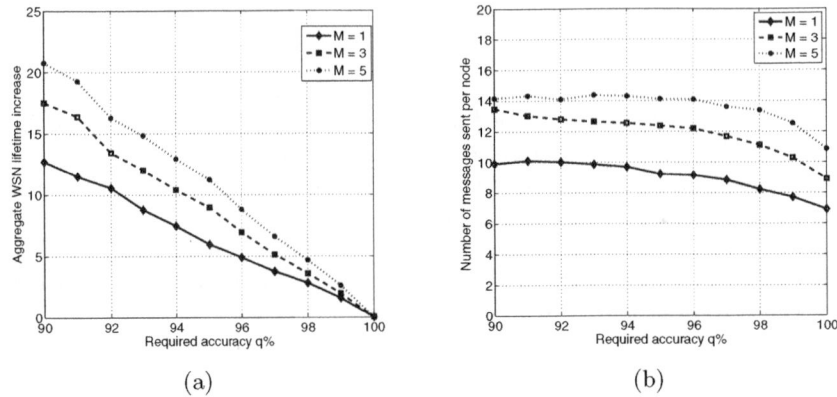

**Fig. 1.** Performance evaluation of the proposed multi-service coalition formation method.
(a) Aggregate WSN lifetime increase. (b) Average number of messages sent per node.

## 5   Conclusions and Future Work

A coalition formation method for efficient multi-service WSN operation has been presented. The coalition formation process optimizes the trade-off between energy efficiency and data accuracy by exploiting the spatial data correlation of the sensed physical phenomena. A set of more powerful nodes perform the optimization of the coalition structure originally achieved by the nodes in a distributive way. The performance and efficiency of the proposed method are assessed through simulations, demonstrating that the proposed architecture can enhance WSN lifetime under multi-modal operation.

The temporal correlation characteristics of the sensed phenomena can also be exploited to further enhance WSN lifetime and make it less susceptible to environmental changes. The design of an adaptive coalition restructuring scheme and the theoretical analysis and evaluation of the respective aggregate WSN performance in the context of multi-service operation are expected to enhance WSN performance with regard to autonomic, self-healing operation.

## References

1. Akyildiz, I.F., Su, W., Sankarasubramaniam, Y., Cayirci, E.: Wireless sensor networks: a survey. Comput. Netw. 38, 393–422 (2002)
2. Vuran, M.C., Akan, O.R.B., Akyildiz, I.F.: Spatio-temporal correlation: theory and applications for wireless sensor networks. Comput. Netw. 45, 245–259 (2004)
3. Meka, A., Singh, A.: Distributed Spatial Clustering in Sensor Networks. In: Ioannidis, Y., Scholl, M.H., Schmidt, J.W., Matthes, F., Hatzopoulos, M., Böhm, K., Kemper, A., Grust, T., Böhm, C. (eds.) EDBT 2006. LNCS, vol. 3896, pp. 980–1000. Springer, Heidelberg (2006)

4. Yoon, S., Shahabi, C.: The clustered aggregation (cag) technique leveraging spatial and temporal correlations in wireless sensor networks. ACM Trans. Sen. Netw. 3 (March 2007)
5. Liu, C., Wu, K., Pei, J.: An energy-efficient data collection framework for wireless sensor networks by exploiting spatiotemporal correlation. IEEE Transactions on Parallel and Distributed Systems 18, 1010–1023 (2007)
6. Pattem, S., Krishnamachari, B., Govindan, R.: The impact of spatial correlation on routing with compression in wireless sensor networks. ACM Trans. Sen. Netw. 4, 1–33 (2008)
7. Dabirmoghaddam, A., Ghaderi, M., Williamson, C.: Energy-efficient clustering in wireless sensor networks with spatially correlated data. In: INFOCOM IEEE Conference on Computer Communications Workshops, pp. 1–2 (March 2010)
8. Cheng, B., Xu, Z., Chen, C., Guan, X.: Spatial correlated data collection in wireless sensor networks with multiple sinks. In: 2011 IEEE Conference on Computer Communications Workshops (INFOCOM WKSHPS), pp. 578–583 (April 2011)
9. Xun, L., Shiqi, T., Merrett, G., White, N.: Energy-efficient data acquisition in wireless sensor networks through spatial correlation. In: 2011 International Conference on Mechatronics and Automation (ICMA), pp. 1068–1073 (August 2011)
10. Aumann, R., Drèze, J.: Cooperative games with coalition structures. International Journal of Game Theory 3, 217–237 (1974)
11. Jindal, A., Psounis, K.: Modeling spatially correlated data in sensor networks. ACM Trans. Sen. Netw. 2, 466–499 (2006)
12. Youssef, M., Youssef, A., Younis, M.: Overlapping multihop clustering for wireless sensor networks. IEEE Transactions on Parallel and Distributed Systems 20(12), 1844–1856 (2009)
13. Rahwan, T., Michalak, T., Elkind, E., Faliszewski, P., Sroka, J., Wooldridge, M., Jennings, N.: Constrained coalition formation. In: The Twenty Fifth Conference on Artificial Intelligence, AAAI 2011, pp. 719–725 (August 2011)

# Autonomous Self-aligning
# and Self-calibrating Capacitive Sensor System

Oscar S. van de Ven, Ruimin Yang, Sha Xia, Jeroen P. van Schieveen,
Jo W. Spronck, Robert H. Munnig Schmidt, and Stoyan Nihtianov

Delft University of Technology, Delft, The Netherlands
o.s.vandeven@tudelft.nl

**Abstract.** An autonomous capacitive sensor system for high accuracy
and stability position measurement, such as required in high-precision industrial equipment, is presented. The system incorporates a self-
alignment function based on a thermal stepping motor and a built-in
capacitive reference, to guarantee that the relative position between the
sensor electrodes is set to 10±0.1 μm. This is needed to achieve the
performance specifications with the capacitive readout. In addition, an
electronic zoom-in method is used to reach the 10 pm resolution with
minimum power dissipation. Finally, periodic self-calibration of the electronic capacitance readout is realized using a very accurate and stable
built-in resistive reference. The performance is evaluated experimentally
and with simulations.

**Keywords:** Capacitive Sensor System, Position Measurement, Self
Alignment, Self Calibration, Thermal Actuator.

## 1 Introduction

For the chosen target application a measurement accuracy better than 100 pm
with a signal bandwidth of 1 kHz is required, while the measurement stability
has to be within 10 pm per minute. Such a measurement can be performed in a
contact-less way, using a proximity type capacitive sensor. The distance between
the parallel sensor electrode and the target electrode is determined based on the
electrical capacitance's between them. With dedicated electronics the ratio of
the total capacitance and the measurable capacitance difference is limited to an
order of magnitude of $10^6$. For high accuracy and stability measurement, the capacitor electrodes should be close and parallel, within 10±0.1 μm. Due to limited
access to the sensor, this cannot be achieved by conventional manual alignment
or by using precision alignment instruments. Also, re-alignment is often required
after transportation. The measurement system should therefore be able to autonomously reposition and realign itself, without compromising the *mechanical
stability* when the system is *at rest*, which is not offered by available, simple systems [2]. Also the readout electronics will need to autonomously (re-)calibrate
periodically to guarantee the measurement precision. Chapter 2 describes the
proposed measurement system and the subsystems thereof. In Chapter 3 the
performance of these sub-systems is evaluated experimentally.

M. Kamel, F. Karray, and H. Hagras (Eds.): AIS 2012, LNCS 7326, pp. 10–17, 2012.
© Springer-Verlag Berlin Heidelberg 2012

## 2   Proposed System

The capacitive displacement measurement system can be split in two main subsystems: (i) the mechanical suspension of the sensor electrode with a motion mechanism and (ii) the electronic system, which measures the capacitance and converts it into a displacement value. Accuracy and stability will be the main design requirements.

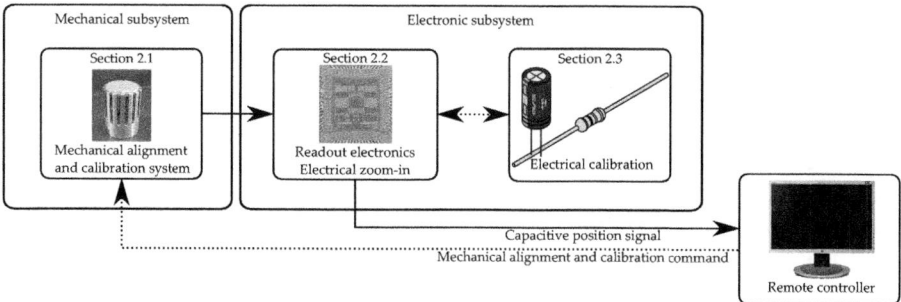

**Fig. 1.** Schematic overview of the measurement system architecture. The solid arrows represent permanent communication, and the dashed arrows represent occasional communication. Only position data and calibration and alignment commands are communicated with the remote controller.

The system accuracy is reached by positioning the measurement electrode (Section 2.1) and using a dedicated capacitance readout (Section 2.2). The output stability is a combination of the mechanical and the electronic stability. The electronic stability is reached by regularly calibrating the readout with a stable reference (Section 2.3). Heat production will also limit the measurement stability and should thus be minimized in the design of the system parts. A final step to improve the stability and accuracy is mechanical calibration of the entire measurement process, from displacement to the reconstructed position, with an accurately known displacement of the motion mechanism.

Implementation of the measurement system will be as follows. First the sensor is mounted with rough position and alignment tolerances. Then the remote controller drawn in Fig. 1 sends an alignment command and the measurement electrode moves towards and aligns with the measurement target. Meanwhile the sensor capacitance is read and sent back to provide feedback for this process. The mechanical calibration will influence the measured position and is thus performed only before or between measurements. Electrical calibration is performed periodically every few minutes during normal operation.

### 2.1   Mechanical Auto-alignment and Calibration Mechanism

The main function of the mechanical subsystem is to generate a motion in three directions (up-down, pitch and roll) to align the electrode, while the position stability at rest is better than 10 pm per minute. The total motion magnitude must be at least 100 μm while the final positioning and alignment accuracy is such that the electrode distance is within 10±0.1 μm. For the mechanical calibration, an accurately reproducible motion is needed. Conventional solutions often cannot meet the stability requirements [2], and therefore the thermal stepper mechanism reported in [3] is used.

 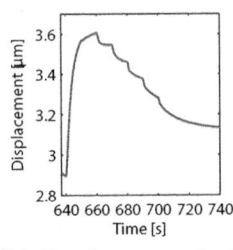 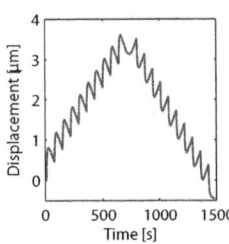

(a) ∅25 mm thermal stepper with integrated switches for the heaters.

(b) Displacement during a single step.

(c) Displacement during a sequence of steps.

**Fig. 2.** Picture of a thermal stepper design and the measured displacement response

The thermal stepper consists of a ring of metallic fingers, a so-called spring nest. (See Fig. 2a.) The fingers clamp the sensor electrode and friction holds it in place. In order to create an overall motion of the measurement electrode, a single finger must be able to slide over the electrode surface. This is accomplished by changing the temperature of one finger while the temperature of the others remains constant, which is done with an electrical heating resistor on each finger. The friction force in this one finger is much larger than the friction force in each of the other fingers, causing only the single finger to start moving. To efficiently actuate the motion directions needed to actively align the electrode, a large number of fingers are used. Using a dedicated heating sequence, an overall net movement can be generated, as is shown in Fig. 3.

To generate this movement, it is essential that the displacement of a single finger, and thus the finger temperature, can be controlled independently. Therefore, the thermal resistance between fingers must be large and the fingers must be thermally connected to a large heat-sink. The preload force of the fingers must be just sufficient for the expected load, because larger contact forces decrease the motion efficiency.

The limited speed due to the thermal time constant (in the seconds range) causes no problems, since the motion is only required during installation. During measurement the actuation will be turned off entirely, reducing its thermal influence to zero.

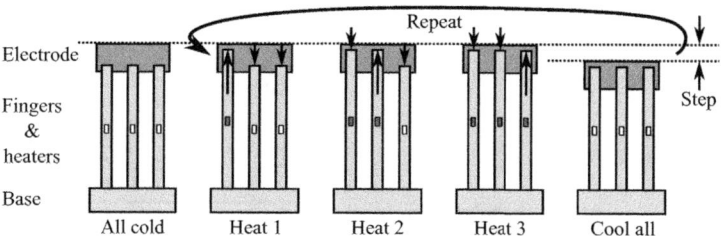

**Fig. 3.** Thermal cycle for a single-dimensional downwards movement. The contact shear forces during motion are indicated. The dark resistors are heating their finger.

To align the electrode surfaces, an upward motion cycle pushes the electrodes towards each other, automatically aligning them when they touch each other. Cooling down all fingers simultaneously from an equal, elevated temperature results in the required electrode distance while maintaining the electrode alignment [3]. When the suspension of the target electrode is too compliant to act as a physical alignment reference, pitch and roll can be actively controlled by the thermal stepper, using a segmented electrode for pitch and roll information. The mechanical calibration uses the calibrated thermal step response of all fingers simultaneously, so that there will be no slip. The measured capacity change is compared to the known displacement to deduce the real sensor gain at the current location [3].

## 2.2 Pico-meter Resolution Capacitive Sensor Interface Circuit

Since the resolution of the capacitance measurement is high, an interface circuit based on the charge balancing principle is a very good candidate for this application. As shown in Fig. 4, this circuit is basically an incremental $\Sigma\Delta$ converter, and can be used to obtain the ratio between the sensor capacitor $C_X$ and the reference capacitor $C_{REF}$ with very high resolution. However, there is one drawback if it is directly applied to the sensor. The measurement range of the circuit is from $-C_{REF}$ to $C_{REF}$, which corresponds to a much larger displacement range than the required $\pm 1$ μm. This causes a waste of system resources.

The solution to this problem is to use electrical zoom-in, as illustrated in Fig. 5. A zoom-in capacitor $C_Z$ is introduced which is driven with an excitation signal opposite to the sensor capacitor ($C_X$) excitation, so that the effective input of the interface becomes $C_X$-$C_Z$. When the value of $C_Z$ is very close the nominal value of $C_X$, the reference capacitor $C_{REF}$ can be largely reduced. In this way the conversion speed of the circuit can be increased.

The realized circuit employs a 3rd order loop filter for sufficient noise shaping with a clock frequency of 5 MHz. It also has a fully differential structure to suppress charge-injection error. The prototype circuit was fabricated in a standard 0.35 μm CMOS technology and consumes 15 mW from a 3.3 V power supply.

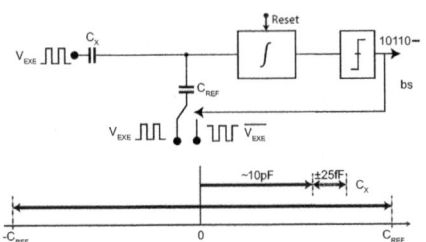

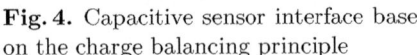

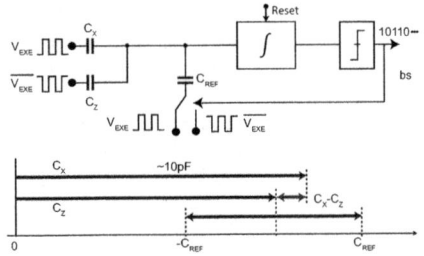

**Fig. 4.** Capacitive sensor interface based on the charge balancing principle

**Fig. 5.** Zoom-in capacitor used to reduce the measurement range of the interface, in order to decrease the conversion time

## 2.3   Electronic Self-calibration with External Stable Reference

As mentioned above, the high resolution capacitive sensor interface uses two extra reference capacitors during operation. One is a zoom-in capacitor $C_Z$ (~10 pF) while the other is a reference capacitor $C_{REF}$ (~100 fF). The performances of the interface, e.g. accuracy and stability, are determined by the quality of these two capacitors.

Unfortunately, capacitive components are not accurate and stable enough for the given application. The best available off the shelf capacitive component is accurate up to 1 % while the thermal stability is in the order of tens of ppm/°C [4]. Therefore, to achieve the required performance, these two reference capacitors have to be calibrated periodically by more stable and accurate references and dedicated electronics.

There are different stable and accurate references available, for instance time-frequency reference, voltage reference and resistor reference. After some investigation and comparison, stable and accurate resistors with 0.005 % accuracy and 0.5 ppm/°C temperature stability [5], were selected as the built-in reference for the capacitance measurement.

The comparison between the capacitor and resistor is based on the charge balancing principle, the charge generated by the capacitor (unknown charge) is balanced by the charge generated by the resistor (reference charge). To improve the energy efficiency and measurement speed, a charge-balancing based $\sum \Delta$ resistance-capacitance comparator is proposed, see Fig. 6.

The circuit works as follows: a reference current source $I_{ref}$ that is generated by a reference voltage $V_{ref}$ and resistor $R_{ref}$ is continuously connected to the input of an integrator. The generated charge is then stored in the integrator, thus making the integrator output non-zero. A comparator monitors the output of the integrator to switch the compensation charge on and off. Only when the output of the integrator is positive, the comparator output is '1', which controls the feedback path to supply a compensation charge. The compensation charge is generated by the unknown capacitor $C_X$ and a reference voltage $V_{ref}$.

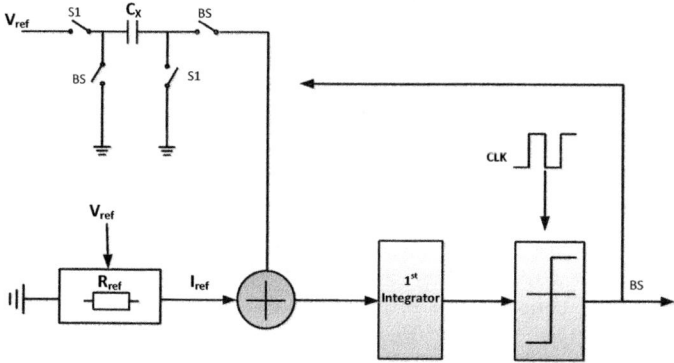

**Fig. 6.** Simplified block diagram of the proposed RC (resistance-capacitance) comparator. The capacitance is charged at every clock by enabling s1 switches. Only when the comparator output (BS) is '1' the charge is applied to the integrator by enabling the BS switches.

Assume the circuit operates for N cycles. The total charge that is supplied by the current source can be calculated as: $Q_{ref} = N \frac{V_{ref}}{R_{ref}} \Delta t$, where $\Delta t$ is the time interval between two decision-making actions of the comparator.

Whenever the output of the comparator is '1', a compensation charge is generated by the capacitor. Therefore, the total compensation charge is: $Q_X = N_1 V_{ref} C_X$, where $N_1$ is the number of '1's in the comparator output during one measurement cycle.

Finally, with sufficient operating cycles $(N)$, the loop will bring the integrator output to zero, meaning that the reference charge balances the unknown charge. Thus the capacitance can be calculated as: $C_X = {}^{N}\!/_{N_1} \Delta t / R_{ref}$. As can be seen from the equation, the effect of the reference voltage is canceled out, thus the final result is a function of a reference resistor and a reference time.

To improve the resolution and speed of the RC comparator, a 3rd order $\sum \Delta$ modulator is implemented. The higher order loop provides better noise shaping, which enhances the resolution with fewer operating cycles required [1]. In addition, the modulator is implemented as a fully differential structure to achieve good immunity to common-mode interferences.

## 3   Experimental Results

### 3.1   Motion Mechanism

Fig. 2a shows a realized motion system. In Fig. 2b the measured sensor displacement of a single upwards step sequence that results in a 0.5 μm net displacement is shown. The series of sequences, in Fig. 2c shows a motion in both directions in the micrometer range with an average speed of approximately $0.5\,\mathrm{\mu m/min}$.

## 3.2   Capacitive Sensor Interface Electronics

Fig. 7 shows the measurement setup for evaluating the performance of the interface with an off-chip capacitive sensor. The sensor is connected to the interface
via shielded coaxial cables, which adds 10 pF parasitic capacitance to load of the
interface. The sensor itself is mechanically actuated by a shaker which is shaking
at 100 Hz, creating a small capacitance variation on the sensor. The measured
response of the sensor is shown in Fig. 8. It can be seen that the measurement
noise floor is 65 aF$_{rms}$, which means that the equivalent displacement resolution
is 65 pm$_{rms}$ if this interface is connected to the designed displacement sensor.
The measurement time for this interface is as low as 20 μs.

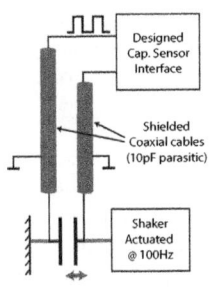

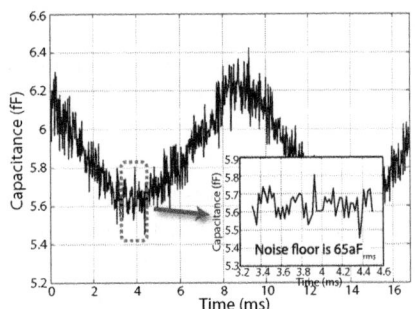

**Fig. 7.** Measurement setup with a mechanically actuated capacitive sensor

**Fig. 8.** Measurement result, the measured noise floor is 65 aF$_{rms}$

## 3.3   Electronic Self-calibration

The RC comparator principle was first implemented and simulated in Matlab.
Fig. 9 shows the simulation result of the comparator with 1500 operating cycles.
In theory, only 500 cycles is sufficient to achieve 20-bit resolution with a 3$^{rd}$ order
$\sum \Delta$ converter. However, this calculation is based on the assumption that only
the quantization noise of the comparator contributes to the error. In reality, the
thermal noise of the system is the limiting factor of the overall performance, thus
more operating cycles are required to suppress the thermal noise. In this design,
the number of operating cycles is selected to be 1500. The result shows that the
conversion error is only a few atto-Farad (aF), while the input capacitance is in
the range of pico-Farad (pF). Therefore more than 20-bit resolution is achieved.
The operating time for each cycle in this design is around 6 μs, which makes the
total measurement time 10ms. The spectrum of the result is shown in Fig. 10,
which clearly shows a 3$^{rd}$ order noise-shaping profile.

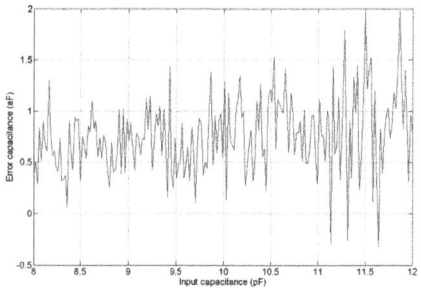

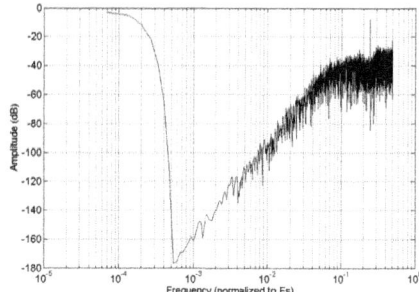

**Fig. 9.** Simulated error versus input capacitance with 1500 operating cycles

**Fig. 10.** Simulated output spectrum of the proposed RC comparator

## 4    Conclusions and Future Work

The developed solution for a stable and high accuracy position measurement is the combination of a mechanical positioning and alignment system and efficient, high resolution readout electronics.

The intelligent operating algorithm of the system first uses the thermal actuator to position the measurement electrode. Secondly, the electronic part is used to zoom in electrically and perform the measurement with high accuracy. Finally a calibrated motion is performed as a mechanical reference and a resistor is used as a stable electrical reference to guarantee the measurement stability over time.

The subsystems that implement these tasks have been designed and their performance is verified with measurements and simulations. The next step consists of further developing these subsystems into an autonomous self-aligning and self-calibrating capacitive sensor element.

This project is funded by the Dutch technology foundation STW.

## References

1. Márkus, J.: Higher-order incremental delta-sigma analog-to-digital converters. Ph.D. thesis, Budapest University of Technology and Economics (March 2005)
2. Ouyang, P., Tjiptoprodjo, R., Zhang, W., Yang, G.: Micro-motion devices technology: The state of arts review. The International Journal of Advanced Manufacturing Technology 38(5), 463–478 (2008)
3. Van Schieveen, J., Spronck, J., Munnig Schmidt, R.: Integrated auto alignment and calibration for high resolution capacitive sensor systems. In: Euspen Conference Proceedings I, pp. 188–191 (2010)
4. Vishay-Precision: High performance, high precision and lowest profile 0402 capacitor, http://www.datasheetcatalog.org/datasheet/vishay/hpc0402b.pdf
5. Vishay-Precision: Ultra High Precision Z-Foil Resistor, http://www.vishaypg.com/docs/63187/zseries.pdf

# Relay Node Positioning in Wireless Sensor Networks by Means of Evolutionary Techniques

José M. Lanza-Gutiérrez, Juan A. Gómez-Pulido,
Miguel A. Vega-Rodríguez, and Juan M. Sánchez-Pérez

Dep. of Technologies of Computers and Communications, University of Extremadura,
Polytechnic School, Campus Universitario s/n, 10003 Cáceres, Spain
{jmlanza,jangomez,mavega,sanperez}@unex.es

**Abstract.** The use of wireless sensor networks (WSN) is a common situation nowadays. One of the most important aspects in this kind of networks is the energy consumption. In this work, we have added relay nodes to a previously defined static WSN in order to increase its energy efficiency, optimizing both average energy consumption and average coverage. For this purpose, we use two multi-objective evolutionary algorithms: NSGA-II and SPEA-2. We have statistically proven that this method allows us to increase the energy efficiency substantially and NSGA-II provides better results than SPEA-2.

**Keywords:** wireless sensor networks, relay node, energy consumption, NSGA-II, SPEA-2, statistic tools, muli-objective evolutionary algorithm.

## 1 Introduction

The use of wireless sensor networks (WSN) is a common situation nowadays. This kind of network has a simple deployment, so it allows us to position devices in places where it would be very expensive or impossible for a traditional network. We can see many examples in civil (industrial control, intensive agriculture...) and military areas (rescue operations, surveillance...) [1].

An important aspect in WSN is the power consumption. Usually, this kind of networks are powered by bateries, thus network lifetime will depend on energetic wear of their elements, like amount of information transmitted by sensors and its scope, among others.

The design of a energy-efficient WSN has been established as a NP-hard optimization problem by some authors [2][3], so we can find some approximations in the literature. Mostly, we can establish two research lines, based on the kind of devices used.

If we focus on the first line, a traditional WSN is composed of sensors that capture information about its environment, and a sink node which collects all information captured. Beginning with heuristics, we can cite the contributions of Cardei et al. [4] (to optimize network lifetime and coverage, splitting the network in disjoint sensor sets and deciding that group must be active at all times) and Xiuzhen et al. [2] (they optimize power consumption by means on assigning

M. Kamel, F. Karray, and H. Hagras (Eds.): AIS 2012, LNCS 7326, pp. 18–25, 2012.

different transmission power levels to sensors). Also, we can find some papers that use evolutionary algorithms (EAs), such as Xiao-Min et al. [5] maximize network lifetime splitting the network in a similar way to Cardei et al. but using a mono-objective EA; Martins et al. [6] optimize lifetime and coverage simultaneously by means of a multi-objective EA (MOEAs), assigning different transmission power levels to sensors, these results are compared with an integer linear programming; and Konstantinidis et al. [7] optimize lifetime and coverage simultaneously, assigning different transmission power levels and maintaining a k-connectivity, using for this purpose a MOEA. Most papers in this reseach line have a common disadvantage: authors use more sensors than necessary to increase network lifetime and this involves a higher cost.

Recently, a new reseach line has appeared, incorporating a third kind of element to WSN: the *relay nodes* (routers), nodes with higher energetic and communicative capabilities. They have been included to minimize communication among sensors, increasing both network speed and sensor lifetime. We can find some references about this topic. Beginning with heuristics, Kenan Xu. et al. [8] study random methods to place routers on a preestablished WSN in order to increase its network lifetime; and Quanhong el al. [9] study how to place the lower number of routers in a random WSN to minimize power consuption. Besides, some authors have used EAs, for example Alfredo J. Perez et al. [10] present a MOEA for the simultaneous optimization of the number of relays and the energy dissipation when deploying a WSN; and Chunhua Zhao et al. [11] study how to deploy the relay nodes to obtain the optimal energy efficiency by minimizing the average path length.

In this work, we study the inclusion of relay nodes in a previously established static WSN to optimize both, average energy consumption and average coverage simultaneously. In order to solve this NP-hard problem, we have used two well-known MOEAs: NSGA-II(Non-dominated sorting genetic algorithm II) [12] and SPEA-2(Strength pareto evolutionary algorithm 2) [13].

The remainder of this paper is organized as follows. In section II, we describe the relay node placement problem, optimizing both average energy consumption and average coverage. The methodology followed to solve this problem appears in section III. In section IV, we present an evaluation of results using statistical procedures. Finally, conclusions and future work are left for section V.

## 2  Relay Node Placement Problem

In this work, we have considered the addition of relay nodes to a traditional static WSN with the least possible number of sensors, optimizing its energy efficiency.

A network, as we use here, is composed of $M$ sensors located at a 2D-scenery of size $D_x \times D_y$ that capture information about its environment with a sensitivity radius $R_s$. Each captured measurement(with a size of $K$ bits) will be sent to a sink node $C$ placed on the center of the scenery. In order to optimice the energy consumption, $N$ devices called relay nodes (routers) are incorporated. This kind of element is only able to establish communications, but it has more energy capacity than sensors (or unlimited). Besides, all network elements can communicate with a radius $R_c$.

We assume a perfect medium access control, such as SMAC defines [14], which ensures that there will be no collisions at any device during data communication and we adopt a simple but relevant path loss communication model as in [15]. In this model, the transmit power level that must be assigned to a sensor $i$ to reach a device $j$ is (1), where $\alpha \in [2,6]$ is the path loss exponent, $\beta = 1$ is the transmission quality parameter and $d_{i,j}$ is the Euclidean distance between $i$ and $j$. This distance must be less than $R_c$, otherwise communication would not be possible. The residual energy of a sensor i in a time t is calculated in (2), where $r_i(t)$ is the traffic load that i receives to relay at time t, " $+1$ " is the data packet captured by i and $amp$ is the energy consumption per bit of the power amplifier. Energy consumed for receipting, sensing and processing are considered negligible and ignored [15].

$$P_i = \beta \times d_{i,j} \tag{1}$$

$$E_i(t) = E_i(t-1) - [(r_i(t) + 1 \times P_i * amp)] \tag{2}$$

In this kind of networks, there is an important concept: the *network lifetime* $(LF)$. LF is the amount of time units (t.u) that a network can provide information about its environment. A coverage threshold is usually used to determine whether the network is valid; if this coverage value is less than threshold, the network will be dead. Initially, all sensors have the same maximum energy charge $(EC)$. Each time a sensor obtains a measure, it is sent to $C$; if the distance is less than $R_c$, sensor sends the measure to $C$ directly, otherwise sensor sends it to other network element (sensor or router) using for this purpose Dijsktra's minimum path [16]. Each transmission involves an energy consumption, so when the battery charge of a sensor is equal to zero, it not will be used again, and then its coverage not will be taken into account.

In this way, given a traditional WSN with its energetic $(\alpha, \beta, amp, EC)$ and physical ($M$ with their fixed positions, $C$, $D_x$, $D_y$, $R_c$, $R_s$) characteristics set, and a number of relay nodes $(N)$, the *relay node placement problem* may be stated as a Multi-objective Optimization Problem, in which the objectives functions are:

- *Average Energy Consumption (AEC)* $(f_1)$: Minimize the network energy consumption along its lifetime, in Joules $(J)$. Note that numSensorsAlive(t) is the number of sensors with a residual energy greater than zero at time t.

$$f_1 = LF^{-1} \times \sum_{t=1}^{LF} \sum_{i=1}^{M} \left( \frac{E_i(t-1) - E_i(t)}{numSensorsAlive(t)} \right) \tag{3}$$

- *Average Coverage (AC)* $(f_2)$: Maximize the network coverage provided by sensors along its lifetime. In order to obtain the percentage of the coverage, there are two possible options [17]. The first one considers that the coverage provided by a sensor is a circumference of radius $R_s$, so the global coverage will be the intersection of all of them. The second one consists of the use of a boolean matrix of $D_x \times D_y$ points over scenery, so for each sensor, points within its radius will be activated; finally we have to count all active points. We have chosen the second option, because although the first one is more exact, it is harder to compute. In this equation $R$ represents the boolean matrix and $R_{x,y}$ its position $(x, y)$.

$$f_2 = LF^{-1} \times \sum_{t=1}^{LF} \sum_{x=1}^{D_x} \sum_{y=1}^{D_y} \left( \frac{R_{x,y}}{D_x \times D_y} \right) \tag{4}$$

## 3  Problem Resolution

As we have said previously, we have used two well-known MOEAs to solve the *relay node placement problem* optimizing two objectives simultaneously.

In an EA, a candidate solution (individual) is encoded as a string named chromosome. These individuals will evolve toward better solutions along the EA generations. In this problem, a chromosome is a 2D-coordinate list $(x, y)$ of $N$ routers.

The evolution starts from a random population and happens in generations. In each generation, several individuals are stochastically selected from the current population, based on their fitness values, and recombined (and probably mutated) to form a new population. Next, best individuals from both populations are selected which will be used in the next iteration of the algorithm. The size of this population is fixed and equal to the size of the initial population. Commonly, the EA ends when either a maximum number of generations or evolutions reached.

In our work, the initial population is generated assigning random coordinates to routers of each chromosome, but with one restriction: all routers must be accessible to $C$. The purpose is to start with an adequate population to facilitate convergence of MOEAs used.

To generate new individuals, we use crossover operator. We take two individuals using the habitual binary tournament and selecting a crossover point randomly. Then, we copy routers from individual 1 to this point, next we copy from individual 2 to the end of the chromosome. The performance of this algorithm is conditioned by the crossover probability. If a randomly generated value is greater than this value, the resulting individual will be a complete-copy of the dominant individual; in other works, crossover will not be performed.

Mutation operator allows us to incorporate random changes in an previously recombined individual, avoiding local minimums and increasing diversity. We perform random changes over coordinates of routers following a mutation probability, then we evaluate the modified individual. If these modifications cause better fitness values, they will be accepted; otherwise mutations will be discarded. The objective is to avoid getting a worse individual than token originally.

To select which individuals must survive at the end of each generation, we use two well-known algorithms: NSGA-II [12] and SPEA-2[13]. The first one is characterized by sorting the population basing on its dominance (Pareto front division), and by using crowding distance to compare individuals in the same front. The second one is based on the file concept, an auxiliary population that saves better solutions over generations. For this case, we follow a strategy that considers, for each individual, the number of individuals that it dominates and the number of individuals dominating to it. Also it uses the density concept as a method of fine assignment.

## 4 Performance Evaluation

The instance data used in this work can be obtained from [18]. The instances represent a couple of scenarios of *100x100* and *200x200* meters, where a set of sensors and a sink node are placed. Our objective is to reduce the energy consumption incorporating relay nodes. Both $R_c$ and $R_s$ take habitual values from [6], *30* and *15* meters respectively. In addition, we have included a new $R_c$ value (*60* meters) to represent devices with higher communication capacities. Energy values ($\alpha=2$, $\beta=1$, $EC=5J$ and $amp=100pJ/bit/m^2$) are taken from [7]. The used information packet size is *128kB*. Collector node is placed in the center of the scenery and the coverage threshold used for lifetime is *70%* .

The number of sensors for both scenarios is the lowest value to cover all the surface: the area covered for a sensor is $\pi \times R_s^2$ and the area of a scenery is $D_x \times D_y$, so it is necessary $\lceil (D_x \times D_y)/(\pi \times R_s^2) \rceil$ sensors. Sensor coordinates have been previously fixed using a mono-objective EA for coverage optimization. In Table 1, for each instance used in this work, we can see dimensions (A), number of sensors (M), $R_c$ for its devices and AEC and AC values for homogeneous conception (without relay nodes).

**Table 1.** Instances used in this work

| Instance | $D_x \times D_y$ | M | $R_c$ | HO-AEC | HO-AC | Ref AEC | | Ref AC | |
|---|---|---|---|---|---|---|---|---|---|
| | | | | | | *ideal* | *nadir* | *ideal* | *nadir* |
| 100x100_15_30 | 100x100 | 15 | 30 | 0.1091 | 89.24% | 0.02 | 0.1 | 100% | 60% |
| 100x100_15_60 | 100x100 | 15 | 60 | 0.1482 | 86.63% | 0.02 | 0.1 | 100% | 60% |
| 200x200_57_30 | 200x200 | 57 | 30 | 0.2791 | 87.10% | 0.10 | 0.30 | 100% | 60% |
| 200x200_57_60 | 200x200 | 57 | 60 | 0.3871 | 82.43% | 0.10 | 0.30 | 100% | 60% |

We follow an habitual strategy to solve this kind of problem [19]. First, we determine the settings that provide the best result for both algorithms (NSGA-II and SPEA-2). Next, we study if any of them provides significatively better performance, using statistical tools. We have set the most common parameters over 30 independent runs in order to determine the best settings: starting on a default configuration, the parameters are adjusted one by one in its optimal

**Table 2.** Obtained hypervolumes for each instance data and routers used. Note that, we use the notation *Instance(x)*, where $x$ is the number of routers used in this instance.

| | NSGA-II | | | | | SPEA-2 | | | | |
|---|---|---|---|---|---|---|---|---|---|---|
| | 50.000 | 100.000 | 200.000 | 300.000 | 400.000 | 50.000 | 100.000 | 200.000 | 300.000 | 400.000 |
| 100x100_15_30(2) | 42.61% | 42.66% | 42.69% | 42.69% | 42.69% | 41.87% | 42.62% | 42.67% | 42.67% | 42.66% |
| 100x100_15_30(3) | 55.17% | 55.30% | 55.59% | 55.66% | 55.69% | 53.62% | 53.90% | 53.83% | 54.01% | 54.01% |
| | | | | | | | | | | |
| 100x100_15_60(2) | 31.57% | 31.84% | 31.92% | 31.94% | 31.94% | 31.47% | 31.56% | 31.83% | 31.84% | 31.91% |
| 100x100_15_60(3) | 59.18% | 59.43% | 59.60% | 59.80% | 60.01% | 58.04% | 58.99% | 59.16% | 59.19% | 59.21% |
| | | | | | | | | | | |
| 200x200_57_30(2) | 35.54% | 37.00% | 37.34% | 37.58% | 37.68% | 33.87% | 33.93% | 34.00% | 34.07% | 34.09% |
| 200x200_57_30(4) | 42.98% | 47.80% | 49.25% | 51.26% | 52.22% | 44.03% | 43.23% | 43.66% | 43.93% | 44.30% |
| 200x200_57_30(6) | 62.42% | 63.77% | 64.57% | 64.65% | 66.03% | 59.35% | 62.15% | 64.25% | 64.37% | 65.05% |
| 200x200_57_30(9) | 75.27% | 76.52% | 77.26% | 78.12% | 78.51% | 69.95% | 74.66% | 75.75% | 76.37% | 76.85% |
| | | | | | | | | | | |
| 200x200_57_60(2) | 23.19% | 23.61% | 23.67% | 24.00% | 24.03% | 22.91% | 23.18% | 23.59% | 23.68% | 23.82% |
| 200x200_57_60(4) | 58.45% | 58.80% | 59.70% | 59.88% | 60.00% | 57.19% | 58.31% | 59.42% | 59.63% | 59.75% |
| 200x200_57_60(6) | 72.75% | 74.49% | 75.14% | 75.72% | 75.88% | 71.34% | 72.57% | 73.84% | 74.23% | 74.75% |
| 200x200_57_60(9) | 87.59% | 88.34% | 89.41% | 89.96% | 89.96% | 84.31% | 86.69% | 89.13% | 89.87% | 89.96% |

value, until all parameters have been adjusted. We have obtained the same settings for both algorithms: a population size of 100 individuals and a crossover and mutation probabilities equal to 0.5.

We have used the hypervolume metric [20] in order to determine the goodness of solutions (Pareto Fronts). This metric needs, for each objective, a couple of reference points called ideal and nadir, representing their best and worst values respectively.

In Table 2, we can observe average hypervolumes for each instance, algorithm and routers used. We have obtained these hypervolumes using different number of evaluations (stop criterium) to study their progression, being these values fixed experimentally. We have used the reference points showed in Table 1, *ref AEC* and *ref AC*, in order to obtain the hypervolumes. The inclusion of relay nodes involves an additional cost; for this reason, we have considered that we should not add more than a 20% of these elements in a network.

With these hypervolumes, we have carry out our statistical study. We have followed an habitual procedure [19]. First, we determine if data obtained for these instances (each of them with 30 runs) follow a normal distribution. For this purpose, we have used Shapiro-Wilk test, obtaining that data do not come from a normal model. To check what algorithm provides better results, we use a non-parametric test: Wilcoxon test. *The result of this study is that NSGA-II provides better solutions than SPEA-2 for these instances.*

To illustrate the advantages provided by the inclusion of relay nodes, in Table 3, for each instance data, we have taken the extreme points of a Pareto Front from NSGA-II. If we compare these results with the homogeneous case in Table 1, we can check how power consumption have been greatly reduced, up to 28 times in some cases.

**Table 3.** Fitness values obtained by means of NSGA-II using 400.000 evaluations

| Routers | 100x100_15_30 | | 100x100_15_60 | | 200x200_57_30 | | | | 200x200_57_60 | | | |
|---|---|---|---|---|---|---|---|---|---|---|---|---|
| | 2 | 3 | 2 | 3 | 2 | 4 | 6 | 8 | 2 | 4 | 6 | 8 |
| AEC | 0,0565 | 0,0425 | 0,0650 | 0,0358 | 0,0195 | 0,0158 | 0,0128 | 0,1000 | 0,0230 | 0,0139 | 0,1000 | 0,1889 |
| AC | 89.52% | 82.37% | 81.88% | 78.42% | 87.16% | 87.92% | 87.60% | 90.38% | 85.58% | 86.70% | 83.41% | 89.39% |
| *Extreme 1* | | | | | | | | | | | | |
| AEC | 0,0673 | 0,0640 | 0,0899 | 0,0618 | 0,0252 | 0,0167 | 0,0137 | 0,9992 | 0,0238 | 0,0160 | 0,0995 | 0,0896 |
| AC | 91.56% | 91.61% | 91.27% | 91.45% | 88.41% | 89.76% | 90.39% | 90.38% | 87.74% | 90.22% | 83.41% | 89.21% |
| *Extreme 2* | | | | | | | | | | | | |

## 5  Conclusions and Future Works

In this work, we have solved the *relay node placement problem* by means of MOEAs, studying how to place relay nodes in a previously defined static WSN with the least possible number of sensors, besides we provide our instance set used. The obtained results have been analyzed using statistical tools, demonstrating for our instance data both, this conception allows us to increase energy efficiency substantially and NSGA-II provides better results than SPEA-2.

As future work, we propose to use other MOEAs and more instance data. Besides, we think that to introduce parallelism would be interesting in order to reduce the execution time due to more complex instances.

**Acknowledgments.** This work has been partially funded by the Spanish Ministry of Education and Science and ERDF (the European Regional Development Fund), under contract TIN2008-06491-C04-04 (the M* project). Finally, thanks to Government of Extremadura for GR10025 grant provided to the group TIC015.

## References

1. Mukherjee, J.Y.B., Ghosal, D.: Wireless sensor network survey. Computer Networks 52, 2292–2330 (2008)
2. Cheng, X., Narahari, B., Simha, R., Cheng, M., Liu, D.: Strong minimum energy topology in wireless sensor networks: Np-completeness and heuristics. IEEE Transactions on Mobile Computing 2, 248–256 (2003)
3. Clementi, A.E.F., Penna, P., Silvestri, R.: Hardness Results for the Power Range Assignment Problem in Packet Radio Networks. In: Hochbaum, D.S., Jansen, K., Rolim, J.D.P., Sinclair, A. (eds.) RANDOM 1999 and APPROX 1999. LNCS, vol. 1671, pp. 197–208. Springer, Heidelberg (1999)
4. Cardei, M., Du, D.Z.: Improving wireless sensor network lifetime through power aware organization. Wireless Networks 11, 333–340 (2005), 10.1007/s11276-005-6615-6
5. Hu, X.M., et al.: Hybrid genetic algorithm using a forward encoding scheme for lifetime maximization of wireless sensor networks. IEEE Transactions on Evolutionary Computation 14(5), 766–781 (2010)

6. Martins, F., et al.: A hybrid multiobjective evolutionary approach for improving the performance of wireless sensor networks. IEEE Sensors Journal 11(3), 545–554 (2011)
7. Konstantinidis, A., Yang, K.: Multi-objective k-connected deployment and power assignment in wsns using a problem-specific constrained evolutionary algorithm based on decomposition. Computer Communications 34(1), 83–98 (2011)
8. Xu, K., et al.: Relay node deployment strategies in heterogeneous wireless sensor networks. IEEE Transactions on Mobile Computing 9(2), 145–159 (2010)
9. Wang, Q., et al.: Transactions papers - device placement for heterogeneous wireless sensor networks: Minimum cost with lifetime constraints. IEEE Transactions on Wireless Communications 6(7), 2444–2453 (2007)
10. Perez, A., Labrador, M., Wightman, P.: A multiobjective approach to the relay placement problem in wsns. In: 2011 IEEE Wireless Communications and Networking Conference (WCNC), pp. 475–480 (2011)
11. Zhao, C., Chen, P.: Particle swarm optimization for optimal deployment of relay nodes in hybrid sensor networks. In: IEEE Congress on Evolutionary Computation, CEC 2007, pp. 3316–3320 (September 2007)
12. Deb, K., Pratap, A., Agarwal, S., Meyarivan, T.: A fast elitist multi-objective genetic algorithm: Nsga-ii. IEEE Transactions on Evolutionary Computation 6, 182–197 (2000)
13. Zitzler, E., Laumanns, M., Thiele, L.: Spea2: Improving the strength pareto evolutionary algorithm. Technical report (2001)
14. Ye, W., Heidemann, J., Estrin, D.: An energy-efficient mac protocol for wireless sensor networks. In: Proceedings Twenty-First Annual Joint Conference of the IEEE Computer and Communications Societies, INFOCOM 2002, vol. 3, pp. 1567–1576. IEEE (2002)
15. Konstantinidis, A., Yang, K., Zhang, Q.: An evolutionary algorithm to a multi-objective deployment and power assignment problem in wireless sensor networks. In: Global Telecommunications Conference, IEEE GLOBECOM 2008, pp. 1–6. IEEE (2008)
16. Cormen, T.H., Leiserson, C.E., Rivest, R.L., Stein, C.: Introduction to Algorithms, 3rd edn., p. 1292. The MIT Press (2009)
17. Wang, B.: Coverage problems in sensor networks: A survey. ACM Comput. Surv. 43, 32:1–32:53 (2011)
18. Lanza-Gutierrez, J.M., Gomez-Pulido, J.A., Vega-Rodriguez, M.A., Sanchez-Perez, J.M.: Instance sets for optimization in wireless sensor networks (December 2011), http://arco.unex.es/wsnopt
19. Lanza-Gutierrez, J.M., Gomez-Pulido, J.A., Vega-Rodriguez, M.A., Sanchez-Perez, J.M.: A multi-objective network design for real traffic models of the internet by means of a parallel framework for solving np-hard problems. In: NaBIC, pp. 137–142 (2011)
20. Knowles, J., Thiele, L., Zitzler, E.: A tutorial on the performance assessment of stochastic multiobjective optimizers. 214, Computer Engineering and Networks Laboratory (TIK), ETH Zurich, Switzerland (2006), revised version

# Autonomous MEMS Inclinometer

F.S. Alves[1], R.A. Dias[1], J. Cabral[1], and L.A. Rocha[2]

[1] ALGORITMI CENTER, Universidade do Minho, Campus de Azurem, Guimarães, Portugal
[2] IPC\I3N, Universidade do Minho, Campus de Azurem, Guimarães, Portugal

**Abstract.** *Pull-in voltage* measurements are used in this work as the transduction mechanism to build a novel microelectromechanical system (MEMS) inclinometer. By successively bringing the microstructure to pull-in while measuring the pull-in voltage allows the detection of external accelerations. Moreover, the availability of asymmetric pull-in voltages that depend on the same mechanical structure and properties enables the implementation of an auto-calibrated thermal compensated inclinometer. The thermal compensation method is described and it relies on the measurement of pull-in voltages only. Both simulations and experiments are used to validate this novel approach and first results show a sensitivity of 50mV/° and a resolution of 0.006°.

**Keywords:** Pull-In voltage, Inclinometer, MEMS.

## 1 Introduction

An inclinometer, or tilt sensor, is a measurement device that uses gravity to measure the inclination of an object. Inclinometers are used in different application areas, such as civil engineering, robotics, automotive and even aviation [1, 2]. The precision of these devices is dependent on the technology employed. State-of-the-art devices can reach resolutions below 0.0025 degrees (°) [2-4].

In the field of electronic inclination sensors, the most common approach consists in using an existing accelerometer that is incorporated into the system [4-7]. Therefore, the characteristics of the accelerometer define the inclinometer characteristics. While inclinometers are devices that do not require large bandwidths, they must have a good stability over time and good resolution. These characteristics call for a dedicated solution that can provide inclinometers with auto-calibration and autonomy from variations in the surrounding environment like temperature and humidity. This is particularly important for structural health monitoring, where autonomy, long term stability and auto-calibration capabilities are desirable [8].

A new approach for an inclinometer that enables thermal compensation and auto-calibration is proposed in this paper. The device relies on a parallel-plate electrostatic actuator that is operated in pull-in mode, i.e., the device is constantly actuated to pull-in, and the measured pull-in voltage is used as a measure of the inclination. Due to the characteristics of the pull-in effect (it only depends on the material properties and dimensions) and using a differential scheme it is possible to have an auto-calibrated and thermal compensated inclinometer.

M. Kamel, F. Karray, and H. Hagras (Eds.): AIS 2012, LNCS 7326, pp. 26–33, 2012.
© Springer-Verlag Berlin Heidelberg 2012

This paper is divided into six sections. After an introduction the pull-in effect is presented followed by the inclinometer working principle. Next, the fabricated micro-structures are introduced and the simulated and experimental results are presented. Finally some conclusions are drawn.

## 2    Pull – In Analysis

Electrostatically actuated parallel-plate MEMS devices have an electrostatic force that is inversely proportional to the square of deflection while the elastic restoring force is proportional to the deflection. Therefore, while for small voltages the electrostatic force is countered by the spring force, when the applied voltage rises to values beyond a critical voltage, the elastic force cannot balance the electrostatic force and there is no equilibrium. In this case, the distance between plates reduces abruptly until the electrodes snap together. This phenomenon is called *pull-in* and the required critical voltage for pull-in to occur is named as *pull-in voltage*.

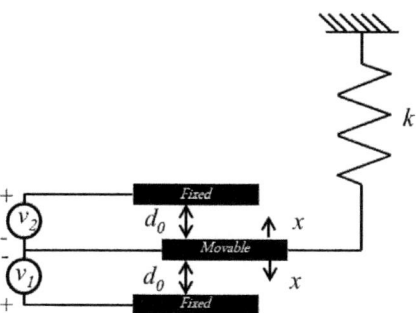

**Fig. 1.** Sketch of the basic device

The simplest structure to study the pull-in phenomenon is a parallel-plate electrostatic actuator with 1- Degree-Of-Freedom (DOF) where the movable electrode is placed equidistant to two fixed electrodes as shown in Fig. 1. The structure enables three different ways to apply voltage into the device and consequently three different *pull-in voltages* can be defined: the asymmetric-right ($V_1 = Vpr$ and $V_2 = 0$), the asymmetric-left ($V_1 = 0$ and $V_2 = Vpl$), and the symmetric ($V_1 = V_2 = Vps$). Assuming ideal conditions the analytical expressions for the three *pull-in voltages* can be found [9]:

$$Vpr = Vpl = \sqrt{\frac{8}{27}\frac{d_0{}^3 k}{\varepsilon_0 wl}} \tag{1}$$

$$Vps = \sqrt{\frac{1}{2}\frac{d_0{}^3 k}{\varepsilon_0 wl}} \; , \tag{2}$$

where $d_0$ is the capacitor initial gap, $k$ is the mechanical spring, $\varepsilon_0 = 8.8546 \times 10^{-12}$ is the air permittivity, and $w$ and $l$ are the capacitor plate width and length, respectively.

Equations (1) and (2) show that the pull-in voltage is dependent on the spring elasticity constant and the capacitor dimensions, and therefore the pull-in voltage depends on the design and process characteristics only.

The previous analysis assumes quasi-static variations and dynamic properties such as mass and damping are neglected. If the displacement due to an external acceleration in the plane of the movable electrode is now considered, the asymmetric-right and asymmetric-left pull-in voltages become:

$$Vpr = \sqrt{\frac{8}{27} \frac{d_{right}{}^3 k}{\varepsilon_0 wl}} \; and \; Vpl = \sqrt{\frac{8}{27} \frac{d_{left}{}^3 k}{\varepsilon_0 wl}} \tag{3}$$

where $d_{right} = d_0 - ma_{ext}/k$ and $d_{left} = d_0 + ma_{ext}/k$ with $m$ being the mass of the movable electrode and $a_{ext}$ the external acceleration. Since the two asymmetric pull-in voltages depend on the external acceleration they can be used as a transduction mechanism to measure external accelerations. Moreover, the two pull-in voltages are easy to measure [11] and they both depend on the same mechanical structure which means that process variations will affect equally the pull-in voltages. This characteristic is used in this work to generate an autonomous inclinometer, i.e., a device that is auto-calibrated and with a zero temperature coefficient.

## 3    Inclinometer Working Principle

Inclinometers rely on the changes of the gravitational force component due to changes in the slope angle to detect different accelerations. A pull-in based inclinometer will be subject to different accelerations, according to (4), and consequently the asymmetric pull-in voltages will change.

$$a_{ext} = 9.8 \sin(\alpha) \; [m/s^2] \tag{4}$$

Since pull-in voltages depend on the external acceleration, and the external acceleration depends on the slope angle $\alpha$, pull-in voltages can be used as a measure of the slope angle. In this work, the difference between the two asymmetric pull-in voltages (5) is proposed as the transduction mechanism for an autonomous inclinometer.

$$\Delta V_{pi}(\alpha) = V_{pl}(\alpha) - V_{pr}(\alpha) \tag{5}$$

A block diagram of the proposed inclinometer is depicted in Fig. 2. The main block of the inclinometer is the microstructure with separated actuators electrodes, for the left and right side. This microstructure is actuated by a digital-to-analog converter (DAC) that generates a voltage ramp with a 30V/s slope for each side of the microstructure. The two electrodes are actuated sequentially and the electrostatic force displaces the mass generating a capacitance change on a set of sensing capacitors. A readout block based on a charge amplifier detects the capacitive changes and sends the values to a microcontroller. The microcontroller is responsible for the control of the entire system, i.e., it controls the DAC to generate the voltage ramp and measures the pull-in voltage based on the output values of the readout circuit. The moment when pull-in occurs is detected by a sudden change on the sensing capacitors. When the readout block detects the sudden large capacitive change, pull-in has occurred and the voltage value currently applied is saved. This procedure is repeated for the two sides and the difference between the two pull-in voltages is calculated. The result value is proportional to the acceleration applied to the microstructure. The acceleration is assumed to be static, and that is why this approach is of special interest for inclinometers. For normal accelerometer operation, this approach would result in a very low bandwidth device.

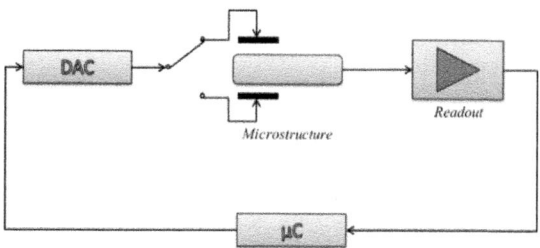

**Fig. 2.** Micro inclinometer block diagram

## 4    Fabricated Devices

The microstructures (Fig. 3) used to experimentally validate the system were fabricated using the SOIMUMPs micromachining process from MEMSCAP [11]. The structures include a symmetrical inertial mass, suspended on four bi-folded springs with parallel-plate capacitors having a 2.25 µm gap (at rest position). The main design parameters are presented in Table 1.

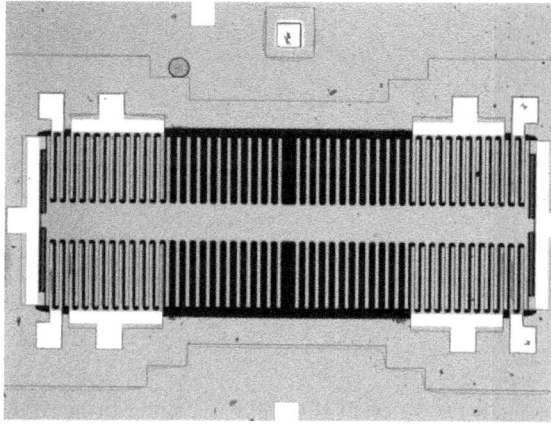

**Fig. 3.** Microscope picture of the fabricated MEMS device

**Table 1.** Device design parameters

| Device Parameters | Value |
|---|---|
| Mass (m) | 0.159 mg |
| Mechanical spring (k) | 4.4873 N/m |
| Zero-displacement gap ($d_0$) | 2.25 µm |
| Natural resonance frequency ($f_0$) | 846 Hz |
| Zero-displacement actuation capacitance | 0.198 pF |
| Damping coefficient (b) | 1.2 mNs/m |
| Zero-displacement sensing capacitance ($C_{d0}$) | 1.1 pF |
| Mechanical-thermal noise | 2.8µg/√Hz |
| Pull-in voltage ($V_{pr}$ /$V_{pl}$) | 5.589V |

## 5    Results

Operation of the device includes applying a voltage ramp and therefore, for a proper study of the system, the dynamic effects must be included. A full system model [12], that includes squeeze-film damping and inertial effects was implemented in Simulink and simulations using the full non-linear system were performed for different tilts. Next, the proposed approach was experimentally verified using the fabricated micro-structures.

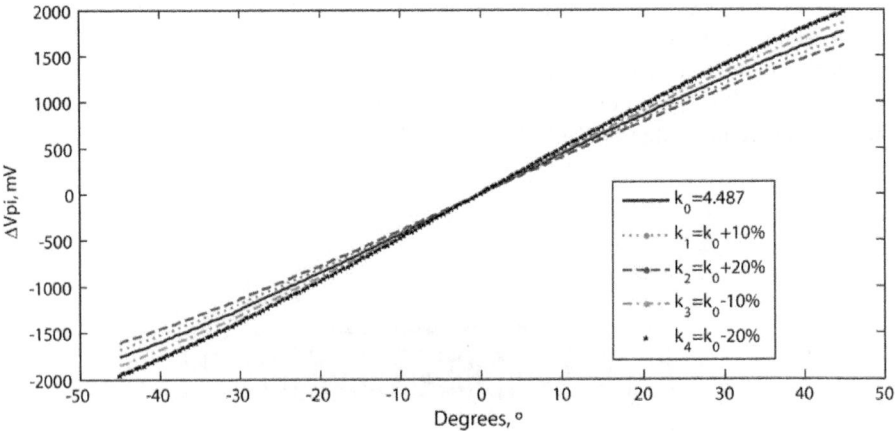

**Fig. 4.** ΔVpi Simulated for different spring constants at different inclinations

### 5.1    Simulations

Simulations were made for different slop angles ranging from -45 ° to 45 ° using steps of 0.5 °. A key characteristic of the proposed transduction method is that pull-in depends on the geometry and material properties only, as already mentioned. Therefore, and since capacitors dimensions are constant, only the mechanical spring is prone to changes due to temperature [13] that will consequently change the behavior of the system. In order to check spring variations on the sensor, simulations were performed for different $k$ (considering changes of ±10% and ±20%) and the resultant $\Delta V_{pi}$ is shown in Fig. 4.

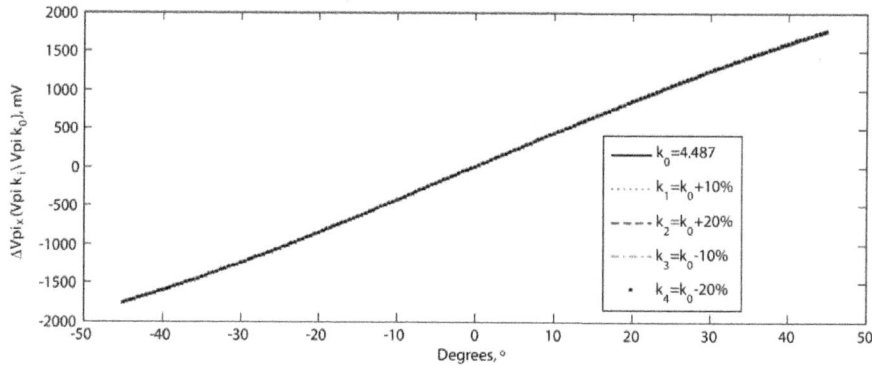

**Fig. 5.** $\Delta Vpi \; \frac{Vpi \; k_i}{Vpi \; k_0}$ Simulated responses for different spring constants at different inclinations

An analysis to Fig. 4 reveals that the sensitivity of the sensor varies for the different spring constants. When $k$ is increased, the $\Delta V_{pi}$ slope is smaller, while for smaller $k$, the $\Delta V_{pi}$ slope increases. Since changes on the spring are affecting the pull-in voltage values with no inclination ($V_{pi0}$) (see table 2), we can multiply the *"ratio"* between the pull-in voltage values with no inclination for each $k$ as shown in table 2 by the simulated $\Delta V_{pi}$ values (6). Fig. 5 shows the new slopes when equation 6 is used and the simulated results show that the sensor response can easily be compensated for changes of the mechanical spring. From Fig. 5 a sensitivity of 45.2mV/° is retrieved. Below -25° and over 25° the response starts to be nonlinear mainly due to the nonlinear characteristics of the pull-in phenomenon.

$$\Delta Vpi \; \frac{Vpi \; k_i}{Vpi \; k_0} \tag{6}$$

## 5.2    Experimental Results

After performing the simulations, experimental tests were made using the fabricated microstructures. A data acquisition board (DAQ) from National Instruments, USB-6251, was used to generate the voltage ramp and the inclination was calculated using a distance measuring device from *SICK* with a resolution of 1mm in order to achieve the most reliable results. The readout circuit output based on a charge amplifier was directly connected to one input of the data acquisition board. A program implemented in Matlab was used to measure the pull-in voltages. The measured $\Delta Vpi$ is depicted in Fig. 6 along with the simulated results. Measurement results show a sensitivity of 50mV/° resulting in a resolution of 0.006° (the DAQ's analog output has a resolution of 300µV).

The differences in the slope between simulated and measured values are due to the differences between real and simulated mechanical spring. MEMS fabrications processes are prone to over-etching resulting in mechanical springs that are less stiff (which explains the larger slope on the experimental results).

**Table 2.** Pull-in Voltages with no inclination

| $k$ | $V_{pi0}$ |
|---|---|
| $k_0$=4.487 N/m | 5.589 V |
| $k_1$= $k_0$+10% N/m | 5.847 V |
| $k_2$= $k_0$+20% N/m | 6.096 V |
| $k_3$= $k_0$-10% N/m | 5.319 V |
| $k_4$= $k_0$-20% N/m | 5.034 V |

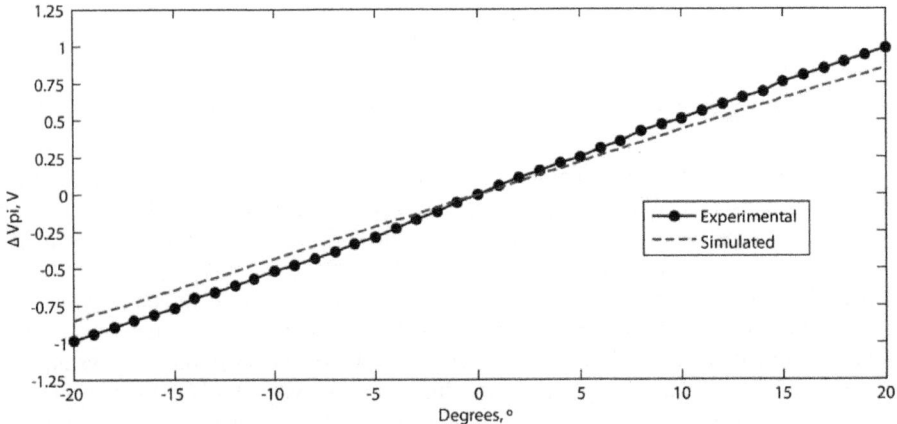

**Fig. 6.** Experimental results for different inclinations

## 6    Conclusions and Future Work

A novel approach for the realization of a high-resolution inclinometer was introduced in this paper. Pull-in operated microstructures have been presented and the proposed approach has the potential for realization of autonomous auto-calibrated inclinometers.

First results are very promising and current device sensitivity is at the same level as state-of-the-art inclinometers. Additional tests are needed to experimentally prove the temperature compensation mechanism and an improved ramp generation circuit (with resolution below 10μV) is being designed in order to improve the system's resolution. A thorough noise analysis study needs to be performed to understand the resolution limitation factors.

**Acknowledgements.** This work is supported by FEDER through COMPETE and national funds through FCT – Foundation for Science and Technology in the framework of the project PTDC/EEA-ELC/099834/2008.

# References

1. Lapadatu, D., Habibi, S., Reppen, B., Salomonsen, G., Kvisteroy, T.: Dual-Axes Capacitive Inclinometer/Low-g Accelerometer for Automotive Applications. In: Proc. 14th IEEE Int'l Conf. on MEMS, Interlaken, Switz., pp. 34–37 (2001)
2. Hoult, N.A., Fidler, P.R.A., Bennett, P.J., Middleton, C.R., Pottle, S., Duguid, K., Bessant, G., McKoy, R.S.: Large-scale WSN installation for pervasive monitoring of civil infrastructure in London. In: Structural Health Monitoring, Naples, Italy, pp. 214–219 (2010)
3. Crescini, D., Baú, M., Ferrari, V.: MEMS tilt sensor with improved resolution and low thermal drift. In: Sensors and Microsystems, pp. 225–228. AISEM (2009)
4. Yu, Y., Ou, J., Zhang, J., Zhang, C., Li, L.: Development of Wireless MEMS Inclination Sensor System for Swing Monitoring of Large-Scale Hook Structures. IEEE Transactions on industrial Electronics 56(4), 1072–1078 (2009)
5. Da-Wei, L., Tao, G.: Design of Dual-axis Inclinometer Based On MEMS Accelerometer. In: Third International Conference on Measuring Technology and Mechatronics Automation, Shangshai, pp. 959–961 (2011)
6. Qinglei, G., Huawei, L., Shifu, M., Jian, H.: Design of a Plane Inclinometer Based on MEMS Accelerometer. In: International Conference on Information Acquisition, Jeju City, Korea, pp. 320–323 (2007)
7. Lin, C.-H., Kuo, S.-M.: High-Performance inclinometer with wide-angle measurement capability without damping effect. In: MEMS, Kobe, Japan, pp. 585–588 (2007)
8. Kottapalli, V.A., Kiremidjian, A.S., Lynch, J.P.: Two-tiered wireless sensor network architecture for structural health monitoring. In: Proceedings of SPIE, vol. 5057, pp. 8–19 (2003)
9. Rocha, L.A., Cretu, E., Wolffenbuttel, R.F.: Analysis and analytical modeling of static pull-in with application to MEMS-based voltage reference and process monitoring. Journal of Microelectromechanical Systems 13(2), 342–354 (2004)
10. Rocha, L.A., Mol, L., Cretu, E., Wolffenbuttel, R.F., Machado da Silva, J.: A Pull-in Based Test Mechanism for Device Diagnostic and Process Characterization. VLSI Design, Article ID 283451, 7 pages (2008)
11. Cowen, A., Hames, G., Monk, D., Wilcenski, S., Hardy, B.: SOIMUMPS Design Handbook. rev. 6.0. MEMSCAP Inc. (2009)
12. Rocha, L.A., Cretu, E., Wolffenbuttel, R.F.: Using dynamic voltage drive in a parallel-plate electrostatic actuator for full-gap travel range and positioning. Journal of Microelectromechanical Systems 15(1), 69–83 (2006)
13. Rocha, L.A., Cretu, E., Wolffenbuttel, R.F.: Compensation of temperature effects on the pull-in voltage of microstructures. Sensors and Actuators A 115, 351–356 (2004)

# Vehicular Ad-hoc Networks(VANETs): Capabilities, Challenges in Information Gathering and Data Fusion

K. Golestan, A. Jundi, L. Nassar, F. Sattar,
F. Karray, M. Kamel, and S. Boumaiza

Department of Electrical and Computer Engineering, University of Waterloo,
Waterloo, ON N2L 3G1, Canada

**Abstract.** Vehicular Ad-hoc Network (VANET) has become an active area of research due to its major role to improve vehicle and road safety, traffic efficiency, and convenience as well as comfort to both drivers and passengers. This paper thus addresses some of the attributes and challenging issues related to Vehicular Ad-hoc Networks (VANETs). A lot of VANET research work have focused on specific areas including routing, broadcasting, Quality of Service (QoS), and security. In this paper, a detailed overview of the current information gathering and data fusion capabilities and challenges in the context of VANET is presented. In addition, an overall VANET framework, an illustrative VANET scenario are provided in order to enhance safety, flow, and efficiency of the transportation system.

**Keywords:** VANETs, Cognitive Information Gathering, Multi-sensor Data Fusion.

## 1 Introduction

Vehicular Ad-hoc Networks (VANETs) have attracted attention in the support of safe driving, intelligent navigation, and emergency and entertainment applications [1–3]. VANET can be viewed as an intelligent component of the Transportation Systems as vehicles communicate with each other as well as with roadside base stations located at critical points of the road, such as intersections or construction sites. The basis of VANET is established on the underlying information that is made available for the 'traveler' to consequently guide the whole VANET system towards the objectives of VANET which aim for Safety, Mobility and Environment issues [4]. In other words, while VANET helps the traveler to have a safe and secure driving experience, by providing them useful mobile information, it implicitly preserves the green environment, e.g. by regularizing fuel consumption rates. A comprehensive well-organized VANET is responsible for extracting, managing and interpreting the information to achieve knowledge, and making it available for travelers.

In most of the research papers on context-aware application, motivating scenario play an important role [5–7]. A scenario is a sequence of events that can

M. Kamel, F. Karray, and H. Hagras (Eds.): AIS 2012, LNCS 7326, pp. 34–41, 2012.

be represented by a directed graph. Scenario description is an acyclic directed graph $G = (V, E)$, where $V$ is the set of events in scenario and $E$ is the set of edges. Each edge in $G$ is an ordered pair $(v_i, v_j)$, where $v_i, v_j \in V$, $v_i \neq v_j$, event $v_i$ is directly followed by event $v_j$. One of the important issue here is to determine the context. Some context, e.g. the context 'raining', can be directly acquired from physical sensors. However, for inferred context, we need to determine the reasoning path and corresponding knowledge. For example, the context 'party' in one room can not be directly sensed. But if the sound, the light and the number of people in the room are sensed by the system, whether there is a party in the room can be deduced. So the sensed contexts that sound, light and number of the people greater than predefined value can determine the inferred context 'party'. The corresponding knowledge can be expressed by a rule: (sound $> \alpha$) and (light=on) and (number-of-people $> \beta$) $\rightarrow$ party, where $\alpha$, $\beta$ are predefined values. In broader sense, a scenario is simply a proposed specific use of the system [5].

The rest of the paper is organized as follows. In section 2, outline of the proposed VANET scheme is shown. An overall VANET framework is presented in Section 3. Section 4 describes in detail the recent advancements as well as the challenges of cognitive information gathering in the context of VANET. Section 5 concludes the paper.

## 2   Overview of the VANET Scheme

A proposed VANET scheme is depicted in Fig. 1(a) which consists of three main components, i) Information gathering and multisensor data fusion, ii) Context aware processing, iii) Communication gateway. The task of the cognitive information gathering scheme is to control the data gathering process and fuses the gathered data in a smart and efficient way to avoid excessive storage and computational requirements. The goal of the context-aware information processing is to disseminate the right information to the appropriate place on right time. The role of the communication gateway is to provide reliable communication link between intelligent vehicles (Vehicle-to-Vehicle(V2V)), Vehicle-to-Roadside Unit(VRU) and Vehicle-to-Infrastructure(V2I). V2V can be used in transferring emergency messages, enhancing GPS accuracy and exchanging related information. VRU can be used in toll collection, transferring localized important information regarding the local terrain and traffic conditions and provide access to some transportation system related services. V2I can be used to connect the vehicle to a collective data bank managed by the service provider of traffic services and it can be used as an internet access medium to provide entertainment services.

## 3   Proposed VANET Framework

An overall VANET framework can be defined by several components coupled together at different levels. Our proposed VANET framework consists of a

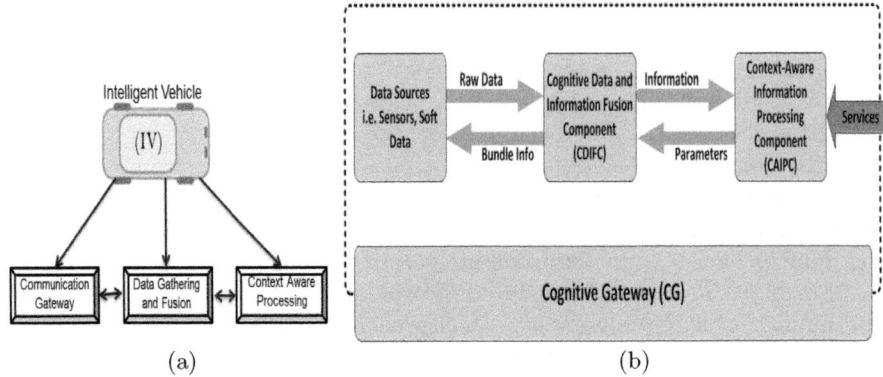

**Fig. 1.** (a) An overview of the proposed VANET scheme,(b) An overall proposed VANET framework

cognitive gateway for hardware, cognitive data and information fusion component, and a context-aware information processing unit. Fig. 1(b) presents the proposed VANET framework showing the links between various components. As depicted in Fig. 1(b), cognitive data and information fusion component(CDIFC) lies in between the two other main components, and therefore plays an important role by connecting the low-level and high-level components of the system.

CDIFC is triggered when a service is asked by an agent, i.e. a driver or the vehicle itself. This service could be of various types such as Safe Driving Service which might be required by a driver, or even an Environmental Green Service that could be requested by the vehicle itself. After finding the required information/context parameters by the Context-Aware Information Processing Component(CAIPC), the CDIFC chooses a bundle of resources based on its various aspects such as available resources, communication bandwidth traffic, degree of emergency, required accuracy etc and gathers the data from lower level, fuses them to exact the meaningful information about the demanded service, and then finally sends the information to preceding level (CAIPC).

Considering the types of services and the selection mechanism of a bundle of resources, as well as the information flow from upper level to lower level and then again to upper level (see Fig. 1(b)), we may face some problems such as the cases when services are asking for the same bundle of resources, or at least bundles with overlapping resources. The other problem could be the incorporation of soft data into the framework, to be used as a potential source of information. Soft data, is the data produced by humans that can be extracted from a wide range of resources such as capturing devices, internet, etc. Such a high level and continuous data needs to be interpreted first by our system to be applicable to the framework. However, some ad-hoc solutions can be adopted such as defining a hierarchy of services and propagating them into the system. Game theory can also be employed when dealing with competition-like requests for acquiring sensor bundles, and solving problems such as defining allocation strategies [8, 9].

The following sections present the cognitive information gathering part, or CDIFC in VANET.

# 4   Cognitive Information Gathering

Cognitive information gathering is essential to provide the required selectivity in VANET. It defines a dynamic process to adaptively gather and process information of interest from the environment [10]. The main obstacle on this path is that the information is amongst a huge amount of data being streamed from various sources in the VANET environment, and dealing with such a huge volume of data is time-consuming and nearly impossible. Furthermore, these data have different meanings and are not necessarily perfect, therefore it is even more challenging to extract information from imperfect data [11]. In this phase, the problem can be defined as a *data fusion* task in which data (in any form) from different sources are processed and fused to make the information meaningful for a particular task. For example, for assisting a traveler with routing service, the data produced by traffic cameras, congestion sensors, and weather data can be fused to provide an optimal route to the traveler in terms of traffic and weather conditions.

Here, we introduce some of the challenges of information gathering in VANET and further propose some of our findings. In the following, let us begin with proposing an illustrative VANET scenario for more clarification.

## 4.1   Proposed VANET Scenario

Here, we propose an illustrative VANET scenario which helps us to effectively describe the information gathering part. Let us assume that there is an intelligent transportation system that is set up in a highway, called VANET Zone, covering a certain area, and aims to help the users in solving their tasks by providing useful information for them. The corresponding data is collected from different types of sensors available in the environment. For example, the data may come from typical on board sensors installed on a vehicle, such as radar, speedometer, odometer etc as well as data being produced by the transportation infrastructures, such as traffic management or weather organizations, or it can be even in the form of a high level data, such as the real-time information produced by a traveler who uses social networks like Facebook and Twitter. For instance, imagine that the information regarding an accident in a highway may be extracted by exploring different tweets. On the other hand, regarding the entertainment application of VANETs, a service can access the information in the Facebook account of the driver to tune the radio to his/her favorite channel.

Considering the fact that the sensors on each vehicle in the VANET zone can also serve as a data source (the idea behind ubiquitous computing) [16], and also by taking into account the data flow from infrastructures and also the travelers, we are dealing with an extremely large number of sensors of different types, formats, accuracy etc, from which we need to choose a set of sensors providing the best possible information for a specific task requested by a traveler.

Imagine that there are a number of services defined and the travelers (or their vehicles autonomously) can ask for a specific one. For instance, in the case of safe driving task, traveler will be advised to maintain a certain speed or lane in order to avoid the risk of collision and also help to traffic flow. For this specific task, this information can be elicited by using a set of data such as from radar and speedometers installed on the vehicles in the vicinity and the data from the traffic management infrastructure. The elicited information can be further sent to the traveler asking for it, and then he/she can follow it using a monitor on the dashboard of his/her vehicle.

Let us assume that besides the current flow of data from different sensors and infrastructures, the data generated by other travelers can be also taken into account. As an example, imagine that a vehicle is driving towards an accident at a specific intersection without a traffic camera (loss of data), and is asking for routing service. Since there is no data showing the accident and probably the congestion caused by it, the system can hopefully take advantage of the data being produced by other travelers present at the accident scene. These data can be acquired by following some specific trends in popular social networks, such as Twitter.

As an illustration, in Fig. 2(a), four vehicles are shown which are driven in a highway. Vehicles 1 and 4 are using the safe driving service and routing service, respectively. Doing so, they are using the data provided by the data sources around them. For instance, Vehicle 1 is using his distance sensors and those of Vehicle 2, to get a more accurate measurement of the distance between them. On the other hand, and in a shared sensor environment, Vehicle 1 might also use the lane deviation data which is measured by some sensors installed on Vehicle 2. The case of using information provided by infrastructures and also including the soft data being created by humans are also depicted in the figure (for Vehicle 4). Finally, music data is another stream of data which is running between Vehicle 1 and Vehicle 4 that as a result imposes too much load of data on both of them.

Here, we introduced a simple VANET scenario and the way the data flow can be used for information elicitation. In this example, new ideas such as ubiquitous computing and soft/hard data fusion are also brought into consideration. The application of different types of data and the way they influence the performance of VANET are presented in Fig. 2(b). Here, at the lower level of the pyramid, we have raw data which often have a huge volume, but provide the least application for VANET. Gradually, as we process the data, and then fuse them to produce information, the volume decreases and applicability increases. On the other hand, deploying soft data at this stage would improve the usefulness of data by including human's high level information. Finally, by fusing the information and gaining knowledge, the more advanced information fusion techniques such as reasoning and prediction can also play a role. In the following, we will discuss the challenges regarding data fusion, as well as briefly investigate the application of ubiquitous computing and soft/hard data fusion.

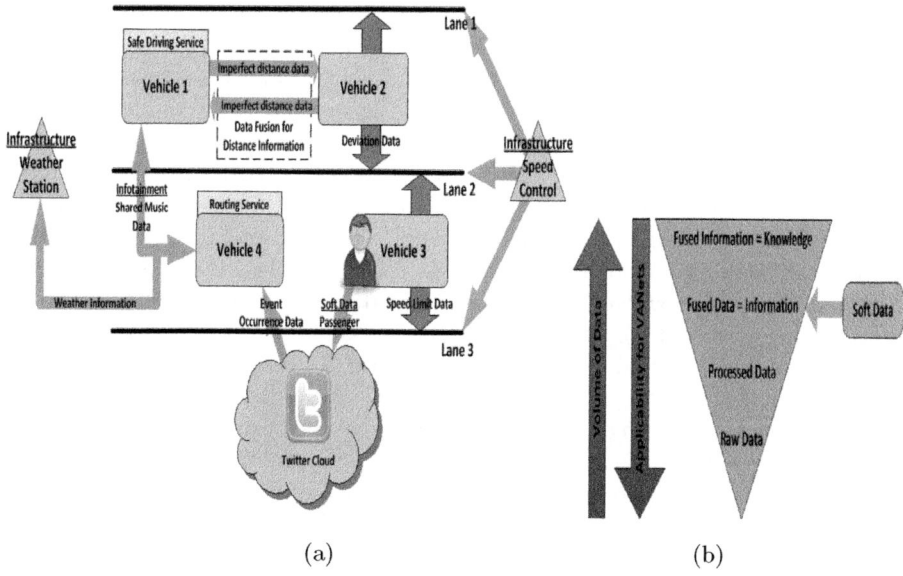

(a)                                         (b)

**Fig. 2.** (a) Proposed VANET scenario,(b) Hierarchy of the corresponding data gathering and fusion process

## 4.2   Multi-sensor Data Fusion and Its Application to VANET

The idea of multi-sensor data fusion is thoroughly studied in [11] where various challenging issues of multi-sensor data fusion and some of the methods to deal with them are discussed. Here we address the data fusion part of VANET and try to match the related problems with the categorization defined in [11]. Various aspects of data fusion can be categorized into three major groups: Imperfection, Correlation, and Inconsistency [12].

Imperfection refers the uncertainty, impreciseness and granularity of the incoming data. Briefly, data is called to be uncertain if the confidence level of the correctness of data falls below 100%, and this can be caused by noises on measurements, errors in readings, etc. In the VANET context, one can imagine certain circumstances in which the data is attributed to be imperfect. For example, a noisy sensor might produce uncertain data, weather station may report ambiguous data about the amount of rain, a traffic camera with blurred lens may provide vague data, and finally we might deal with incomplete data when facing disconnections in data transmission.

In the correlated case, the sensors are exposed to some external noises being made by each other which have consequently biased their measurements. In such circumstances, the data from all of these sensors are correlated and not dealing with it may result in errors regarding data confidence. Referring to VANET, for example the speed of the vehicle can be measured by a speedometer, a GPS device, or even a traffic camera. All of these data sources produce the same information about vehicle speed, but in different modalities, i.e. a speedometer, GPS device and traffic camera show the vehicle's speed by counting wheel rotations, measuring displacement based on the longitude and latitude of the vehicle, and assessing displacement based on imaging in different time slots, respectively.

According to [12], inconsistent data can be observed to have conflicts, outliers, or disorder. When we say data from two different sources have conflict with one another, it means that the information is contradicting each other. For example, in VANET environment and at the traffic intersection, a congestion meter may produce data claiming heavy traffic, whereas a traffic camera reports something completely opposite, and notifies a light traffic. On the other hand, the data could be the outlier when the information produced by them is irrelevant to the event being observed. For instance, a radar sensor may produce data showing a car behind, while actually there is no car behind. Moreover, data disorder could happen when the data produced by the source have different rates and fusing them may produce wrong information. Therefore, such problem which can easily be occured in VANET, is another challenge to solve. Making use of all these data and constructing a coherent global view of the current event is another challenge in data fusion, which can be obviously seen in VANET area.

The arising ideas of soft/hard data fusion [13–15] and ubiquitous computing [16], which lead the idea of shared sensors, coupled with more advanced method of data fusion, such as distributed data fusion, can be considered as an extension to the definitions of information gathering part of VANET. Referring to our VANET scenario, ubiquitous computing can play an important role by defining a cloud of data source from which the vehicles or travelers can take advantage and employ to accomplish a task by collecting useful information. On the other hand, as mentioned previously, high level data flow from the traveler's hand held devices such as laptop or cell phone or social network like Twitter, can be fused with some low level data, which is originated from physical sensors, such as traffic camera, in order to generate rich or meaningful information and make it available for the travelers or other users. Finally, splitting the data fusion to the distributed systems, rather than doing it in a centralized way, is one of the major task that is known as *distributed data fusion* [17], and in-parallel increasing the robustness of the system, relaxes the computational load, and thereby increases the processing speed making the system available for real-time tasks. However, dealing with distributed systems and solving the data fusion problem in a distributed way is a difficult task.

## 5   Conclusion

The cognitive information gathering and data fusion play a significant role for a comprehensive well-organized VANET which is responsible for extracting, managing and interpreting the information to achieve knowledge, and making it available for travelers. Therefore, in this paper we have discussed the important role of information gathering in vehicular ad-hoc network (VANET), some of the challenging issues arising from multi-sensor data fusion and information elicitation in VANET. We have further introduced the overall VANET scheme and its framework, an illustrative VANET scenario as well as some ideas of data fusion in the context of VANET based on soft data, ubiquitous computing and distributed data fusion.

# References

1. Xu, Q., Mak, T., Ko, J., Sengupta, R.: Vehicle-to-vehicle safety messaging in DSRC. In: Proc. of ACM Workshop on Vehicular Ad Hoc Networks, VANETs (2004)
2. Fazio, M., Palazzi, C.E., Das, S., Gerla, M.: Facilitating real-time applications in VANETs through fast address auto-configuration. In: Proc. of the IEEE CCNC Int. Workshop on Networking Issues in Multimedia Entertainment (2007)
3. Fuler, H., Fuler, H., Hartenstein, H., Mauve, M., Käsemann, M., Vollmer, D., Mauve, M., Asemann, M.K.: Location-Based Routing for Vehicular Ad-Hoc Networks (2002)
4. Boukerche, A., Oliveira, H.A.B.F., Nakamura, E.F., Loureiro, A.A.F.: Vehicular Ad Hoc networks: A new challenge for localization based systems. Comput. Commun. (2008)
5. Carroll, J.M.: Five reasons for scenario-based design. Interacting with Computers 13(1), 43–60 (2000)
6. Dssouli, R., Some, S., Vaucher, J., Salah, A.: A service creation environment based on scenarios. Information and Software Technology 41(11), 697–713 (1999)
7. Kim, J., Kim, M., Park, S.: Goal and scenario based domain requirements analysis environment. Journal of Systems and Software 79(7), 926–938 (2006)
8. Mullen, T., Hall, D., Pennsylva, T.: A market-based approach to sensor management. Information Fusion 4(1) (2009)
9. Brynielsson, J.: An Information Fusion Game Component. Information Fusion 1(2) (2006)
10. Wen, D., Yan, G., Zheng, N.-N., Shen, L.-C., Li, L.: Towards cognitive vehicles. IEEE Intelligent Systems (2011)
11. Khaleghi, B., Khamis, A., Karray, F.O.: Multisensor data fusion: A review of the state-of-the-art. Information Fusion (2011) (in press)
12. ITS Strategic plan,
    http://www.its.dot.gov/strategic_plan2010_2014/index.htm
13. Sambhoos, K., Llinas, J., Little, E.: Graphical methods for real-time fusion and estimation with soft message data. In: Proc. Int. Conf. on Information Fusion, pp. 1–8 (2008)
14. Pravia, M.A., Babko-Malaya, O., Schneider, M.K., White, J.V., Chong, C.Y.: Lessons learned in the creation of a data set for hard/soft information fusion. In: Proc. Int. Conf. on Information Fusion, pp. 2114–2121 (2009)
15. Premaratne, K., Murthi, M.N., Zhang, J., Scheutz, M., Bauer, P.H.: A Dempster-Shafer theoretic conditional approach to evidence updating for fusion of hard and soft data. In: Proc. Int. Conf. on Information Fusion, pp. 2122–2129 (2009)
16. Challa, S., Gulrez, T., Chaczko, Z., Paranesha, T.N.: Opportunistic information fusion: A new paradigm for next generation networked sensing systems. In: Proc. Int. Conf. on Information Fusion, pp. 720–727 (2005)
17. Memorandum, T.: Distributed Information Fusion and Dynamic (2008)

# Vehicular Ad-hoc Networks(VANETs): Capabilities, Challenges in Context-Aware Processing and Communication Gateway

L. Nassar, A. Jundi, K. Golestan, F. Sattar, F. Karray,
M. Kamel, and S. Boumaiza

Department of Electrical and Computer Engineering, University of Waterloo,
Waterloo, ON N2L 3G1, Canada

**Abstract.** Vehicular Ad-hoc Networks (VANETs) have attracted attention in the support of safe driving, intelligent navigation, and emergency
and entertainment applications. VANET can be viewed as an intelligent
component of the Transportation Systems as vehicles communicate with
each other as well as with roadside base stations located at critical points
of the road, such as intersections or construction sites. In this paper, we
provide an overview of the context-aware processing and communication
gateway associated with Vehicular Ad-hoc Network (VANET). The concept of context-awareness, the recent advances and various challenges involved in context-aware processing are discussed. Some arising ideas such
as based on context ontology, relevancy, hybrid dissemination, service oriented routing are also presented. This paper further briefly describes the
communication gateway in VANET which includes its functional view
together with the standards and their detailed preliminary specifications
applicable to VANET.

**Keywords:** VANETs, Context-Awareness, Contextual Information Processing, Communication Gateway.

## 1 Introduction

Vehicular Ad-hoc Network (VANET) has become an active area of research, stan-
dardization, and development because it has tremendous potential to improve
vehicle and road safety, traffic efficiency, and convenience as well as comfort to
both drivers and passengers. Therefore, recent research efforts have placed a lot
of emphasis on VANET context-aware processing as well as VANET communication gateway.

The goal of the context-aware processing is to disseminate the right information to the appropriate place on right time. Context-awareness has been a popular research area for a number of years [1] defines context as "any information
that can be used to characterize the situation of an entity. An entity is a person,
place, or target that is considered relevant to the interaction between a user
and an application, including the user and applications themselves". Context
can be target's identity, location, posture, past or current activities, or intent.

M. Kamel, F. Karray, and H. Hagras (Eds.): AIS 2012, LNCS 7326, pp. 42–49, 2012.

The secondary context-types could be for example birth date, email address, nearby people, forecasted weather. Context-aware mobility is the ability to dynamically capture and use the surrounding contextual information of a mobile entity to improve the performance of the system. Context-awareness is therefore a key factor in any ubiquitous system and provides intelligence to the system, allowing computing devices to make appropriate and timely decisions on behalf of users. One of the most important context information in context-aware systems is location. The majority of early context-aware systems utilized location-awareness as the exclusive type of sensed context for application adaptability. Increased mobility of users in ubiquitous computing environments made knowledge of the user location an essential element in adapting to the users needs. Location is divided into two categories: Comparative and Non-Comparative locations. The former used to describe object positions in relation to other objects. The latter described exact location of objects in terms of GPS coordinates, addresses, and other forms irrelevant of the location of nearby objects.

On the other hand, the communication gateway facilitates the interfacing between the intelligent vehicles, road-side units and current infrastructure linking to a collective data bank to manage traffic and provide services. The role of the communication gateway is thus to provide reliable communication link between intelligent vehicles (Vehicle-to-Vehicle(V2V)), Vehicle-to-Roadside Unit(VRU) and Vehicle-to-Infrastructure(V2I). V2V can be used in transferring emergency messages, enhancing GPS accuracy and exchanging related information. VRU can be used in toll collection, transferring localized important information regarding the local terrain and traffic conditions and provide access to some transportation system related services. V2I can be used to connect the vehicle to a collective data bank managed by the service provider of traffic services and it can be used as an internet access medium to provide entertainment services.

The rest of the paper is outlined as follows. In the next section, we give an overview of the context-aware processing as well as some challenges in the context of VANET. We then address briefly the communication gateway in VANET. Finally, the conclusions are drawn in the last section.

## 2  Context-Aware Processing

There are challenges currently being experienced in knowledge querying and information dissemination in VANET. The first is that important messages are not delivered on time to interested nodes due to network overload. The second is the fact that the current network addressing schemes which are used to define groups for information dissemination, like multicasting and broadcasting, do not consider the contextual characteristics of the nodes for optimizing the network traffic [2]. To handle these challenges context modeling and processing accompanied by smart dissemination mechanisms should assure efficient use of network resources and on time delivery of relevant services to interested nodes. Consequently, context aware systems in Ad-hoc Networks like VANET should

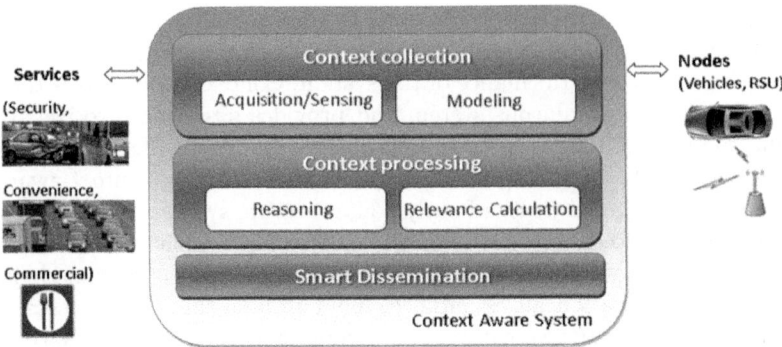

**Fig. 1.** VANET context aware processing

have a middleware responsible for modeling the context, processing it and disseminating information using routing protocols that adapt dynamically to the current context, see Fig. 1.

## 2.1   Context Collection and Modeling

According to [3], context may have four categories: 1) computing context, which refers to the system capabilities such as communication bandwidth and available resources, 2) physical context which is accessible to the nodes using sensors such as location, 3) time context and 4) user context such as user profiles and preferences. Context can also be classified according to its continuous change overtime to static and dynamic context [4]. Static context is acquired directly from users or services and stored in a central repository, while dynamic context can be accessed by sensors and it is processed right away. A third classification is the low level context, i.e. the data from HW/SW sensors, and high level context which is deduced from reasoning over low level context [4]. In order to process all above mentioned context, a context model is needed. The model should define and store context data in a machine processable form. Ontology based context models are usually preferred since they allow knowledge sharing, reuse, validation and aggregation. Usually static context is saved in the ontology while dynamic context is handled programmatically [11].

## 2.2   Context Processing

The rule based reasoning over the ontology facilitates the inference of high level implicit context from low level explicit context [4]. This is one example of context processing while relevance calculation is another crucial stage in context processing. In Ad-hoc Networks both the context of the service provider as well as that of the interested nodes should be considered for matching nodes with relevant services. For example, in VANETs services provided are generally classified as security services (e.g. post crash notification), convenience/efficiency services (e.g. congested road notification) or commercial services (e.g. shopping) [5–7]. The nodes interested in these services can either be vehicles or roadside unites (RSU). To achieve efficient use of available resources, services should only be delivered to nodes when they are highly relevant to the nodes' context. This relevance can be divided into two types, binary relevance as in [8] or partial relevance as in [9]. In case of binary relevance the service context can either be relevant or irrelevant to the node context and as a result a node can either be a member or a non-member of the service contextual virtual group. On the other hand, the partial relevance discussed in [9] is based on using the vector space model and each context parameter can be represented as a vector. The partial relevance provide a degree of relevance measured by calculating the weighted distance between the optimal relevance vector and the node's context vector.

## 2.3   Smart Dissemination

Finally, the communication/dissemination mechanisms used to deliver the relevant services to interested nodes should dynamically adapt to the services provided in addition to their relevance to nodes' context and available network resources at runtime in order to maximize network utilization. The first factor that helps in choosing an appropriate communication as well as routing mechanism for the current situation is the type of provided services. VANET can be seen as a special case of MANET especially when it utilizes V2I communication. Nevertheless, VANET is differentiated by its higher processing power and its high dynamic nature [10]. These special features of VANET necessitate that the V2I communication be complemented by the decentralized V2V communication [11]. However, a criteria is still needed to decide when is it best to use each of these communication methods. Therefore, the type of services provided can be seen as one factor in deciding whether V2V or V2I communication is more suitable to use. One view of how the type of provided services can affect the chosen communication method was presented in [12]; V2V communication was recommended for safety services such as vehicle breakdown warning while V2I was chosen to serve as a better communication for commercial services like locating attractions for tourists. On the other hand, the work done in [5] showed how the type of service offered can also be used to infer the most appropriate dissemination and routing mechanism. As a demonstration of how the routing method should adapt to the type of services provided, broadcast/GeoCast was recommended for safety services, beaconing for convenience services and unicast communication for commercial services.

The other factor that can affect the chosen addressing mechanisms as well as dissemination and transmission techniques is the relevance of the provided services to the node context and interest. In fact, the specific identities of the service receivers do not concern most of VANET applications; they are more concerned about the characteristics of those intended receivers [13]. Therefore a shift in addressing mechanisms from node addressing to message addressing, where message content is considered in the dissemination process, was found to be more suitable for VANET [10]. This allows disseminating the service based on its relevance to the node's contextual parameters. One example is the use of content based addressing model where the intended receiver set of nodes can be defined by parameters included in the content of the data packet, for instance speed and vehicle types. The parameters should be chosen based on their usefulness for the services provided by the upper application layer [13]. On the other hand, a move towards conditional transmission was recommended in [14], where the receiving nodes dynamically evaluate application dependent conditions, such as the distance from the sender or time, to decide whether to retransmit the message or not. Work in [13] shows that the decision to rebroadcast is not only affected by the receiver's context but also by a recorded interest table that holds the interest of neighboring vehicles. Finally in [11], the proposed V2I communication architecture utilized the inference techniques to adapt information disseminated to vehicles according to user point of interest so that only interested vehicles can receive relevant services to user preferences using inference rules.

## 3    Communication Gateway

We require a communication interface to link between intelligent vehicles, roadside nodes and current infrastructure linking to a collective data bank to manage traffic and provide services such as safety, data and wireless connectivity as

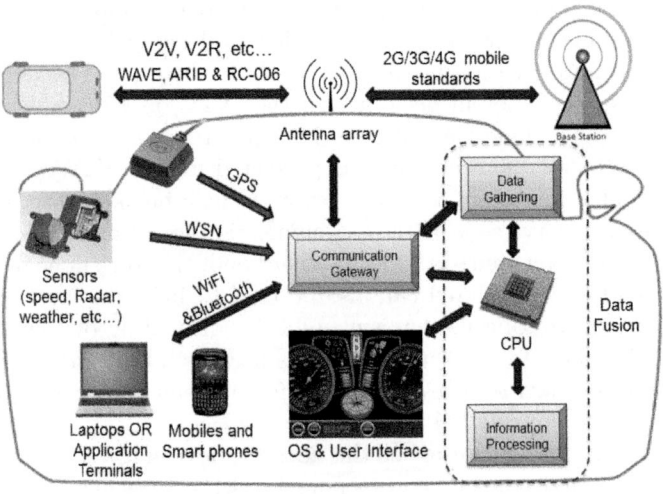

**Fig. 2.** Proposed functionality diagram

illustrated by Fig. 2. The objectives here are 1) providing reliable connectivity between nodes(vehicles), 2) securing high data rate connectivity upon request 3) being adaptive and future proof in terms of design, 4) providing global hardware with (software adaptation) [15–20]. In order to determine the required specifications for the gateway, we need to define the set of standards to include in our design. They are mainly three categories based on the type of communication which includes Direct Short Range Communication (DSRC), Mobile i.e. Wide Area Network (WAN), WiFi based on IEEE 802.11 standard (WLAN) [21–24].

Fig 3 demonstrates a challenging set of specifications to meet with a single transceiver frontend. Our recommendation is to limit the specifications to reduce the complexity of the design of the VANET communication gateway. This decision does not necessarily give negative impact to the user. For instance, in North America the vehicle would not operate with frequencies below 700MHz since all

| | DSRC | | | WAN | | | WLAN | | | |
|---|---|---|---|---|---|---|---|---|---|---|
| Standards | Wave | ARIB-T75 | RC-006 | GSM | UMTS | LTE | 802.11 b | 802.11 g | 802.11 n | 802.11 ac |
| Freq. Bands (MHz) | 5850-5925 | 5770-5850 | 715-725 | 380-500, 698-960, 1710-1991 | 410-500, 716-960, 1710-2690 | 698-960, 1427-2200, 2500-2690 | 2400–2485 | 2400–2485 | 2400-2485, 4910-5835 | 5170-5835 |
| Channel Bandwidth | 10MHz, 20MHz | 5MHz | 10MHz | 200KHz | 5MHz-10MHz | 5-20MHz | 22MHz | 22MHz | 20MHz, 40MHz | 20MHz, 40MHz, 80MHz, 160MHz |
| Maximum Avg EIRP | 33dBm | 14.8dBm | 20dBm | 33dBm | 33dBm | 23dBm | 20dBm | 17dBm | 30dBm | 30dBm |
| Max PAPR | ~12dB | ~6dB | ~12dB | ~0 | ~3.5dB | ~8dB | ~2.5dB | 10dB< | 10dB< | 10dB< |
| EVM/ACPR | -25dB/- | -/30dB @2.5MHz | -25dB/- | -/-60dB @400KHz | -/54dB @400KHz | -22dB/- | -/-30dB @20MHz | -25dB/- | -25dB/- | -25dB/- |
| Modulation | OFDM | ASK, QPSK | OFDM | GMSK | QPSK-64QAM | OFDM (DL), SC-FDMA(UL) | DSSS (DBSK, DQPSK) | OFDM | OFDM | OFDM |
| Multiple Access | CSMA | CSMA/ TS–CSMA | CSMA | TDMA | CDMA | SC-FDMA | CSMA | CSMA | CSMA | CSMA |

DSRC: Dedicated Short Range Communication
ARIB: Association of Radio Industries and Business
WAVE: Wireless Access in Vehicular Environments
EIRP: Effective Isotropic Radiated Power
PAPR: Peak-to-Average Power Ratio
ASK: Amplitude Shift Keying
PSK: Phase Shift Keying
OFDM: Orthogonal Frequency Division Multiplexing
SC-FDMA: Single-Carrier Frequency Division Multiple Access
DSSS: Direct-Sequence Spread Spectrum
GMSK: Gaussian Minimum Shift Keying
QPSK: Quadrature Phase Shift Keying
GSM: Global System for Mobile Communications
UMTS: Universal Mobile Telecommunications System
LTE: Long Term Evolution

**Fig. 3.** Specifications suggested for VANET communication gateway

the services needed are assigned at higher frequencies. Therefore, a vehicle that is operating in North America does not need to be equipped for European or Japanese standards. Such recommendation for geographical based design could downscale the scope and reduce the complicacy of the design with double edged outcome in the long run. On the other hand, due to the rapid development of wireless communication, it becomes important to leave room for adaptability which directly conflicts with limiting the specification's set. We therefore need to decide which standards are to follow and their operating zone while keeping in mind the future proof of the design. This decision will shed light on some unknowns which will be used to define our set of specifications. These decisions pose an immediate challenge that requires a firm stand early on in the design stage as it helps in deciding the skeleton of the VANET gateway, for example the number of frontend transceivers that are needed.

## 4    Conclusion

In this paper, we have explored the context aware processing in VANET in which middleware is the main data repository in a vehicle. It provides the dynamic representation of the vehicle environment and maintains the fused sensor data, information exchanged with other vehicles. Different stages in context-aware processing are described in detail covering the important concepts of context ontology, smart dissemination, service oriented routing, etc. Further the standards and their detailed preliminary set of specifications for the communication gateway are presented with the aim for geographical based design in order to downscale the scope and reduce the complications in the VANET gateway design despite some limitations for long term perspective.

## References

1. Dey, A.K., Abowd, G.D.: Towards a better understanding of context and context-awareness. GVU Technical Report GIT-GVU-99-22, College of Computing, Georgia Institute of Technology (1999)
2. Yasar, A., Preuveneers, D., Berbers, Y.: Adaptive context mediation in dynamic and large scale vehicular networks using relevance backpropagation. In: Proc. Int. Conf. on Mobile Technology, Applications, and Systems, Taiwan, pp. 1–8 (2008)
3. Chen, G., Kotz, D.: A Survey of context-aware mobile computing research. Technical Report TR200-381. Dept. of Computer Science, Dartmouth College (2000)
4. Yilmaz, O., CenkErdur, R.: iConAwa - An intelligent context-aware system. Elsevier (2011)
5. Bako, B., Weber, M.: Efficient Information Dissemination in VANETs, Advances in Vehicular Networking Technologies. In: Almeida, M. (ed.). InTech, http://www.intechopen.com/articles/show/title/efficient-information-dissemination-in-vanets

6. Karagiannis, G., Heijenk, G., Altintas, O., Ekici, E., Jarupan, B., Lin, K., Weil, T.: Vehicular networking: A survey and tutorial on requirements, architectures, challenges, standards and solutions, University of Twente, Enschede, The Netherlands, TOYOTA InfoTechnology Center, Tokyo, Japan, Ohio State University, Columbus, OH, USA, Booz Allen Hamilton, McLean, VA, USA, Raytheon Polar Services, Centennial, Colorado, USA

7. Hartenstein, H., Laberteaux, K.: VANET Vehicular Applications and Inter-Networking Technologies. John Wiley & Sons (2010)

8. Yasar, A.U.H., Vanrompay, Y., Preuveneers, D., Berbers, Y.: Optimizing information dissemination in large scale mobile peer-to-peer networks using context-based grouping. In: IEEE Intelligent Transportation Systems Conf. (ITSC), pp. 1065–1071 (2010)

9. Preuveneers, D., Berbers, Y.: Architectural backpropagation support for managing ambiguous context in smart environments. In: Stephanidis, C. (ed.) UAHCI 2007 (Part II). LNCS, vol. 4555, pp. 178–187. Springer, Heidelberg (2007)

10. Kukliski, S., Wolny, G.: CARAVAN: A Context-AwaRe Architecture for VANET, Mobile Ad-Hoc Networks: Applications. In: Wang, X. (ed.). InTech, http://www.intechopen.com/articles/show/title/caravan-a-context-aware-architecture-for-vanet

11. Santa, J., Gómez-Skarmeta, A.F.: Sharing context-aware road and safety information. IEEE Pervasive Computing 8(3), 58–65 (2009)

12. Santa, J., Muñoz, A., Skarmeta, A.F.G.: A Context-Aware Solution for Personalized En-route Information Through a P2P Agent-Based Architecture. In: Gervasi, O., Gavrilova, M.L. (eds.) ICCSA 2007, Part III. LNCS, vol. 4707, pp. 710–723. Springer, Heidelberg (2007)

13. Li, J., Chigan, C.: Achieving robust message dissemination in VANET: Challenges and solution. In: IEEE Intelligent Vehicles Symposium (IV), pp. 845–850 (2011)

14. Ducourthial, B., Khaled, Y., Shawky, M.: Conditional transmissions: performances study of a new communication strategy in VANET. IEEE Transactions on Vehicular Technology 56(6), 3348–3357 (2007)

15. Olariu, S., Weigle, M.C.: Vehicular Networks: FromTheory to Practice. Chapman & Hall, CRC (2009)

16. Emmelmann, M., Bochow, B., Kellum, C.: Vehicular Networks: Automative Applications and Beyond. John Wiley (2010)

17. Mizui, K., UChida, M., Nakagawa, M.: Vehicle-to-vehicle communications and ranging system using spread spectrum techniques. Electronics and Communications in Japan, Part 1 79(12) (1996)

18. Hung, C., Yarali, A.: Wireless services and intelligent vehicle transportation systems. In: Canadian Conf. Electrical and Computer Engineering (CCECE), Niagara Falls, Canada (2011)

19. Papadimitratos, P., et al.: Vehicular communication systems: Enabling technologies, applications, and future outlook on intelligent transportation. IEEE Communication Magazine 47(11), 84–95 (2009)

20. Bishop, R.: Intelligent Vehicle Technology and Trends. Artech House, Norwood (2005)

21. http://standards.ieee.org/findstds/standard/802.11p-2010.html

22. http://www.itsforum.gr.jp/Public/J7Database/p35/ITSFORUMRC006engV1_0.pdf

23. http://www.arib.or.jp/english/html/overview/doc/5-STD-T75v1_0-E2.pdf

24. Research and Innovation Technology Administration (RITA) Combined DSRC Workshop, http://www.its.dot.gov/presentations/DSRC_Workshop.pptx

# Multi-Cue Based Place Learning
# for Mobile Robot Navigation

Rafid Siddiqui and Craig Lindley

Department of Computing,
Blekinge Institute of Technology, Karlskrona, Sweden
{rsi,cli}@bth.se

**Abstract.** Place recognition is important navigation ability for autonomous navigation of mobile robots. Visual cues extracted from images provide a way to represent and recognize visited places. In this article, a multi-cue based place learning algorithm is proposed. The algorithm has been evaluated on a localization image database containing different variations of scenes under different weather conditions taken by moving the robot-mounted camera in an indoor-environment. The results suggest that joining the features obtained from different cues provide better representation than using a single feature cue.

**Keywords:** place learning, visual cues, place recognition, robot navigation, localization.

## 1    Introduction

Learning and recognizing the already visited places is an important feature of autonomous navigation and it remains a fundamental challenge in robotic vision research. The evidence that biological organisms utilize information in the surroundings (termed as "*allothetic cues*") [1] in addition to internal path integration mechanisms ("*idiothetic cues*") [2] for multiple navigation tasks, motivate the use of visual cues for mobile robotic navigation tasks. Some important sub-tasks of navigation for which visual cues are helpful are localization, cognitive mapping, place recognition, loop closing and kidnapped recovery. Place recognition in this respect plays an important role not only in answering the fundamental question of where the robot is in the environment but also serving as a bridge between other navigation tasks. For example, knowing the places in a robot's traversed environment provides a way to abstract low level sensor information into a cognitive map (e.g. a topological tree) where each node provides localization information and links between multiple nodes provide an input for path planning and reasoning tasks [3].

Place recognition can use either global features or local features extracted from scene images for the representation of places. In [4] a probabilistic learning approach using a Bayesian classifier is used for scene recognition while a non-linear dimensionality reduction serves as a global representation of the images. Similarly, another automatic place detection and localization approach is presented in [5] in which an

M. Kamel, F. Karray, and H. Hagras (Eds.): AIS 2012, LNCS 7326, pp. 50–58, 2012.

incremental statistical learning approach has been employed for modeling the probability distribution of extracted features from images. Such place learning methods that utilize global descriptions of the images usually have higher efficiency due to the lower dimensionality of features. However, they lack the ability to encapsulate detailed information that is otherwise provided by local cues. Moreover, such global feature-based learning techniques contradict biological visual representation mechanisms found in most organisms that are believed to use more general cues to identify places in spite of using a snapshot representation.

The use of visual cues for obtaining important information from the surroundings has been used either for the representation and extraction of landmarks or for the representation of whole visited places, although the relationship between these two tasks is one-to-many where each scene representation may or may not use multiple landmark representations. A landmark can be extracted and represented in many ways due to its unique characteristic in the environment. The representation could take the form of point features [6] or can be based on the saliency of a visual region [7]. The former usually uses point based feature extraction methods [8][9] that employ a pyramidal *Difference Of Gaussians* (DOG) approach toward representation of landmarks. Such landmarks provide a way to build cognitive maps when arranged in topological order, where each landmark acts as a node and each link between two neighboring nodes expresses the similarity between these nodes.

After the representation of individual cues, there is a need to integrate multiple cues together in order to form a general description of a visited place. Although the idea of using multiple cues is not new, there is very little that has been done for its use in the recognition of places. From this perspective, a multi-cue based place recognition technique has been presented in [10] in which a weighted *Discriminative Accumulation Scheme* (DAS) [11] has been used to integrate multiple cues. The results suggested that integration improved the overall class accuracy of the recognition task. In [12] similar cue integration-based place recognition is proposed, although it also provides a confidence level for each recognition decision in order to provide a self-improvement mechanism for generating more precise actions. A semantic description of visited places has been presented in [13]. The semantic descriptor is formed by integration of color and global image features using a multi-layer *Hull Census Transform* (HCT) [14]. In all such cue-integrated place recognition approaches, improvements have been observed compared to single-cue approaches. This provides an indication of the need for more investigation to be performed in this direction. In this article we propose a new multi-cue-based feature description technique that integrates three cues (corners, color and edges) for representing a given place. An evaluation is presented using multiple learning techniques performed on the COLD database [15] that contains a 48GB of mono-camera images under multiple weather conditions.

The paper is organized as follows: section 2 describes the individual cue representation methods and then provides the multiple cue feature description that is obtained by integrating multiple feature cues. In section 3 the experimental setup is explained and results are presented. Finally, in section 4 conclusions and future work are discussed.

## 2     Multi Cue Place Recognition

This section describes the multi-cue based place recognition approach. The method presented is supervised and needs to be trained on multiple instances of each single place with a nominal label. Each place instance provides an intrinsic representation of the place obtained by extracted multiple cues from input images and integrated in a coherent feature description. Unlike usual integration schemes, this approach doesn't use an individual classifier for each cue. Instead a single classification stage is used based upon obtained integrated features. The system architecture is represented in figure 1. Given an input image '$i$', a set of patches $\delta_j^i$ are obtained having a fixed window size $s = (w/div, h/div)$ where $w, h, div$ are width, height and number of divisions of image respectively. The overall performance can vary with the number of divisions being taken. If $div$ is too large, it might not be possible to extract meaningful cues; on the other hand taking it too small may result in decrease of localized representation. For each patch three feature cues are extracted and represented in a separate feature vector. The feature extraction mechanism and their integration is explained in the following subsections.

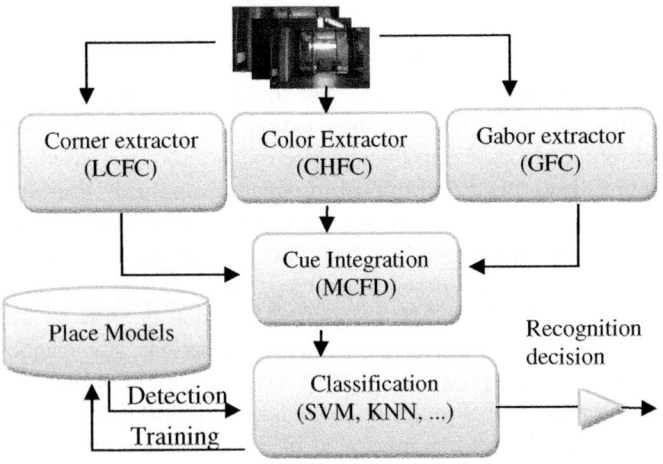

**Fig. 1.** Architecture of Multi-cue-based place recognition system

### 2.1     Local Corner Feature Cues (LCFC)

Corner points provide important cues for identification of a place. A corner in an image is a pixel that is surrounded by pixels whose intensity is in high contrast to itself. For each image patch a set of corners $C_k^j$ are obtained using a FAST corner detector [16]. The feature vector $f_k^j$ is formed by the pixels in the $m \times n$ neighborhood of the corresponding corner $c_k^j$ as given in equation 1.

$$f_k^j = \{c_{k(x-m)(y-n)}^j, \dots, c_{k(x-1)(y-1)}^j, c_{kxy}^j, c_{k(x+1)(y+1)}^j, \dots, c_{k(x+m)(y+n)}^j\} \qquad (1)$$

These corner features are combined to give a coherent corner feature cue representation $\zeta$ as follows:

$$\zeta_j^i = \frac{1}{m \times n} \sum_c^{m \times n} (f_k^j - \mu_k)^2 \qquad (2)$$

where $\mu_k$ is the mean of the intensities of a corner's surroundings. The variable number of corners detected gives variable sized features, so a fixed size feature vector is used with zero padding.

## 2.2    Color Histogram Feature Cues (CHFC)

The color cue also plays an important role in the identification of a place when other cues are insufficient or when intensity is effected by lightening variations. Although defining a general color histogram of an image patch based on color is tricky, a good degree of invariance to affine transformations can be obtained by weighting the pixels in two different channels. Hence, any deformation in an image region will also change the corresponding weights in the image channels. We use an affine color histogram approach [17] for the representation of prominent color features. The color histogram feature cue $\eta$ is given in equation 3.

$$\eta_j^i = \sum_l h_l^j \qquad (3)$$

where $h_l^j$ is a '$l$' level invariant histogram. The color histograms used for the representation of color cue information are shown in figure 2 along with other visual cues.

## 2.3    Gabor Feature Cues (GFC)

Another important cue that captures frequency and orientation can be obtained by transforming the image into Gabor feature space. This feature space has the potential to represent and discriminate textures in scenes. There is also evidence that a Gabor space representation is performed by simple cells of the visual cortex of mammalian brains [18] which makes such a representation important and interesting to study for object perception and autonomous navigation tasks. A 2D representation of a Gabor filter is a Gaussian-based filter that can be defined as:

$$\omega(x, y, f, \Theta) = \frac{f^2}{\pi \gamma s} e^{-\left(\frac{f^2}{\gamma^2} x'^2 + \frac{f^2}{s^2} y'^2\right)} e^{\iota 2\pi f x'} \qquad (4)$$

$$G_j^i(x, y, f, \Theta) = \omega(x, y, f, \Theta) * \delta_j^i(x, y) \qquad (5)$$

where '$f$' is the frequency of a sinusoidal wave, '$\gamma$' is the spatial width of the wave along the horizontal direction, '$s$' is the spatial width along the perpendicular to the wave, and '$\theta$' is an anti-clockwise rotation. The response of the filter '$G$' is obtained by convolving the image function $\delta_j^i$ as presented in equation 5. In order to obtain a general representation of a scene, the corresponding feature should be invariant to changes in the environment. This requires the feature space to be invariant to scale, rotation, translation and illumination changes. A simple feature space of Gabor features that are invariant to changes is given by [19]:

$$\eta_j^i(x,y,f,\Theta) = c\, G\left(x,y,\frac{f}{a},\theta - \phi\right) \;*\; \delta_j^i(x,y) \tag{6}$$

where '$c$' is the illumination constant, '$\Phi$' is the rotation angle and '$a$' is the scale normalization factor. We represent the input image in Gabor space by convolving an individual image patch and forming an invariant feature vector with a low resolution version of the original feature using interpolation in order to reduce the dimension space of the learning classifier.

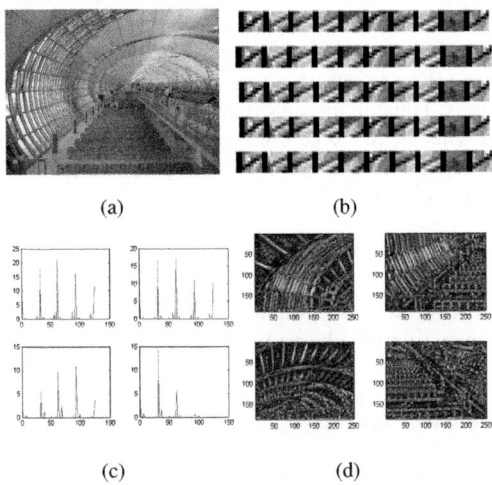

(a)        (b)

(c)        (d)

**Fig. 2.** Visual cue extractions. (a) Original image. (b) Some corner features where each row has 10 corner features. (c) Color histograms are shown for multiple sections of the image. (d) High resolution Gabor features are shown.

## 2.4    Multi-Cue Feature Description (MCFD)

All of the important cues are obtained above, so now there is a need to integrate them in order to form a concise representation of a given place. The integration can be done in many ways; one way is to use a weighted Discriminative Accumulation Scheme (DAS) [11]. This integration method requires a classifier to be trained for each cue, which becomes computationally expensive. We integrate multiple cues linearly by joining the individual features to obtain a multi-cue feature   descriptor $\psi$ which is given by:

$$\psi_j^i = \left\{\zeta_j^i\, \eta_j^i\, \rho_j^i\right\} \tag{7}$$

These integrated feature representations along with place labels are used as inputs to machine learning classifiers in order to get a place model that allows recognition to be performed on a test image.

Corridor     Printer     Office     Stairs     Bathroom

Sunny

Night

Cloudy

**Fig. 3.** Example pictures of COLD database following a path around different places

## 2.5    Learning

The integrated representation of the cues that is discussed in previous section provides an input to the learning classifier. In this study we are using supervised learning methodology for the learning of feature cues associated with a certain place. A training dataset is build where each attribute is represented by each coefficient of $\psi_j^i$ and is assigned a nominal class describing the place where the image was taken. The training stage is done offline and models are built which are then used later for detection of new unseen images.

## 3    Experiments and Results

The place learning method has been evaluated in order to determine which classification method works best with the feature representation used and to determine how it performs for different variations of a scene. Since the algorithm has potential application in robotic localization tasks, the COLD [15] image localization database provides good evaluation data. COLD has thousands of images taken under varying weather conditions (Sunny, Night, Cloudy) as shown in figure 3, and also provides labels for images corresponding to places they belong to. There are a total of 8 places that are visited by the robot while taking the images. Some of these places form a set of labels depending on the path the robot has followed. These place labels are used to build training data where each individual instance of MCFD has been assigned a class label associated with the label of that image. A total of 4377 images have been used with 1459 for each different type of weather condition. The whole dataset is divided into a training set and a test set, with 60 percent training and 40 percent test images. The optimal parameters for the classifiers have been manually selected based on their performance using 10-fold cross fold validation on the training samples.

Among the selected classifiers (SVM, KNN, J48) which are to be linearly attached to the place learning algorithm, the performance of RBF kernel used for SVM based classification is sensitive to different parameter values. So, in order to determine the best possible parameters, a Grid search is used for finding radial field and cost parameter of the kernel function. Similarly, a suitable number of neighbors for KNN and the confidence level for the decision tree model is also determined. The classification performance of selected classifiers with cross-fold validation performed on the selected datasets is given in figure 4 (a) while a performance comparison with other place recognition methods is depicted in figure 4 (b).

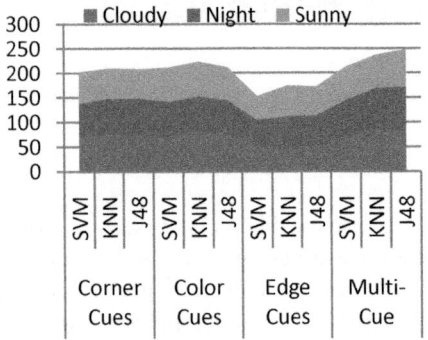 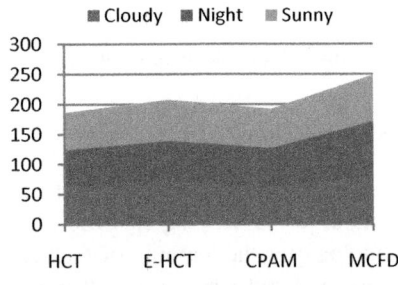

**Fig. 4 (a).** Classification rates of different classifiers with single or multiple-cues in different weather conditions

**Fig. 4 (b).** Comparison of MCFD with other methods; HCT (Hull Census Transform), E-HCT (Extended HCT), CPAM (Color-Pattern Appearance Model)

The single feature cues can perform better in recall rate, for example color cues perform better than other single feature cues on average classification rate. However, the increased false positive rate for single features increases the classifications error. This can be reduced by the addition of other features for better discriminative representation, although inclusion of multiple features into a multi-cue feature representation increases the dimensionality, and hence increases the training and testing times of the classifier. This is one of the reasons why low resolution Gabor features are used; although low resolution has resulted in low classification rate when used alone, however, they contribute towards an overall better classification rate obtained by a multi-cue representation.

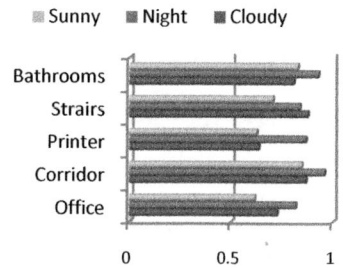

■ Sunny    ■ Night    ■ Cloudy

**Fig. 5.** Classification rates of participating classes

**Table 1.** Confusion matrix for the best performing classifier, 2PO (2-Person Office), Cor (Corridor), PA (Printer Area), ST (Staris), BT (Bathroom)

| classes | 2PO | Cor | PA | ST | BT |
|---------|-----|-----|-----|-----|-----|
| 2PA | 154 | 27 | 21 | 5 | 4 |
| Cor | 44 | 897 | 70 | 16 | 4 |
| PA | 15 | 60 | 177 | 12 | 12 |
| ST | 11 | 18 | 7 | 135 | 25 |
| BT | 3 | 3 | 2 | 22 | 240 |

It is important to see whether the classifier has performed well for each participating class to verify that its good performance is not merely a result of good performance for one class. A classification rate for all participating classes against the different datasets used in the experimentation is given in figure 5.

After obtaining a suitable classifier, which is the decision tree in this case, a place identification framework according to figure 1 has been established. Final testing is performed using the best performing classifier on a dataset formed by randomly choosing instances from weather categories. The results of this are presented in Table 1 in the form of a confusion matrix.

## 4    Conclusions and Future Work

In this work we have presented a multi-cue-based feature representation method that can be used for place recognition. Individual cue features extracted from images of a given scene are combined linearly in order to obtain a joint feature with more discrimination ability than individual features when used with different classifiers. The results suggest that joining the feature cues can be useful and can increase the overall accuracy of a recognition algorithm. Care must be taken in forming individual features as the dimensionality of the final feature can exponentially increase training and testing times. It can also be noted that simple feature cues with a low number of dimensions (e.g. color cues used in the experiments) are also very useful and are robust than other local features (e.g. corners). However, such feature representations are very susceptible to environmental variations and hence they need to be joined with other cues in order to have less false positives and negatives. In future we aim to extend the work to build a topological mapping of an environment which would be used in localization of a robotic platform; moreover, better integration scheme is also intended to be devised in later studies in order to increase processing time.

# References

1. Nardini, M., Jones, P., Bedford, R., Braddick, O.: Development of cue integration in human navigation. Current biology 18, 689–693 (2008)
2. Mittelstaedt, M.L., Mittelstaedt, H.: Homing by path integration in a mammal. Naturwissenschaften 67, 566–567 (1980)
3. Yeap, W.K.: How Albot0 finds its way home: A novel approach to cognitive mapping using robots. Topics in Cognitive Science 3, 707–721 (2011)
4. Ramos, F., Upcroft, B., Kumar, S., Durrant-Whyte, H.: A Bayesian approach for place recognition. Robotics and Autonomous Systems 60, 487–497 (2012)
5. Chella, A., Macaluso, I., Riano, L.: Automatic place detection and localization in autonomous robotics. In: IEEE International Conference on Intelligent Robots and Systems, pp. 741–746 (2007)
6. Luke, R.H., Keller, J.M., Skubic, M., Senger, S.: Acquiring and maintaining abstract landmark chunks for cognitive robot navigation. Presented at the (2005)
7. Ouerhani, N., Hügli, H., Gruener, G., Codourey, A.: A Visual Attention-Based Approach for Automatic Landmark Selection and Recognition. In: Paletta, L., Tsotsos, J.K., Rome, E., Humphreys, G.W. (eds.) WAPCV 2004. LNCS, vol. 3368, pp. 183–195. Springer, Heidelberg (2005)
8. Lowe, D.G.: Distinctive image features from scale-invariant keypoints. International Journal of Computer Vision 60, 91–110 (2004)
9. Bay, H., Tuytelaars, T., Van Gool, L.: SURF: Speeded Up Robust Features. In: Leonardis, A., Bischof, H., Pinz, A. (eds.) ECCV 2006, Part I. LNCS, vol. 3951, pp. 404–417. Springer, Heidelberg (2006)
10. Xing, L., Pronobis, A.: Multi-cue discriminative place recognition (2010)
11. Nilsback, M., Caputo, B.: Cue integration through discriminative accumulation. In: Proceedings of the 2004 IEEE Computer Society Conference on Computer Vision and Pattern Recognition, CVPR 2004, p. II-578 (2004)
12. Pronobis, A., Caputo, B.: Confidence-based cue integration for visual place recognition. In: IEEE International Conference on Intelligent Robots and Systems, pp. 2394–2401 (2007)
13. Wang, M.-L., Lin, H.-Y.: An extended-HCT semantic description for visual place recognition. International Journal of Robotics Research 30, 1403–1420 (2011)
14. Lee, S.W., Kim, Y.M., Choi, S.W.: Fast scene change detection using direct feature extraction from MPEG compressed videos. IEEE Transactions on Multimedia 2, 240–254 (2000)
15. Pronobis, A., Caputo, B.: COLD: The CoSy localization database. The International Journal of Robotics Research 28, 588 (2009)
16. Rosten, E., Drummond, T.: Fusing points and lines for high performance tracking. In: Tenth IEEE International Conference on Computer Vision, ICCV 2005, pp. 1508–1515 (2005)
17. Domke, J., Aloimonos, Y.: Deformation and viewpoint invariant color histograms. In: British Machine Vision Conference, pp. 509–518 (2006)
18. Daugman, J.G.: Complete discrete 2-D Gabor transforms by neural networks for image analysis and compression. IEEE Transactions on Acoustics, Speech and Signal Processing 36, 1169–1179 (1988)
19. Kyrki, V., Kamarainen, J.K., Kalviainen, H.: Simple Gabor feature space for invariant object recognition. Pattern Recognition Letters 25, 311–318 (2004)

# Bio-inspired Navigation of Mobile Robots

Lei Wang, Simon X. Yang, and Mohammad Biglarbegian

University of Guelph, School of Engineering,
Guelph, Ontario, N1G 2W1, Canada
{lwang02,syang,mbiglarb}@uoguelph.ca

**Abstract.** This paper presents a bio-inspired neural network algorithm for mobile robot path planning in unknown environments. A novel learning algorithm combining Skinner's operant conditioning and a shunting neural dynamics model is applied to the path planning. The proposed algorithm depends mainly on an angular velocity map that has two parts: one from the target, which drives the robot to move toward to target, and the other from obstacles that repels the robot for obstacle avoidance. An improved biological learning algorithm is proposed for mobile robot path planning. Simulation results show that the proposed algorithm not only allows the robot to navigate efficiently in cluttered environments, but also significantly improves the computational and training time. The proposed algorithm offers insights into the research and applications of biologically inspired neural networks.

**Keywords:** Path Planning, Collision Avoidance, Mobile Robot, Robot learning, Biological Inspiration, Neural Network, Neural Dynamics.

## 1 Introduction

The concept of bio-inspired artificial intelligence has been introduced in past decade [1]. A biologically inspired neural network is a type of neural network that can control a robot in dynamic environments while avoiding obstacles and approaching target. In the past fifty years, many researchers have focused on algorithms imitating complex neuron networks in cognitive functioning in humans and animals. Designing these algorithms is a crucial step and forms a complexes issue in mobile robot navigation [2],[3],[4],[5] and [6]. Neural systems have been developed in two main areas, computational neuroscience and neural engineering. The basic theory has been summarized in [1]. In computational neuroscience, Yang and Meng [7] proposed a new real-time shunting path planning models based on the Hodgkin and Huxley's membrane neural network; through simulations they demonstrated that an online control can be achieved in mobile robot navigation. In neural engineering and robotic learning, Gutnisky and Zanutto [5] adopted an operant learning model in mobile robot obstacle avoidance. They concluded that reinforcement learning is better than a Q-learning algorithm. A key model, based on Grossberg'e neural model of conditioning for robot approach and avoidance behaviors was presented by Chang and Gudiano [4]. Further, Aren, Fortuna and Patané [6] introduced a network of spiking neurons with classical conditioning

M. Kamel, F. Karray, and H. Hagras (Eds.): AIS 2012, LNCS 7326, pp. 59–68, 2012.
© Springer-Verlag Berlin Heidelberg 2012

learning devoted to navigation control. Despite these advances, there is a continuing difficulty in applying the algorithms and adapting them to real robots. They also have a feasibility problem associated with high computational costs.

This paper introduces a bio-inspired local path-planning approach that is a combination of two algorithms proposed by Grossberg in [2] and [10]. The biological learning theory, i.e., Skinner's operant conditioning [8] in Grossberg's first algorithm is combined with the membrane ionic mechanisms be used, adapting them from Yang and Meng [7]. After training, the robot builds its own knowledge, learns to track the target and avoid the obstacles.

## 2   The Proposed Approach

### 2.1   Model of Mobile Robots

A model of a mobile robot with respect to the global coordinate is shown in Fig. 1. The position and posture of the mobile robot is determined by $x$, $y$ and $\theta$. Two inputs are used to control the motion of the robot-translational velocity $v$, and angular velocity $\omega$. The kinematic equations of the robot are expressed as

$$\theta_r(t+1) = \theta_r(t) + \Delta t \omega_r(t), \tag{1}$$

$$x_r(t+1) = x_r(t) + v\Delta t \cos(\theta_r), \tag{2}$$

$$y_r(t+1) = y_r(t) + v\Delta t \sin(\theta_r), \tag{3}$$

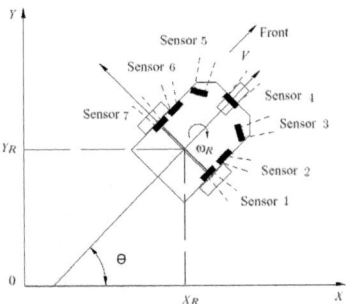

**Fig. 1.** A nonholonomic mobile

where $\Delta t$ is the time interval, $\theta_r$ is the orientation angle with respect to the robot mass center, and $x_r$, $y_r$ and $\omega_r$ are the coordinates and angular velocity in the global frame, respectively.

## 2.2   Biological Inspiration

An outside stimulus will cause an inner reaction in animal and human bodies, the communication of neurons depend on electrical signals traveling in one direction along a connection. Neurons come in two categories: excitatory and inhibitory. Excitatory neurons launch the connections between previous and post neurons and become compact; the inhibitory neurons build the connections which tend to decrease or block the activation between previous and post neurons [1]. Similarly, there are two kinds of stimulus: conditioned stimulus (CS) and unconditioned stimulus (UCS). Through learning,neurons accept a positive feedback after a stimulus as a reinforcement signal; a negative feedback is, however, avoided. Much research is based on these phenomena, such as classical conditioning and operate conditioning. Our main implementation along follows Grossberg [2], Chang and Gaudiano [9]. A description of the complete neural network involved can be found in [2] and [9].

## 2.3   Model Algorithm

The proposed algorithm has two main parts: angular velocity map that maps the sensors output to angular velocity that is required to drive the robot, and a modified shunting model to ensure the angular velocity obtained lies in a safe boundary.

### Angular Velocity Map

An angular velocity map is a 1-D layer of identical neurons that is used to control the movement of the mobile robot, as shown in Fig. 2. The input to the angular velocity map comes from sensors, and the output appears as the robot's angular velocity. In the map, only one neuron is active at a time, with most active neuron being "the winner" of this layer. The map activates this winner neuron maximally and suppresses all other neurons. At any given time, the robot's angular velocity is chosen based on the winner neuron, meaning that the index of the active neuron is more important than its activation level. For example, activating the leftmost neuron means that robot turns left with angular velocity $-\omega_{max}$ as shown in Fig. 2 (a). Activating the central neuron represents straight movement, and activating the rightmost neuron causes the robot to turn right with angular velocity of $\omega_{max}$.

The transformation from angular velocity to the angular velocity map is performed via a sigmoidal transformation, which ensures a smooth transition between angular velocity and the map. In the proposed algorithm, given a certain angular velocity value $\alpha(t)$, one neuron in the map will be activated as [4].

$$n_i(t) = \begin{cases} \dfrac{N}{2} + \dfrac{N\left(\alpha_\omega + 0.5\omega_m\right)\alpha(t)}{\omega_m(1.0 + \alpha(t))}, & \text{if } \alpha(t) > 0 \\[3mm] \dfrac{N}{2} + \dfrac{N\left(\alpha_\omega + 0.5\omega_m\right)\alpha(t)}{\omega_m(1.0 - \alpha(t))}, & \text{otherwise} \end{cases} \tag{4}$$

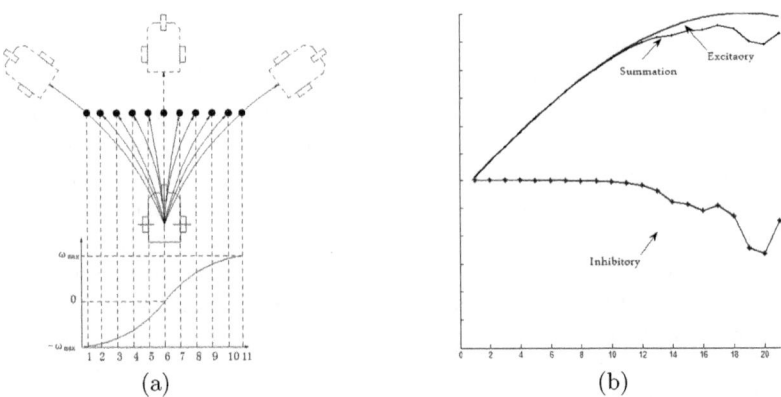

**Fig. 2.** Angular velocity map modified from [4]. (a) Angular velocity map; (b) The peak shift property.

where $n_i(t)$ is the most active neuron activated at time $t$, $N$ is the number of the neurons in the angular velocity map, and $\omega_m$ is the maximum of angular velocity of mobile robot, and $\alpha_w$ is a constant that could controls the slope of a sigmoid function. A Gaussian distribution around the desired node in the angular velocity map is given as

$$G_{x_j}(t) = \exp\left[\frac{-(j - n_i(t))^2}{\sigma^2}\right] \qquad (5)$$

where $j$ is the neuron index in the map, and $\sigma$ is the standard deviation of the Gaussian distribution.

The angular velocity map consists of two important parts, excitatory and inhibitory. The excitatory input comes from the target (a positive Gaussian distribution) that leads the robot to choose the direction to approach the target by generating the shortest path. The inhibitory part is generated directly from the sensory neurons (a negative Gaussian distribution), decreasing the collision chance of the robot once it senses obstacles and moves away. The summation of two Gaussian distributions generates the robot's final moving direction through a process of peak shifting. The winning neuron's activation level indicates the angular velocity $\omega(t)$ used to drive the robot [4]

$$\omega(t) = \begin{cases} \dfrac{\omega_m\left[J(t) - \dfrac{N}{2}\right]}{N(\alpha_w + 0.5\omega_m) - \omega_m\left[J(t) - \dfrac{N}{2}\right]}, & \text{if } J(t) > \dfrac{N}{2} \\[4ex] \dfrac{\omega_m\left[J(t) - \dfrac{N}{2}\right]}{N(\alpha_w + 0.5\omega_m) + \omega_m\left[J(t) - \dfrac{N}{2}\right]}, & \text{otherwise} \end{cases} \qquad (6)$$

where $J(t)$ is the most active node at time $t$. Note that (6) is the inverse versions of (4): (4) is mapping a given angular velocity value into the map, while (6) shows how a winner neuron's activation is used to generate the angular velocity.

## A Modified Shunting Neural Dynamics Model

Grossberg's shunting model [10] is based on Hodgkin and Huxley's [7] model which describs the contingencies of individuals' real-time adaptive behavior in complex and dynamic environments. By analyzing Grossberg's neural network theory of classical and operant conditioning for mobile robots used in Chang and Gudiano's model [4], explains why the sharp angular velocity jumps and abates an oscillation. This caused by the single sensory neuron input from the layer S while the robot is moving parallel to an obstacle or a wall. A shunting model angular velocity controller is proposed to solve the oscillation problem. To produce the model an angular velocity dynamics equation with respect to the target angular velocity in the driving direction is obtained

$$\frac{d\theta_r(t)}{dt} = -A\theta_r(t) + (B - \theta_r(t))f(\theta_t(t)) - (D + \theta_r(t))g(\theta_t(t)), \qquad (7)$$

where $A$, $B$, and $D$ are positive constants that represent the passive decay rate in membrane potential, the upper and lower bounds of the neural activity; $f(x)$ is a linear-above-threshold function defined as $f(x)=\max\{x,0\}$ and the non-linear function $g(x)$ is defined as $g(x)=\max\{-x,0\}$. (7) ensures that a sudden single increasing input generates an angular velocity output that is bounded in the upper and lower bounds of the neural activity.

## 2.4    The Navigation Algorithm

The proposed algorithm has two parts-the modified Grossberg operant conditioning model using in [2] and a shunting model. Robot velocity is assumed to be constant; the neural network inputs the sensor data and outputs the torque required to run the robots. The initial position of the robot is facing straight toward the target, and it is natural that it moves to the target directly. Any sensor measurement that is smaller than its minimum value indicates a a collision will occur. When the robot detects an obstacle, it rotates based on the result of the calculation of (6). While the robot is moving toward the target, if there are no obstacles along the path, the path is a straight line from the robot's initial position as it points the target.

**Step 1:** Calculate the activation of the layer S and update the environmental information from the sensors. With a simple initialization of sensors information, assigning sensor measurement in $[0, 1]$ as

$$x_i(t) = \frac{I_i(t)}{|I(t)|}, \tag{8}$$

where $I_i(t)$ is the measurement of sensor $i$, and $I(t)$ is the summation of all sensor measurements.

**Step 2:** Calculate the activation $y(t)$ of drive node D: process input sensor measurements as $y(t) = 1$, if the collision has occurred, and $y(t) = 0$ otherwise. Any sensor measurement smaller than its minimum value indicates a collision will occur. Drive node D tests whether the robot moves to the obstacle domain area or not. This step has been included to improve calculation and ensure the training period and computational time is shorter than Chang-Gudiano algorithm.

**Step 3:** Calculate the activation $K_i(t)$ of the layer P

$$K_i(t) = x_i(t)y(t) \tag{9}$$

**Step 4:** Adjust connection weights $z_{ij}(t)$ between layer P and angular velocity map. Mapping the desired angular velocity leads the robot to the target and generates an excitatory Gaussian distribution, with the desired neuron in angular velocity based on (4) and (5). The learning weight between layer P and the angular velocity map is described by

$$z_{ij}(t) = z_{ij}(t-1) - Fx_i(t)\left[\frac{G_{x_i}(t)}{1+(j-J(t))^2} + z_{ij}(t-1)\right] \tag{10}$$

where $z_{ij}(t)$ is the adaptive weight from the node $i$ in the sensory layer to the node $j$ in the angular velocity map, $F$ is the learning rate, $G_{x_i}(t)$ is the Gaussian function value around the winning node that indicates the robot's moving direction, and $J(t)$ is the node index with the maximum value in the map.

**Step 5:** Generate the activation $m_j(t)$ of the angular velocity map, that includes one excitatory component from the target given by $G_{x_j}(t)$, and an inhibitory component derived from the sensed obstacles, given by $H(t)$

$$m_j(t) = G_{x_j}(t) + H(t) \tag{11}$$

$$H(t) = \sum_i K_i(t)z_{ij}(t-1) \tag{12}$$

where $m_j$ is the activity of the node $j$ in the angular velocity map. If the index of the maximum values node in $H(t)$ in much different than the index of the maximum values node in $G_{x_j}(t)$, $n_{inew}(t) = N - n_i(t) + 1$, return to Step 4 and recalculate $z_{ij}(t)$ using (5) and (10); if it is not, go on to Step 6.

**Step 6:** Determine the angular velocity that the robot might perform in its next movement before the final adjustment according to (6).

**Step 7:** Generate the angle for the robot motion using

$$\theta_r(t+1) = \theta_r(t) + \Delta t f(y(t)), \qquad (13)$$

where $f(y(t))$ is defined as

$$f(y(t)) = \begin{cases} -A\theta_r(t) + (B - \theta_r(t))f(\theta_t(t)) - (D + \theta_r(t))g(\theta_t(t)), & \text{if } Ty(t) > 0 \\ \omega(t), & \text{otherwise} \end{cases}$$
$$(14)$$

where $Ty(t) = \xi(\mu - \theta_e(t))$ works as a vigilance parameter that determines when the neural dynamics-based shunting model is activated. $\theta_e(t) = |\theta_r(t) + \Delta t \omega(t) - \theta_t(t)|$ is the calculation error between the angle generated from the angular velocity map and the target heading angle. The term $\rho(\mu - \theta_e)$ indicates that the vigilance parameter $Ty$ recovers to its maximum value $\mu$ at the rate $\xi$.

**Step 8:** Update the distance between the robot's current position and the target position. If it is smaller than a constant value, stop the robot; if not, return to Step 1.

The proposed algorithm structure is shown in Fig. 3.

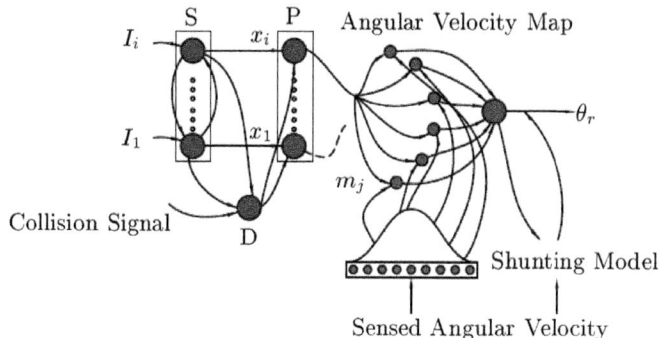

**Fig. 3.** The proposed bioinspired neural model

## 3    Simulation Results

This section presents simulation results for a mobile robot using the Chang-Gudiano and the proposed algorithms. Two simulations were conducted in Matlab, with fair comparison between the algorithms, ensured by using the same environments during each simulations. The parameters required for simulations are kept the same for both algorithms. Also for both, repeated training was conducted before the simulation. The learning weight between layer P and the angular velocity map is fully developed, as shown in Figs. 4(a) and (b). We let the robot randomly moves at a simple environment cluttered with obstacles. This robot performance will lets the model develop a set of weights which increasing the connection between different angular velocities and motion neurons,

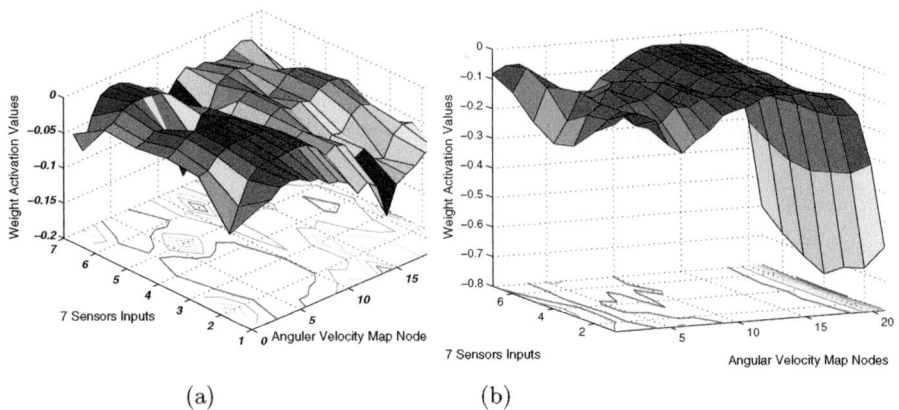

**Fig. 4.** The weight between sensory layer $P$ and the angular velocity map $m_j$. (a) The initial weight before training; (b) The weight after training.

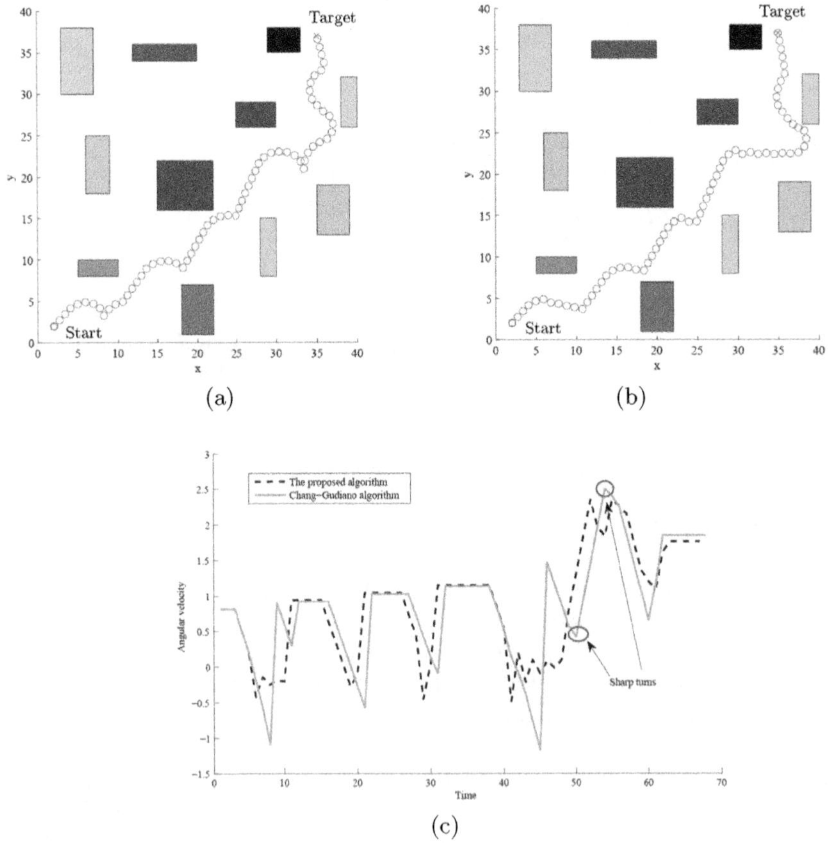

(a)

(b)

(c)

**Fig. 5.** Simulation 1 results. (a) The Chang-Gudiano algorithm; (b) The proposed algorithm; (c) Angular velocities.

by activating contiguously the corresponding nodes in angular velocity map. The original weight between the two layers is randomly selected in the range of [0, 0.5], to makes the robot turn right or left when the opposite sensory neuron is activated.

## 3.1   Simulation 1: Simplified Scenarios

In the first simulation, Figs. 5 (a) or 5 (b) shows the simulation environment, which has 40 by 40 squares in the presence of obstacles. The performances of the Chang-Gudiano algorithm and the proposed algorithm are shown in Figs. 5 (a) and (b), Fig. 5 (c) compares the angular velocities of the algorithms. As can be see, the algorithm proposed in this paper outperforms the Chang-Gudiano algorithm. The reason appears to be that Chang-Gudiano method is based only on the activation of the side sensors' measurement. Hence, the robot can never navigate parallel to walls or obstacles a size are significantly larger than itself, which could cause the robot to make sharp turns. For example, sharp turn appears when robot approaches the last obstacle.

We have also compared the performance of the two algorithms in terms of the completion time, based on the training time. The training time and runtime about the Chang-Gudiano algorithm and the proposed algorithm are 56, 39; 69 and 67, respectively. The proposed algorithm requires shorter time to be trained, as a result of the Step 2 modification of the Chang-Gudiano algorithm, which ensures faster running time for the our methodology.

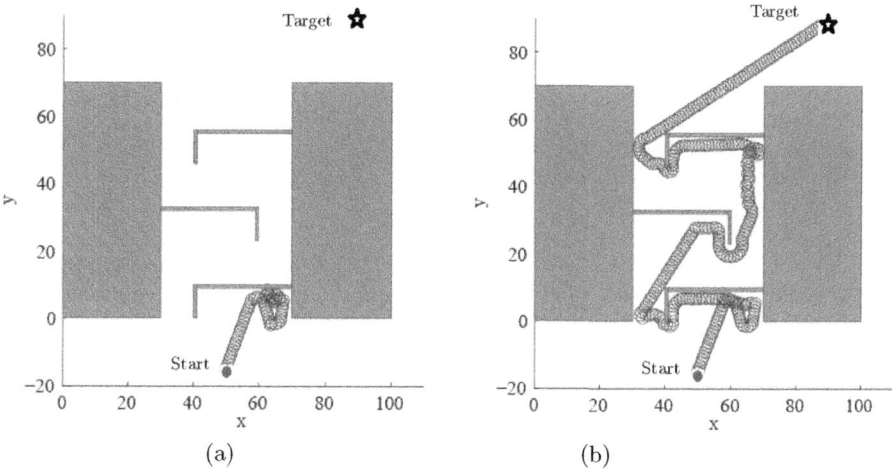

**Fig. 6.** Simulation 2 results. (a) The Chang-Gudiano algorithm; (b) The proposed algorithm.

## 3.2    Simulation 2: Complicated Scenarios

In the second simulation, Figs. 6 (a) and (b) show the simulation environment, which has 120 by 120 squares in the presence of obstacles. The performance of both algorithms were computed and presented. Our algorithm successfully led the robot to reach the target; while the robot using the Chang-Gudiano algorithm failed. Because the Chang-Gudiano algorithm causes the robot to move in one direction, it dose not adapt to complex environments.

In the proposed algorithm an opposite peak shifting appears in the opposite direction. The most active inhibitory node in the angular velocity map generated by the sensors input, differ significantly from the excitatory node in the map, generated form the target. The modification implemented the Chang-Gudiano algorithm in Step 5 ensures the adaptability of the proposed algorithm.

## 3.3    Conclusion

This paper proposed a bio-inspired algorithm for mobile robot path planning. By combining two biologically inspired neural networks, the developed algorithm can drive the robot move smoothly in an unknown environment. Simulation results shows that the proposed algorithm has less computational cost, less training time and more adaptable in complex environments.

# References

1. Floreano, D., Mattiussi, C.: Bio-Inspired Artificial Intelligence Theories, Methods, and Technologies. MIT Press, Cambridge (2008)
2. Grossberg, S.: On the dynamics of operant conditioning. Theoretical Biology 33, 225–255 (1971)
3. Yang, S.X., Lou, C.: A Neural Network Approach to Complete Coverage Path Planning. IEEE Trans. Systems 33, 718–724 (2004)
4. Chang, C., Gudiano, P.: Application of biological learning theories to mobile robot avoidance and approach behaviours. Complex Systems 1, 79–114 (1998)
5. Gutnisky, D.A., Zanutto, B.S.: Learning Obstacle Avoidance with an Operant Behavior Model. Artificial Life 10, 65–81 (2004)
6. Aren, P., Fortuna, L., Patané, L.: Learning Anticipation via Spiking Networks: Application to Navigation Control. IEEE Trans. Neural Netw. 20(2), 202–216 (2009)
7. Yang, S.X., Meng, M.: An efficient neural network approach to dynamic robot motion planning. IEEE Trans. Neural networks 13, 143–148 (2000)
8. Saksida, D.S., Sariff, L.M.: Operant Conditioning in Skinnerbots. Adaptive Behavior 5, 1–28 (1997)
9. Gaudiano, P., Chang, C.: Adaptive obstacle avoidance with a neural network for operant conditioning: experiments with real robots. In: Computational Intelligence in Robotics and Automation, pp. 13–18 (1997)
10. Grossberg, S.: Nonlinear neural networks: principle, mechanisms, and architecture. Neural Networks 1, 17–61 (1988)

# Market-Based Framework
# for Mobile Surveillance Systems

Ahmed M. Elmogy[1], Alaa M. Khamis[2], and Fakhri Karray[3]

[1] Tanta University, Egypt
[2] GUC, Egypt
[3] University of Waterloo, Canada

**Abstract.** The active surveillance of public and private sites is increasingly becoming a very important and critical issue. It is therefore, imperative to develop mobile surveillance systems to protect these sites. Modern surveillance systems encompass spatially distributed mobile and static sensors in order to provide effective monitoring of persistent and transient objects. The realization of the potential of mobile surveillance requires the solution of different challenging problems such as task allocation, mobile sensor deployment, multisensor management, and cooperative object tracking. This paper proposes a market-based framework that can be used to handle different problems of mobile surveillance systems. Task allocation and cooperative target-tracking are studied using the proposed framework as two challenging problems of mobile surveillance systems. These challenges are addressed individually and collectively.

**Keywords:** Surveillance systems, task allocation, market-based techniques, Target tracking.

## 1  Introduction

One of the hot topics is how to automate surveillance tasks based on mobile and fixed sensors platforms [1]. Many benefits can be anticipated from the use of multisensor systems in surveillance applications, such as decreasing task completion time, and increasing mission reliability. Advanced surveillance systems include a vast array of cooperative (static and mobile) sensors with varying sensing modalities that can sense continuously the volume of interest [2]. The main goal of the surveillance system is to adjust the sensing conditions for improved visibility, and thereby improve performance [3]. In such setting, surveillance is a complex problem posing many challenging problems.

This paper presents a market-based framework for mobile surveillance systems. The goal is to develop a framework that efficiently distributes tasks among the mobile sensor team to achieve the surveillance mission. Such a framework will support the operation of the mobile sensors so that they can collaboratively perform tasks such as detecting and tracking moving targets. In order to maximize the effectiveness of the mobile sensor team collaborating as a group, the action of every mobile sensor should consider the contribution of its teammates

M. Kamel, F. Karray, and H. Hagras (Eds.): AIS 2012, LNCS 7326, pp. 69–78, 2012.

towards the mission objectives. How to accomplish this is a complex problem, which is currently an active area of research [4, 5].This requires tackling some of the challenging problems of mobile surveillance that have not been investigated collectively in the past. These problems include, but are not limited to, task allocation, mobile sensor deployment, cooperative object detection and tracking, and decentralized data fusion. This paper tackles two of these problems: task allocation, and cooperative detection and tracking. These problems are tackled individually and collectively.

The rest of this paper is organized as follows. Section 2 presents the components of the proposed mobile surveillance framework. A market based approach is introduced in section 2.2. Section 3 summarizes the conducted experimental results tackling complex task allocation, and target detection and tracking. Finally, conclusion and future work are summarized in section 4.

## 2   Proposed Mobile Surveillance System Framework

This section highlights the major components of the proposed mobile surveillance framework as shown in Fig. 1 tackling two main challenges- complex task allocation, and target detection and tracking.

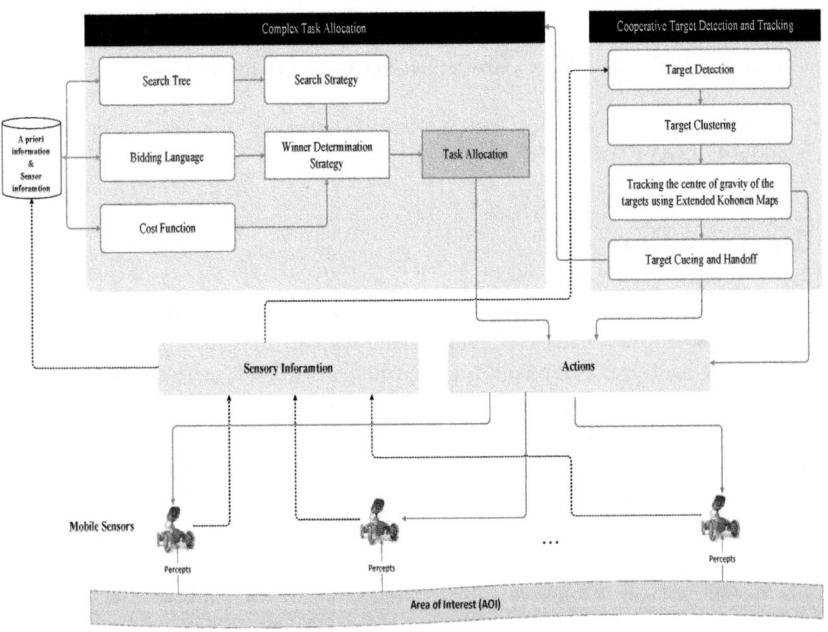

**Fig. 1.** Market-based framework for mobile surveillance systems

## 2.1   Complex Task Allocation

This section provides problem formulation for complex task allocation which is one of the main focuses of this paper. The goal is to assign sensors to tasks so as to maximize overall expected performance, taking into account the priorities of the tasks and the skill ratings of the sensors.

For the general case, the problem of task allocation [6] is to find the optimal allocation of a set of tasks $T$ to a subset of sensors $S$, which will be responsible for accomplishing it [7]:

$$A : T \to S \tag{1}$$

Each mobile sensor $s \in S$ can express its ability to execute a task $t \in T$ , or a bundle of tasks $G \subseteq T$ through bids $b_s(t)$ or $b_s(G)$.

The most common global objective is to minimize the sum of the team member costs, which can be described mathematically as follows:

$$C(A) = \sum_{s=1}^{n} b_s(G_s) \tag{2}$$

where $C(A)$ is the total required cost for executing the allocation $A$, and $G_s$ is the bundle of tasks that is won by sensor $s$.

In this work, we are using a shortest sequence planning algorithm (SSP) [8] in order to find the minimum cost path for each mobile sensor given the tasks locations.

## 2.2   Proposed Market-Based Approach

Market-based approaches have received significant attention and are growing very fast in the last few decades especially in multi-agent domains [5, 9]. These approaches are considered as hybrid approaches that combine the centralized and distributed strategies (i.e. market-based approaches have elements that are centralized and distributed). The decision to use auctions in this paper comes from the existence of several desirable properties of auction approaches [10, 11] such as efficiency, robustness, and scalability.

### A. Auction Design

The task allocation approach proposed in this paper imitates the auction process of buying and selling services through bidding. Sellers or auctioneers are responsible of processing the bids sent by buyers or bidders and determining the winning bidder. In order to design a auction, the main focus will be based on maximizing a utility function. For task allocation problem, utility is a satisfaction (value of profit) derived by a mobile sensor $s$ from accomplishing a task $t$.

Given a mobile sensor $s$ and a task $t$, if $s$ is capable of executing $t$, utility can be defined [7] on some standardized scale as:

$$u = p(t) - d(t) \tag{3}$$

Where $p(t)$ is the total payment it receives after executing the task $t$, and $d(t)$ is the total distance it travels to reach the task. The priorities of tasks to be executed should be taken into account while designing the task allocation framework. Our objective is to find the optimal assignment of tasks $T$ to sensors $S$ in order to minimize cost and thus maximize the overall utility. Consequently, system performance is ideally optimized.

## B. Search Tree

Most of task allocation approaches treated tasks as atomic units [11, 12, 13, 14]. Thus allowing only static description for each task and so the only degree of freedom is determining to which sensor the task will be assigned. While this description is fine in case of simple tasks, it is not with complex tasks. Given the bid submitted, search over all possible allocations can be used as a winner determination strategy. In this case, a search tree can be used as a better description for the tasks. In this tree, mobile sensor team members are permitted to bid on nodes representing varying levels of task abstraction, thereby enabling hierarchical planning, task allocation, and optimization among the team members.

## C. Fixed Tree Task Allocation

Consider a team of mobile sensors assembled to perform a particular task. Consider further, that each mobile sensor is capable of executing one task at once, and each task can be accomplished by one sensor. The task information is continuously available to the mobile sensors team. The goal of the team is to perform the task efficiently while minimizing costs. In the context of fixed task tree allocation, a set of constrains dictates that the whole auction mechanism is based only on one task tree, which is proposed by the operator or the auctioneer. The proposed algorithm allows using only one auctioneer from the start to the end of auctioning, and so considered as a centralized task allocation. It also allows changing the auctioneer during auctioning while considering only the plan of the original operator. In this case, our proposed mechanism can be seen as a hierarchal task allocation mechanism. Another constraint dictates that at most one node can be sold to each bidder per auction. This is because upon awarding one node to a bidder the bid prices on other nodes become invalid due to the fact that bid prices are conditioned on the current commitments of each participant.

## D. Dynamic Tree Task Allocation

The proposed fixed task tree allocation described in the previous section could be seen as an instance of decompose-then-allocate approach. The main drawback of this approach is that the cost of the final plan cannot be fully considered because the complex task is decomposed by the auctioneer without knowledge of the eventual task allocation. Also, backtracking is not allowed in this approach, and so any costly mistakes in the auctioneer decompositions cannot be rectified. Generally, the allocate-then-decompose method tries to avoid the drawbacks of the decompose-then-allocate method. However, there are still some disadvantages. Motivated by the drawbacks of both methods, we are proposing dynamic tree allocation to allow backtracking in order to recover the bad plans made by the auctioneers. The algorithm allows auctioning on all levels of abstraction of the mission task implemented by the task tree from the top to the bottom. Each mobile sensor evaluates its ability to execute the required task based on its plan not on the plan of auctioneer. Our proposed dynamic algorithm is either executed by allowing only one auctioneer (centralized allocation) or allowing different auctioneers (hierarchical allocation).

## 2.3 Target Detection and Tracking

The main objective of this section is to give a detailed description of the developed methodology for tracking multiple objects, which will be incorporated into the proposed mobile surveillance framework.

The proposed mobile surveillance framework requires knowledge of the current targets' positions as well as their future positions, in order to be able to determine the sensor assignments and states. In this context, the overall problem of tracking starts by Detecting every target within a given scene to be categorized as obstacles or objects of interest to be clustered using a hybrid subtractive-K-means clustering technique. The next step is track the center of gravities of the clustered targets using Extended Kohonen Maps. the final step is to track the exit targets by the best suitable trackers using a cueing/handoff market-based approach.

The goal is to develope an algorithm that maximizes the following metric:

$$AC = \sum_{tt=1}^{t_e} \sum_{j=1}^{n} \frac{g(OM(tt), j)}{t_e} \tag{4}$$

where

$$g(OM(tt), j) = \begin{cases} 1 \text{ when there exist } i \text{ such that } om_{ij}(tt) = 1 \\ 0 \qquad\qquad \text{otherwise} \end{cases} \tag{5}$$

In other words, the problem requires maximizing the number of targets that are observed by the mobile sensors. $t_e$ is the execution time of the algorithm. It is assumed that:

$$\bigcup_{s_i \in S} SC(s_i, tt) << AOI \qquad \text{for any tt} \tag{6}$$

## A. Target clustering

In order to make our approach dependent on the distribution of the targets rather than the density of targets as in [15, 16], we chose to cluster the targets in the environment and then track the clusters centers instead of tracking each target separately. Clustering the targets makes our approach energy efficient one because not all the trackers will be active at all the time. Out of the numerous available clustering techniques, two were selected; subtractive clustering, and K-means [17] clustering techniques. The reason behind using K-means clustering technique is its high level of accuracy [18]. However, K-means clustering technique has the problem of selecting the initial locations of clusters and the number of clusters, which will affect the speed of convergence and the accuracy of the algorithm. So, we propose using the subtractive clustering technique to find the initial number and locations of the clusters centers to be fed into the K-mean clustering technique in order to find the final exact locations of clusters centers.

### B. EKM-based Tracking Algorithm

After clustering the detected targets which lie within the tracker sensing range. The center of gravity of the detected targets is reached by adopting a representation of the sensory input vector: $u_p = (\theta, d)$, where $\theta$, and $d$ are the angle difference and distance between the tracker and the target respectively. Each tracker uses an Extended Kohonen Map (EKM) [19] in order to reach the target. The Extended Kohonen Map is an extension of Kohonen Map [20], and is considered as one of the most famous unsupervised learning neural network. Extending the Kohonen Map, which is done by adding fan-in input weights to its input layer gives the map the ability to learn by supervised learning. However, this is not always the case; unsupervised learning could still be used. Each neuron in the EKM has a sensory weight vector $w_i = (\theta_i, d_i)$.

### 2.4    The Proposed Cooperative Multi-target Tracking Approach

The mobile sensors used in this paper are assumed to be equipped with high-bandwidth communications and an array of sensors and actuators, which give the sensors the ability to achieve cooperative behavior at the group level. A cueing/handoff market-based method like the one in [4] is used to guarantee that there is only one mobile sensor that will respond to the help call coming from the sensor that detects a target about to exit its sensing range (or in its predictive tracking range). The mobile sensor that detects an exiting target will do the task of an auctioneer (i.e., every tracker can do the function of the auctioneer) This makes our algorithm more robust than methods that use only one coordinator because there is no central point of failure in this case. Using a cueing/handoff market-based algorithm will guarantee that the most suitable sensor will track the exit target. In other words, there is no need to explore the environment or to check the answered help calls as in [15]. This makes the proposed method a more energy-efficient one than the method in [15]. The cueing/handoff market-based method proceeds as follows:

1. **Help cueing:** the mobile sensor that needs help (detects an exiting target) broadcasts a help call to its teammate. One sensor can issue multiple help calls according to the number of targets about to be lost. Each help call includes the position of the target to be lost.
2. **Bid submission:** after each sensor receives the help call, they send their bids to the auctioneer (the sensor that issued the help call). The bids include the distances of the bidders to the target.
3. **Close of auction:** the auctioneer processes the bids, determines the winner, and notifies the bidders with a message which indicates who is the winner.

4. **Task handoff:** the winner will proceed by orienting itself to face the target and move with its maximum speed to put the target inside its sensing range.

The existence of the auctioneer does not mean that the proposed system is completely centralized. The system is still distributed and the function of the auctioneer is only to start the action of tracking.

# 3   Results and Decision

As mentioned previously in this paper, two main phases should be considered in order to accomplish the surveillance mission: task allocation, and target detection and tracking. In this section, the simulations and results of these phases are presented in detail.

## 3.1   Complex Task Allocation Simulations and Results

In order to evaluate the proposed approach, we consider an area surveillance application where the goal is to monitor some areas in a public place with a team of mobile sensors, each equipped with a vision system, and laser ranger sensor. An example of this is the use of a team of mobile sensors to survey an indoor environment, such as malls or airports.

The proposed system has been tested considering Waterloo airport in the city of Waterloo, Ontario, Canada, which consists of six main areas, and so the goal of the proposed system is to track targets within these areas, such as people, in order to secure the airport. In order to accomplish this, the airport areas (areas of interest (AOIs)) should first be allocated to the available mobile sensors. Each mobile sensor will scan the allocated area, if any, looking for targets to track. Not all sensors will have areas to scan. In other words, the proposed task allocation approach may allocate more than one area to one mobile sensor in order to minimize the traveling cost. The results of this scenario, considering both fixed and dynamic tree task allocation algorithms using different numbers of mobile sensors are shown in Fig. 2, and Fig. 3. The experiments are performed within a 2D simulation environment using Player/Stage simulator.

The average cost is computed by calculating the cost of executing the mission task using 50 runs and then taking the average. In terms of this average cost, the results in Fig. 2, and Fig. 3 show that both fixed and dynamic tree allocations (centralized and hierarchical) consistently outperform the other algorithms. It is also seen that the dynamic tree allocation outperforms fixed tree allocation, which was expected as the replanning ability is added to the sensors in the dynamic tree allocation.

On average, the hierarchical task tree algorithm is better than the centralized task tree algorithm besides its good feature of relying on different auctioneers compared to one auctioneer in the centralized algorithm. This is because the hierarchical auctioning allows more auction rounds to happen. Thus the hierarchical auctioning increases the possibility of improving the system performance than the centralized auctioning.

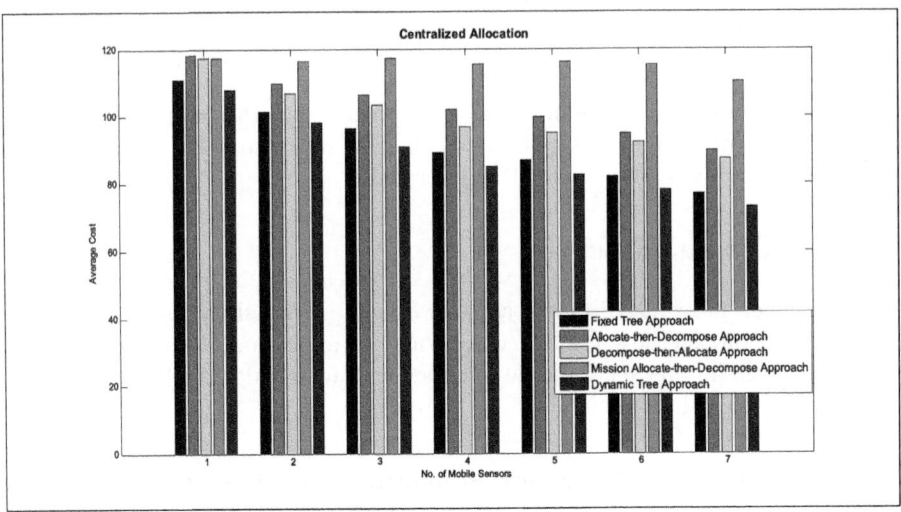

**Fig. 2.** Comparison of the average cost for centralized allocation mechanism

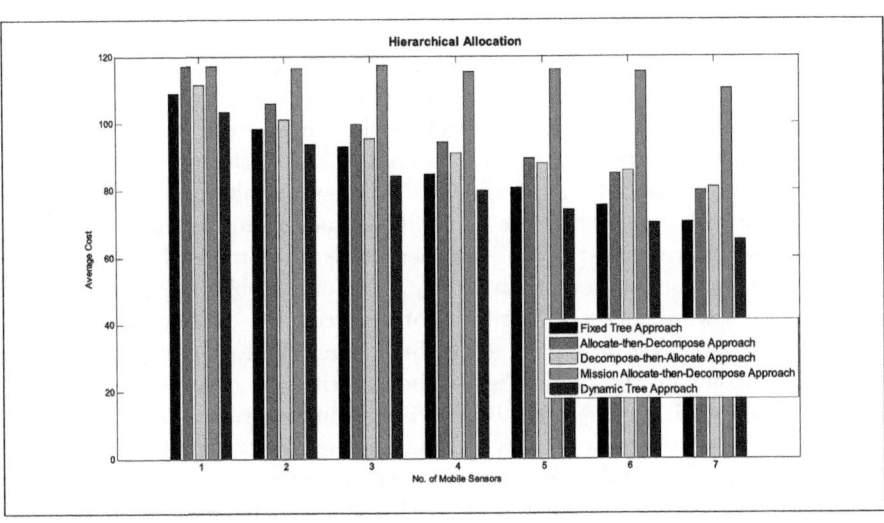

**Fig. 3.** Comparison of the average cost for hierarchical allocation mechanism

## 3.2   Target Tracking Simulation and Results

As mentioned above, the target tracking algorithm for an individual target is decoupled from the cooperative tracking algorithm for a multi-tracking system. So, the simulation results for single target tracking system, as a basis layer of the cooperative multi-target system, are presented first. As seen from in Fig. 4, the unsupervised learning EKM has provided a good tracking performance in terms of accuracy and tracking speed.

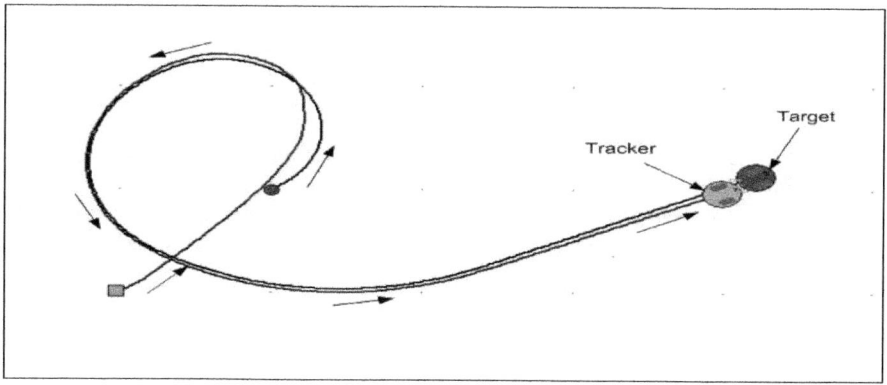

**Fig. 4.** Single target tracking using unsupervised learning EKM

## 4   Conclusion

A market-based framework for mobile surveillance systems has been presented in this paper. The proposed framework capitalizes on the strengths of market economies that enable mobile sensing agents to collectively execute complex tasks efficiently and reliably. Task allocation and cooperative target-tracking have been studied in this paper using the proposed framework as two challenging problems of mobile surveillance systems. These challenges are addressed individually and collectively. The results of the conducted experiments showed that hierarchical dynamic tree task allocation outperforms all other techniques. In the future, we consider extending the proposed algorithms so that constrained and tight tasks can be handled. An example for constrained tasks is two tasks that cannot be done independently as the same sensor would obviously have to do both of them. Tight tasks cannot be decomposed into further single sensor tasks.

## References

1. Delle Fave, F.M., Canu, S., Iocchi, L.: Multi-objective Multi-robot Surveillance. In: Proceedings of the 4th International Conference on Autonomous Robots and Agents (2009)
2. Elmogy, A., Karray, F., Khamis, A.: Auction-based Consensus Mechanism for Co-operative Tracking in Multi-sensor Surveillance Systems. In: Proceedings of 4th (HNICEM) International Conference, Manila, Philippines, pp. 149–158 (2009)

3. White, H.D.: Sensor Models and Multisensor Integration. Int. Journal of Robotics Research 7(6), 97–113 (1998)
4. Gerkey, B.P., Mataric, M.J.: Sold!: Auction Methods for Multirobot Coordination. IEEE Transactions on Robotics and Automation 18(5), 758–768 (2002)
5. Dias, M.B., Stentz, A.: A market approach to multi-robot coordination. Tech. Rep., Carnegie Mellon University, USA (2001)
6. Sayyaadi, H., Moarref, M.: A Distributed Algorithm for Proportional Task Allocation in Networks of Mobile Agents. IEEE Transactions on Automatic Control 56(2), 405–410 (2011)
7. Zlot, R., Stentz, A.: Market based multirobot coordination for complex tasks. Int. Journal of Robotics Research 25(1), 73–101 (2006)
8. Wurll, C., Henrich, D., Worn, H.: Multi-Goal Path Planning for Industrial Robots. In: International Conference on Robotics and Application (RA 1999), Santa Barbara, USA (1999)
9. Martinez-Jaramillo, S., Tsang, E.P.K.: An heterogeneous, endogenous and coevolutionary GP-based financial market. IEEE Trans. on Evol. Comput. 13(1), 33–55 (2009)
10. Zlot, R.M.: An Auction-Based Approach to Complex Task Allocation for Multirobot Teams. PhD Thesis, Robotics Institute, Carnegie Mellon University, USA (2006)
11. Dias, M.B.: TraderBots: a new paradigm for robust and efficient multi-robot coordination in dynamic environments. PhD Thesis, Robotics Institute, Carnegie Mellon University, USA (2004)
12. Gerkey, B.P., Mataric, M.J.: Sold!: auction methods for multi-robot control. IEEE Trans. Robot. Autom (Special Issue on Multi-Robot Systems) 18(5), 758–768 (2002)
13. Botelho, S.C., Alami, R.: M+: A scheme for multi-robot cooperation through negotiated task allocation and achievement. In: Proc. IEEE Int. Conf. Robot. Autom (ICRA), pp. 1234–1239 (1999)
14. Tang, F., Saha, S.: An anytime winner determination algorithm for time-extended multi-robot task allocation. In: Proc. Int. Conf. on Automation, Robotics, and Control Syst., pp. 123–130 (2008)
15. Kolling, A., Carpin, S.: Cooperative Observation of Multiple Moving Targets: an Algorithm and Its Formalization. Int. Journal of Robotics Research 26(9), 935–953 (2007)
16. Parker, L.E.: Distributed Algorithms for Multi-robot Observation of Multiple Moving Targets. Int. Journal of Autonomous Robots 12(9), 231–255 (2002)
17. MacQueen, J.B.: Some Methods for Classification and Analysis of Multivariate Observations. In: Proceedings of 5th Berkeley Symposium on Mathematical Statistics and Probability, vol. 12(9), pp. 281–297 (1967)
18. Kanungo, T., Mount, D.M., Netanyahu, N.S., Piatko, C.D., Silverman, R., Wu, A.Y.: An Efficient k-Means Clustering Algorithm: Analysis and Implementation. IEEE Transactions on Pattern Analysis and Machine Intelligence 24(7), 881–892 (2002)
19. Ritter, H., Martinetz, T., Schulten, K.: Neural Computation and Self-Organizing Maps: an Introduction. Addison-Wesley Longman Publishing Co., Inc., Boston (1992)
20. Kohonen, T.: Self-Organization and Associative Memory. Series in Information Sciences, Heidelberg (1989)

# MobiFuzzy: A Fuzzy Library to Build Mobile DSSs for Remote Patient Monitoring

Flavio Frattini[1], Massimo Esposito[2], and Giuseppe De Pietro[2]

[1] Università degli Studi di Napoli Federico II, Naples, Italy
flavio.frattini@unina.it
[2] National Research Council of Italy,
Institute for High Performance Computing and Networking (ICAR), Naples, Italy
{esposito.m,depietro.g}@na.icar.cnr.it

**Abstract.** Recently, a new mobile generation of decision support systems (DSSs) is appearing to seamlessly and ubiquitously support the monitoring of patients' health status during the activities of daily living. This work proposes MobiFuzzy, a Java Micro Edition fuzzy library characterized by a light-weight and update-versatile implementation for resource-limited mobile devices. The library eases the design process of fuzzy DSSs for Remote Patient Monitoring by providing the user with a wide range of fuzzy connectives, membership functions, implication, aggregation and defuzzification methods. MobiFuzzy has been evaluated on different smart-phones in terms of time-processing with respect to a home-monitoring scenario, proving its capability to proficiently build fuzzy mobile DSSs for healthcare applications where real-time performance demands have to be met.

**Keywords:** Fuzzy Logic, Decision Support Systems, Mobile Computing, Pervasive Healthcare.

## 1 Introduction

During the last decade, Remote Patient Monitoring (RPM) technologies have shown their effectiveness in monitoring population affected by chronic diseases [1]. Recently, RPM is sensibly changing due to radical advances in pervasive technologies applied to health care. Indeed, several factors, such as the miniaturization of integrated circuits and the improvement and reduction of the costs of communications, have increased quality and performances of tiny mobile devices, so as to enable a new long-term care option based on both modern sensor infrastructures and Decision Support Systems (DSSs) for monitoring patients' health status during the activities of daily living, so as to promote individual independence and well-being [2].

Constantly developed during the years, DSSs for RPM are now designed as knowledge-based (KB), i.e. they support the analysis and interpretation of a number of monitored parameters by simulating the process followed by the physicians, in order to early detect critical conditions and apply appropriate countermeasures, such as actions and recommendations.

M. Kamel, F. Karray, and H. Hagras (Eds.): AIS 2012, LNCS 7326, pp. 79–86, 2012.

This typology of DSSs, typically associated with desktop systems, is recently migrating on mass-market smart phones with more and more increasing capabilities to perform a wide variety of activities thanks to advanced CPUs, memories, and displays. In particular, mobile health DSSs for remote patient monitoring are increasingly appearing, with the aim of facilitating self care and communications with doctors by offering reasoning facilities for processing data gathered by sensor devices directly on mobile devices [3]. Existing examples of mobile DSSs essentially rely on rule-based programming, and, in particular, they model the knowledge in terms of crisp rules and exploit rule engines for reasoning purposes [4, 5]. However, they do not reproduce the real physician's decision-making process often pervaded by uncertainty and vagueness, and, thus, they can represent an unrealistic oversimplification of reality, leading to possible wrong interpretations when compared to a direct observation.

Fuzzy Logic [6] has been proposed as the most suitable approach for profitably tackling uncertainty and vagueness and providing for enhanced DSSs [7, 8]. Building advanced mobile DSSs based on Fuzzy Logic for monitoring patient's health status requires software tools and libraries able to reduce the development time and effort and ease the exploration of new theoretical aspects. In the last few years, several tools tailored to the fuzzy paradigm have been developed [9, 10, 11], but they are all dedicated to desktop computer applications and have also constraints on the set of fuzzy operations they support and on the complexity of the systems they can design. Furthermore, to the best of our knowledge, no fuzzy tool has been deployed for mobile devices until now, neither system-oriented researches regarding fuzzy libraries for mobile DSSs have been developed in that direction.

In such a sense, this work proposes a Java Micro Edition fuzzy library, named MobiFuzzy, characterized by a light-weight, easy-to-develop, and update-versatile implementation suitable for resource-limited mobile devices. The library eases the design process of Fuzzy DSSs by providing the user with a wide range of fuzzy connectives, membership functions, implication, aggregation and defuzzification methods.

It is entirely programmed in Java, in accordance with the object-oriented paradigm, which makes it easy to maintain and extend. Besides, it is extremely portable since executable on any mobile device with a compatible Java Virtual Machine installed. As a proof of concept, MobiFuzzy has been evaluated on different smart-phones in terms of time-processing with respect to a home-monitoring scenario.

The rest of the paper is organised as follows. Section 2 depicts the library functionalities while, in Section 3, its implementation is described. Finally, Section 4 discusses results and concludes the work.

## 2     Library Functionalities

MobiFuzzy allows to define single-input-single-output (SISO), multi-input-single-output (MISO), or multi-input-multi-output (MIMO) fuzzy systems on the basis of the number of input and output variables. It implements a FITA (First Infer Then Aggregate) approach for the inferential process, i.e. it first applies the compositional inference rule at each rule and, then, it combines the conclusions of all rules. It supports the building of both Mamdani and Sugeno systems [12, 13] depending on the used reasoning mechanism.

Moreover, several membership functions, fuzzy operators, implication, aggregation, and defuzzification methods are provided to allow users to create a Fuzzy DSS best suited to their needs. All these issues are fully described below.

**Membership Functions:** MobiFuzzy enables the construction of membership functions characterized by different shapes, as reported in the following:

- *Trapezoidal, triangular, and step:* they are the most common membership functions adopted above all in Mamdani-type fuzzy systems. Such functions are diffusely used since they satisfy the orthogonality requirement, i.e. for each element of the universe of discourse the sum of all its membership values is equal to one.
- *Singleton*: it is provided for supporting Sugeno fuzzy systems, commonly applied to classification problems.
- *Bell curve, Cosine, Gaussian, and Sigmoidal:* they are derivable membership functions introduced to support the automatic tuning of their shapes. Indeed, after formalizing the physician's knowledge under the form of rules, the designer has to choose the shape and location of membership functions for all the linguistic values related to all the linguistic variables involved. This requires both medical expertise and technical intervention along with great effort to identify which among the design choices are suited to the given problem. An adaptive fuzzy system allows to construct a set of rules and membership functions on the basis of knowledge provided by physicians and to adapt its behaviour thanks to additional knowledge extracted by data. In particular, in order to apply neural network approaches usually based on gradient-descent back-propagation, only the reported derivable membership functions can be used.
- *Piecewise:* it allows to define a function with linear pieces, point by point.

**Fuzzy Operators:** typical implementation foreseen in Mamdani systems is provided, i.e. the fuzzy operators *And* and *Or* defined by using the minimum (*AndMin*) and the maximum (*OrMax*), respectively [12]. Furthermore, MobiFuzzy enables the definition of the probabilistic operators; in this case the *And* is defined as a product (*AndProd*) and the *Or* by the algebraic sum (*OrAlgebraicSum*). The *Not* operator is implemented as the simple standard complement.

Nevertheless, in many situations, the use of these fuzzy operators can generate a scarcely intuitive behaviour for physicians. For instance, let us consider the case illustrated in Fig. 1, which refers to two linguistic terms, namely *medium* and *high*, associated the linguistic variable *fever*.

The membership functions, named $\mu_A$ and $\mu_B$ in figure, associated to the terms *medium* and *high* are partially overlapped; the disjunction achieved by using the *OrMax* operator provides the fuzzy set depicted in light gray in the figure. This implies that, for each input value between 37 and 39, the degree of membership will be lower than the ones in the ranges [35.8; 37] and [39; 40.6]. But, since such values are centigrade degrees associated to the linguistic variable *fever*, this kind of behaviour may be incorrect. Indeed, the concept of *medium-high fever* expressed as disjunction of the two terms should have an additive behaviour so that the degrees of membership are the same in the area where the two fuzzy sets are overlapped.

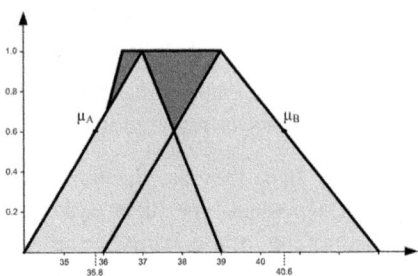

**Fig. 1.** An example of application of Lukasiewicz operator Or

To overcome these issues, MobiFuzzy supports the operators defined by the *Lukasiewicz* logic. Let us consider Fig. 1 again. The *Lukasiewicz* operator *Or* also considers the dark gray part as result of the disjunction, generating a different fuzzy set. In such a way, the result of the disjunction will be 1 also between 37 and 39, so achieving a behaviour in accordance with the physicians' common sense.

**Implication:** There are many typologies of implication function defined in literature to perform different inferences. A binary operator I:[0, 1] x [0, 1] $\rightarrow$ [0, 1] is an implication function if it is non-increasing in the first variable, non-decreasing in the second variable and I(0, 0)=I(1, 1)=1 and I(1, 0)=0 [14].

The most diffused inference rules are *modus ponens* and *modus tollens* [14]. MobyFuzzy supports implication methods, such as *Kleene-Dienes* and *Lukasiewicz*, which generalize the classical implication since they take value 1 when antecedents take value 0, and, in addition, also the Mamdani implication, which does not respect the properties of the classic implication operator since I(0,0)=min(0,0)=0 and it is not equal to 1. Moreover, MobiFuzzy also provides the *product* (*Prod*) and *BoundedDifference* methods.

**Aggregation:** Similarly to the fuzzy operators for aggregating antecedents when evaluating a rule, also the reasoning process requires to unify the outputs of each rule into a single fuzzy set by means of an aggregation operator. In particular, MobiFuzzy implements the most common method, based on the use of the *maximum* (*Max*) and the *BoundedSum* in order to support the Lukasiewicz logic. Other aggregation methods provided are also *Sum*, *AlgebraicSum* and *NormedSum*.

**Defuzzification:** In literature there are several Defuzzification methods, each of them being suitable to face the requirements of a particular application.

Indeed, in some cases the computational efficiency of the defuzzification method is the main parameter for the choice, since, for instance, in real-time systems the number of operations required to evaluate a defuzzified value should be strongly reduced to concretely achieve the maximum efficiency. Another criteria for choosing the defuzzification method is the transparency, i.e. a defuzzification method should be easy to understand rather than based on a complicated mathematical formula [15].

MobiFuzzy provides *CenterOfArea* and *CenterOfGravity* defuzzification methods that are the most diffusely used since their output is calculated taking into account the trend of the function in all its domain [15]. Moreover, the library also provides *CenterOfGravitySingleton* method, which evaluates the center of gravity of a set of singletons, and maximum based methods, such as *LeftMostMax*, *RightMostMax*, and *MeanOfMaximum*.

# 3    MobiFuzzy Implementation

Both the design and implementation of MobiFuzzy deal with some interesting development and packaging issues raised by the extremely small footprints (memory, CPU, etc.) of mobile devices, consisting in how to fit this complex library into resource-constrained devices.

MobiFuzzy was developed for the J2ME [16] according to its minimum Java libraries, technology APIs (Application Programming Interfaces) and Virtual Machine capabilities, by allowing the creation of a mobile DSS based on a Fuzzy Inference System offering the functionalities stated in the previous Section 2. The choice to use Java is due to its platform independence, i.e. it is extremely portable on any mobile device equipped with an operating system supporting a compatible java VM. Moreover, Java is always supported by Symbian and Android mobile devices, which in April 2011 have a market share of 19.2% and 38.5%, respectively [17].

However, since J2ME does not include a complete Math library for resource-limited devices, a java floating point library, named Real [18], for J2ME mobile devices has been used and, more precisely, opportunely expanded, for instance to allow to return numbers of double type. The resulting library has been compared to the Java Standard Edition implementation on the basis of 1 million randomly generated numbers to prove its correctness. In the 87.6% of tests, differences on the results have been picked out only from the eighteenth significant digit; in the 100% of cases, differences have been picked out only from the sixteenth significant digit.

From an architectural perspective, a Fuzzy Inference System (FIS) buildable by means of MobiFuzzy is made of the set of components depicted in Fig. 2.

The system has $n$ input variables, so there are $n$ values as input, namely $x_1, ..., x_n$, one for each variable. Each variable has one or more terms, each one characterized by a membership function, $\mu_i$.

The *Fuzzifier* allows to define such membership functions; moreover, input values are crisp and there is the necessity to translate them as values for a fuzzy term, hence such a module also provide methods for fuzzification.

The component *Rules* simply contains the set of $p$ rules of the system, $r_1, ..., r_p$. In more detail, the following general representation form is proposed for a rule:

| | |
|---|---|
| (i) | IF antecedent THEN consequent; |
| (ii) | IF antecedent$_1$ connector antecedent$_2$ THEN consequent; |
| (iii) | IF antecedent$_1$ connector$_1$ antecedent$_2$ connector$_2$ ... antecedent$_N$ THEN consequent. |

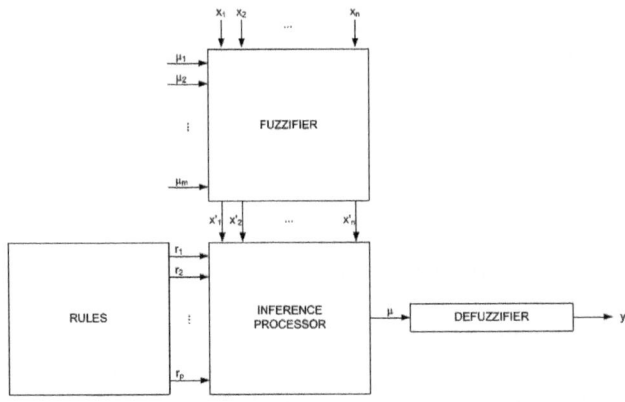

**Fig. 2.** Fuzzy Inference System - FIS

Each antecedent may be composed of (i) a simple expression of only one term (e.g., A is large), (ii) by two terms (e.g., A is large AND B is small), or (iii) by composite expressions (e.g., A is large AND (B is small OR C is high)). In the latter case, priorities have also to be managed. Indeed, when describing such rules on paper it is possible to use brackets to define expressions that have to be first evaluated.

The *Inference Processor* evaluates each rule by aggregating antecedents, performing implications, and aggregating results; $\mu$ denotes the membership function of the output variable opportunely cut during the inference.

It is worth highlighting that the inference process has been optimized in order to face the limited resources the mobile devices are equipped with. As a matter of fact, the aggregation is performed in MobiFuzzy at the end of the implication of each rule and not after the evaluation of all rules. Usually, an inference process evaluates all rules and subsequently aggregates their results. However, such a way, the result of each rule should be stored as a fuzzy set opportunely clipped by the rule, i.e. an instance of the output variable and its terms with modified membership functions should be used for each rule. Since mobile devices do not have huge memories, this solution is to be avoided. Instead, performing aggregation after the evaluation of each rule enables to operate on only one instance of the output fuzzy set, avoiding the use of additional memory.

Finally, the *Defuzzifier* module provides as output a single crisp value, y.

As a proof of concept, the behaviour of the library has been evaluated on three mobile devices, namely Nokia N97 mini [19], Nokia N8 [20], and HTC Desire HD [21] in terms of time required for the inference process, with respect to a home monitoring scenario, where a chronic patient has to be monitored in real time by means of a pulse oximeter to detect a severe exacerbation of a chronic obstructive pulmonary disease.

In the worst case considered for such a scenario, i.e. 40 rules with 10 antecedents, the time required by MobiFuzzy to provide a crisp value for the output variable was 33.50 ms, 29.38 ms, and 46.75 ms for Nokia N97 mini, Nokia N8, and HTC Desire HD, respectively. In order to equip a mobile device with a DSS built on the top of MobiFuzzy and able to infer whether a critical situation is happening, the vital signal parameters gathered and transmitted by the pulse oximeter should be evaluated no more than two times per minute [22]. The results achieved on three different smart phones prove that MobiFuzzy can be proficiently used in the experimental scenario since it largely meets real-time performance demands. Generally speaking, these preliminary tests performed on MobiFuzzy gives a proof of its feasibility, suggesting that it can be applied to face new challenging RPM scenarios.

# 4    Discussion and Conclusions

MobiFuzzy is a Java Micro Edition library characterized by a light-weight, easy-to-develop and update-versatile implementation suitable for resource-limited mobile devices so that it can be used to build fuzzy DSSs for remote patient monitoring. MobiFuzzy is entirely programmed in Java in accordance with the object-oriented paradigm which makes it easy to maintain and extend.

The library eases the design process of Fuzzy DSSs by providing the user with a wide and self-contained range of fuzzy connectives, linguistic hedges, membership functions, implication, aggregation and defuzzification methods. Indeed, the current version of MobiFuzzy enables the user to build mobile DSSs embedding different types of FIS, from classical examples such as Mamdani, Sugeno, and Lukasiewicz, to customized solutions characterized by a variegate and ad-hoc combination of fuzzy operators and implication, aggregation and defuzzification methods.

Moreover, MobiFuzzy is extremely portable since executable on any mobile device equipped with an operating system supporting a compatible Java Virtual Machine. Indeed, in addition to Symbian and Android operating systems used in the experimental evaluation, also Windows Mobile and Blackberry OS could be adopted since equipped with a compliant Java VM. As a result, no further adaptation for different operating systems is necessary, i.e. the library is designed and developed just once and can be re-used independently of the actual platform and device it is running on.

Since, to the best of our knowledge, no fuzzy implementation has been deployed for mobile devices until now, neither system-oriented researches regarding fuzzy libraries for mobile DSSs have been developed in that direction, a comparative evaluation between the proposed library and other similar approaches in terms of performance has been impossible to plan and carry out.

In conclusion, the presented library offers an innovative and valuable tool to build a new mobile generation of fuzzy decision support systems for remote patient monitoring. Since it has a general basis, it can be profitably applied to face a variegate and heterogeneous set of new challenging pervasive healthcare scenarios, where information must be supplied, received, and/or used anywhere for supporting patients or physicians seamlessly and ubiquitously in their decision-making tasks.

# References

1. Guy, P., Mirou, J., Claude, S.: Systematic review of home telemonitoring for chronic diseases: The evidence base. J. Am. Med. Inform. Assoc. 14, 269–277 (2007)
2. Eren, A., Subasi, A., Coskun, O.: A Decision Support System for Telemedicine Through the Mobile Telecommunications Platform. J. Med. Syst. 32(1), 31–35 (2008)

3. Minutolo, A., Sannino, G., Esposito, M., De Pietro, G.: A rule-based mHealth system for cardiac monitoring. In: 2010 IEEE EMBS Conference on Biomedical Engineering and Sciences, pp. 144–149 (2010)
4. Albuquerque, R., Guedes, P., Filho, C.F., Robin, J., Ramalho, G.: Embedding J2ME-based Inference Engine in Handheld Devices: The KEOPS Case Study. In: The Workshop on Ubiquitous Agents on Embedded, Wearable, and Mobile Devices (2002)
5. Hall, L., Gordon, A., Newall, L., James, R.: A Development Environment for Intelligent Applications on Mobile Devices. Expert Systems Application 27(3), 481–492 (2004)
6. Zadeh, L.: Fuzzysets. Inform. Control. 8, 338–353 (1965)
7. Alayón, S., Robertson, R., Warfield, S.K., Ruiz-Alzola, J.: A fuzzy system for helping medical diagnosis of malformations of cortical development. J. Biomed. Inf. 40(3), 221–235 (2007)
8. Dong-Her, S., Hsiu-Sen, C., Binshan, L., Shih-Bin, L.: An Embedded Mobile ECG Reasoning System for Elderly Patients. IEEE Trans. Inf. Technol. Biomed. 14(3), 854–865 (2010)
9. jFuzzyLogic: Open Source Fuzzy Logic, http://jfuzzylogic.sourceforge.net/html/index.html
10. Mathworks, Fuzzy Logic Toolbox, http://www.mathworks.com/products/fuzzylogic
11. Velo, F.J.M., Baturone, L., Solano, S.S., Barriga, A.: Rapid design of fuzzy systems with Xfuzzy. The 12th IEEE International Conference on Fuzzy Systems 1, 342–347 (2003)
12. Mamdani, E.H., Assilian, S.: An experiment in linguistic synthesis with a fuzzy logic controller. International Journal of Man-Machine Studies 7(1), 1–13 (1975)
13. Takagi, T., Sugeno, M.: Fuzzy identification of systems and its applications to modeling and control. IEEE Trans. Syst. Man Cybern. 15(1), 116–132 (1985)
14. Mas, M., Monserrat, M., Torrens, J., Trillas, E.: A Survey on Fuzzy Implication Functions. IEEE Trans. Fuzzy Syst. 15(6), 1107–1121 (2007)
15. Roychowdhury, S., Pedrycz, W.: A survey of defuzzification strategies. Int. J. Intell. Syst. 16(6), 679–695 (2001)
16. Oracle, Java Micro Edition, http://www.oracle.com/technetwork/java/javame/overview/index.html
17. Gartner: Gartner Says Android to Command Nearly Half of Worldwide Smartphone Operating System Market by Year-End 2012 (2011), http://www.gartner.com/it/page.jsp?id=1622614
18. Lauritzsen, R.: Real – Java floating point library for MIDP devices (2011), http://real-java.sourceforge.net/Real.html
19. Nokia, N97 mini device details, http://developer.nokia.com/Devices/Device_specifications/N97_mini (retrieved February 2011)
20. Nokia, N8 device details, http://developer.nokia.com/Devices/Device_specifications/N8-00 (retrieved February 2011)
21. HTC, HTC unveils Desire HD, http://www.htc.com/www/press.aspx?id=144266&lang=1033 (retrieved February 2011)
22. Nonin, Onyx II 9550, http://www.nonin.com/documents/5196-000-04_9550_Spec_Sheet.pdf (retrieved February 2011)

# Tactical Resource Planner for Workforce Allocation in Telecommunications

Ahmed Mohamed[1], Hani Hagras[1], Sid Shakya[2], and Gilbert Owusu[2]

[1] School of Computer Science and Electronic Engineering,
University of Essex, Wivenhoe Park, Colchester, CO4 3SQ, UK
[2] British Telecommunication Adastral Park, Martlesham, Ipswich, UK

**Abstract.** Resource planning is one of the most important operational issues for many companies. This is especially crucial for telecommunications companies. Resource planning aims to provide a high quality of service to the customers while trying to keep the cost as low as possible. This is done by trying to utilize the available resources (workforce) as much as possible so that they can match the expected demand for services. Tactical resource planning looks at medium-term planning periods, i.e. weeks to months, and aims to establish coarse-grain resource deployments. This paper focuses on fuzzy based resource planning approach in British Telecom (BT). We will present a hierarchical based fuzzy logic system which calculates the compatibility between resources (technicians) and the allocated tasks, and then matches the most compatible tasks and technicians to each other. The proposed hierarchical fuzzy logic based system in an experimental setting was able to achieve very good results in comparison to the original system, where the proposed system was able to achieve 12.2% improvement in utilization, 34% increase in technician deployment ,10.8% decrement in travel time and 116.2% improvement in number of important tasks being completed. The proposed system is being incorporated in the workforce planning system in BT.

**Keywords:** fuzzy logic systems, hierarchical fuzzy logic systems, tactical resource planning and telecommunications.

## 1    Introduction

Resource planning is an integral component of service chain management as it ensures that customer commitments are met, that a high quality service is maintained and that operational costs are kept as low as possible. Service enterprises with varied manifold service offerings often employ large, diverse workforces to deliver their services. Such resource planning scenarios are particularly challenging as different types of services often require individual planning approaches but, at the same time, the aim is to resource plan for the entire workforce to guarantee an optimal overall balance [1].

The main aim of resource planning is to flex the available resources so that they match the expected demand for services as closely as possible. The challenge is to

M. Kamel, F. Karray, and H. Hagras (Eds.): AIS 2012, LNCS 7326, pp. 87–94, 2012.

strike the right balance between the quality of service delivered to customers on one hand and the resource costs incurred by the business on the other hand [1].

Resource planning can be divided into three main stages (as shown in Figure 1a): strategic, tactical and operational. Strategic resource planning is long-term planning, usually over time horizons of several months to several years, and is concerned with the overall balance between service demand and resource capacity. Tactical resource planning looks at medium-term planning periods, i.e. days to months, and aims to establish coarse-grain resource deployments. Such deployment plans are refined during the operational resource-planning phase, i.e. the scheduling phase, which sees the allocation of individual resources to specific tasks at a very detailed schedule (for example, for each technician, what task to be executed at what specific time of the day). Such schedules are generated for short periods. Information flows from strategic to tactical to operational resource planning (as shown in Figure 1a) as the resource utilization becomes more and more refined [1].

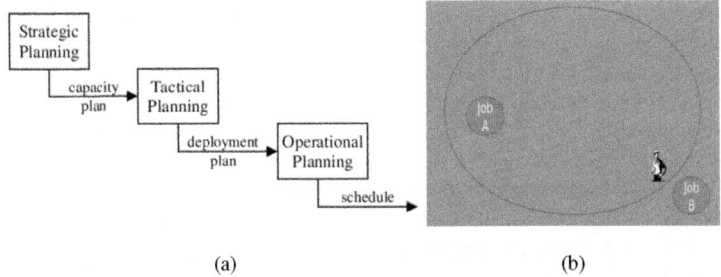

(a)                                             (b)

**Fig. 1.** (a) The stages of resource planning [1] (b) First limitation

This paper mainly focuses on solving the tactical resource planning problems faced by the original BT system, which is based on using a crisp based approach. In the telecommunication domain, the resource refers to the technician who is allocated to resolve various tasks (the tasks include fixing telephone lines, installing an internet connection, etc.).

On one hand, tactical resource planning focuses on planning in the medium-term, i.e. Similar to strategic planning, it provides resource managers with an overall, condensed view of the balance between demand and supply as well as capabilities to influence this balance. On the other hand, and interlinked with operational planning, tactical resource planning often involves the anticipatory deployment of individual resources and thus has to provide a detailed plan about the expected utilization of the workforce. The requirement to generate both coarse-grain and fine-grain plans pose certain challenges to the development of tactical resource planning systems [2] where:

- The tactical planning has to address the balance between quality of service and cost reduction objectives of the business.
- The tactical planning must provide an automated approach for generating both coarse-grain and fine-grain plans. The more accurate and optimal these plans, the quicker they will be implemented.

Several approaches were introduced to solve the resource planning problems such as particle swarm optimization [3] and data envelopment analysis [4]; however these approaches employed crisp sets that couldn't deal with allocating resources to over-lapping areas. Other systems employed type-2 fuzzy logic based system optimized by evolutionary algorithms [5], however these systems suffered from the fuzzy logic curse of dimensionality and hence it could only deal with limited problem sizes. In [6], an analytic hierarchy process which included fuzzy mathematics comprehensive evaluation was employed for human resource allocation, however this system couldn't be deployed in real time and it could be only used for the system overall evaluations. In [7], a fuzzy mixed-integer linear programming was employed for re-source allocation; however, it takes a long time to do the allocation, which is not fea-sible for resource allocation in large geographical areas.

In this paper, we will present a hierarchical based fuzzy logic tactical resource planner for the workforce allocation. The proposed system uses a hierarchical fuzzy logic system to avoid the curse of dimensionality. The system calculates the compati-bility between technicians and the allocated tasks and then the system assigns the most compatible tasks and technicians to each other. The proposed hierarchical fuzzy logic based system was able to achieve very good results in comparison to the original system.

Section 2 will present an overview of the existing resource-planning tool employed in BT and its problems. Section 3 will present the proposed hierarchical fuzzy logic system. Section 4 will present the proposed fuzzy logic tactical resource planner. Section 5 will present the experiments and results. Finally the conclusions and future work will be presented in Section 6.

# 2    Overview of the Resource Planning Tool in BT

BT uses the Field Optimization Suite (or FOS), which is a collection of tightly inte-grated components for service chain management [1].

FOS is developed by BT's Intelligent System Lab and is targeted to companies that provide their services through a field force such as BT's large field engineering work-force or transportation companies with large workforces of drivers or businesses, which rely on a workforce of salespeople [1].

## 2.1    Limitations of the Original Approach

One of the resource planning engines does not take into account distance based factors when deploying technicians. It assumes that a technician can do any jobs within a dep-loyed geographical working area in a domain, giving equal weight to each job in terms of the travel distances. This was one of the key limitations that had to be overcome, i.e. an update to original logic was required.

Let's list the limitations of the original logic and describe them with examples

1.    Technicians are limited to work inside their assigned geographical working area.
2.    Taking the distance between tasks and technicians was limited to a certain level
3.    Technicians' finish locations weren't taken into account when assigning tasks to the technicians towards the end of the day.

To illustrate the first limitation (as shown in Figure 1b), let's assume the following:

- Job A is far, has low importance and requires skill that is not very compatible with technicians skill preference, but it is inside the technician's assigned working area
- Job B is close, has high importance and has good skill preference but is outside the technician's assigned working area and there is no available technician to do it

With the original crisp based system, the technician will be assigned to job A although job B is closer to the technician, has higher importance than job A, has better skill and area preference.

Concerning the second limitation, the original system doesn't take distance between the task and technician into consideration and it assigns tasks to the technicians based on task's importance, skill preference and area preference while ignoring distance, which is a very important factor.

In the third limitation, the original crisp based system doesn't take distance to technician's finish location (home location) into consideration when it assigns tasks to him so at the end of the day. Hence, at the end of the day, the technician might be assigned to a task, which is far from his home, and then the technician might waste time travelling back home at the end of the day. On the other hand, if the distance to finish location was taken into consideration, the technician could have been assigned to a task closer to his home and hence we could have saved on his time consumed to travel back home.

## 2.2    Improving the Original System

Improving the original system is a challenge and requires implementing a new system that takes the same inputs as the old system in addition to the new inputs (distance between the technician and the task, distance between the technician and his finish location (home location), tasks' density and technician's expected working time left) and generates a new output or a new deployment plan that looks like the old one so it can be passed to the operational planning to provide a more refined schedule. In order to improve the original system, various challenges were faced such as different kinds of uncertainty in the technician's current, start and finish locations, size of the assigned working area and tasks density. Fuzzy logic systems are known to provide a very good performance when dealing with uncertainties. However, due to the large number of inputs involved, traditional fuzzy systems will not be practical due to the big number of needed rules to be generated as a result of the curse of dimensionality problem. Hence, we will present an improved hierarchical fuzzy logic system to allow using the benefits of fuzzy logic systems while avoiding the curse of dimensionality problems.

## 3      The Proposed Hierarchical Fuzzy Logic System

The proposed hierarchical fuzzy logic system mainly concentrates on the input with the highest priority (in this case, it is the distance between the technician and the task) by taking it as a common factor with all the other inputs (as shown in Figure 2). In this case, in the lowest level of the hierarchical fuzzy system, there will be fuzzy subsystems (FLSs) where each FLS receives the input with the highest priority (distance in our case) and another input from the other available inputs. Each FLS then produces the fired rules by the incoming two inputs. The system then takes all the fired rules from all the fuzzy subsystems and passes them to the defuzzifier to calculate an overall compatibility score between the task and the technician.

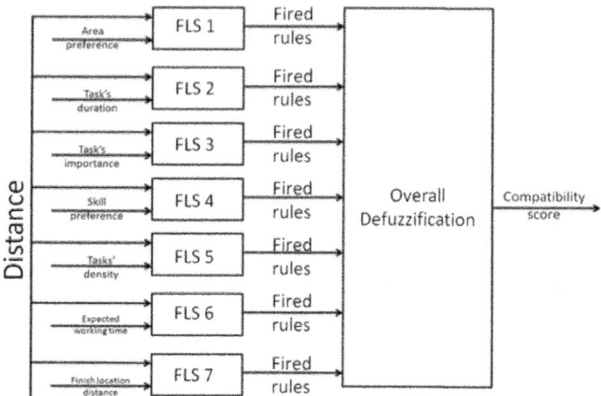

**Fig. 2.** 1 Level Hierarchical Fuzzy Logic System

In the proposed hierarchical structure, the same number of fuzzy sets will represent all the inputs in the lowest level of the hierarchical structure. Hence, the number of rules increases linearly ($m^2$ will be a constant number) with the number of inputs as follows:

$$\sum_{i=1}^{N-1} m^2 \tag{1}$$

Where $m$ is the number of fuzzy sets used to represent each input and $N$ is the total number of inputs.

In our application we have 8 inputs and hence there will be 7 FLSs where each FLS will have as input the distance and one of the other remaining 7 inputs as shown in Figure 2. Each of the inputs is represented by three fuzzy sets. Hence, each FLS in the lowest level of the hierarchical structure will have 9 rules. Hence, the number of rules in the lower level of the hierarchical structure will be 63 rules. In addition, in our proposed hierarchical structure, there is no need for high-level arbitration or blending between the low level behaviors as was the case in [8].

In Figure 2, we can see that each low level FLS outputs its fired rules that go into the overall defuzzification phase. The overall defuzzification uses the center of sets defuzzification to calculate the final output, which is written as follows:

$$y(x) = \frac{\sum_{l=1}^{M} y^{-l} w^l}{\sum_{l=1}^{M} w^l} \tag{2}$$

Where $M$ is the total number of fired rules, $y^{-l}$ is the center of gravity of the output set of the $l^{th}$ rule and $w^l$ is the firing strength of the $l^{th}$ rule irrespective, which FLS is firing this given rule. The crisp output of the hierarchical fuzzy system represents the compatibility score between the given task and the given technician.

The hierarchical fuzzy logic systems in [9] address the "curse of dimensionality" problem but they face the intermediate inputs and rules problem where it is not easy to define rules for them.

On the other hand, the proposed hierarchical fuzzy logic system shown in Figure 2 addresses the "curse of dimensionality" problem where its rules grow linearly with the number of inputs and it has no intermediate inputs and rules. In addition, the proposed hierarchical fuzzy system can be easily designed and it is modular where low levels FLSs could be easily added or deleted.

## 4    The Proposed Hierarchical Fuzzy Logic Based Tactical Planner

As shown in Figure 3, the proposed system works as follows, an expert configures the system; the system uploads today's data (resources and tasks) from the database, then the system sorts the tasks between the resources and vice versa. The system then calculates the compatibility scores between each resource and his tasks and between each task and its resources using the hierarchical fuzzy logic system mentioned in the previous section, after that the system tries to find the best matching couple to start the work simulation using the compatibility scores and a matching algorithm that takes the point of view of the tasks and the resources to match them together. If the system failed to find the best matching couple; this means that it is done and the system will output the results.

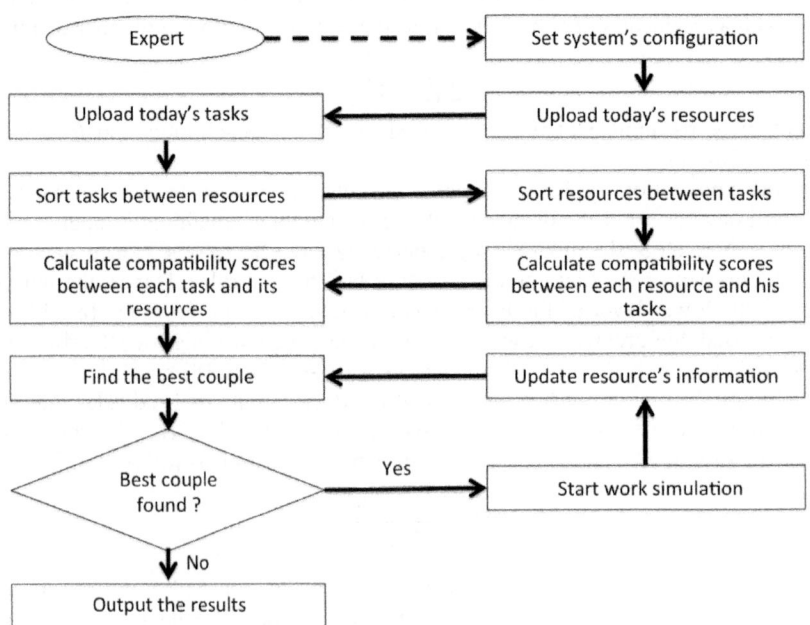

**Fig. 3.** An Overview of the proposed hierarchical fuzzy logic based tactical resource planner

The proposed system uses the following 8 inputs to calculate the compatibility score between a task and a resource:

1. Distance between the resource and the task
2. Area preference
3. Skill preference
4. Task's duration
5. Task's importance
6. Tasks' density
7. Resource's expected working time
8. Distance between the resource and his finish location

After calculating the compatibility scores, a matching algorithm takes these compatibility scores and starts a matching procedure to match the best resources and tasks together. The behaviour of this matching algorithm can be configured by changing the rule base of the hierarchical fuzzy logic system.

# 5    Experiments and Results

The proposed system is being incorporated in the workforce planning in BT. The experiments were carried in a controlled environment and the results presented below are the expected improvements from deployment planning point of view. The output of the deployment plan is then passed to the operational planner which optimally schedules the technicians and dispatches the jobs within the deployed area.

In this section; we will show experimental average results based on five UK geographical areas over 4 days:

These domains were chosen to exemplify domains, which are densely populated, medium populated and low populated domains.

**Table 1.** Comparison between the original system and the fuzzy system

| Measure | Average of all domains | | % Difference |
|---|---|---|---|
| | Current system | Fuzzy System | |
| Deployed techs | 15.9 | 21.3 | 34.0% |
| Total techs | 23.5 | 23.5 | 0.0% |
| Covered Tasks | 65.22 | 94.72 | 45.2% |
| Tasks per resource | 3.92 | 4.40 | 12.2% |
| Deployed techs utilized hours | 121.78 | 163.36 | 34.1% |
| Deployed techs total hours | 134.37 | 178.85 | 33.1% |
| Tech's utilization % | 62.01 | 83.19 | 34.1% |
| Average travelling distance | 72.96 | 65.11 | -10.8% |
| Average time needed to travel home | 0.8 | 0.69 | -13.3% |
| Number of important tasks | 10.89 | 23.55 | 116.2% |

As shown in Table 1, the experimental results from the proposed hierarchical fuzzy system deployed more resources by 34%, covered more tasks by 45.2%, did more tasks per resource by 12.2%. In addition, our system worked more by 34.1% and technician's utilization was more by 34.1%. Our system resources average traveling distance was less by 10.8% and the average time needed to travel home was less by 13.3% and the number of important tasks done was more by 116.2%.

# 6    Conclusions and Future Work

In this paper, we presented a hierarchical fuzzy logic based approach for tactical resource planning.

We have introduced an improved hierarchical fuzzy logic system, which is able to handle the curse of dimensionality problem in fuzzy systems where the rules grow linearly rather than exponentially. In addition, there is no need to define the intermediate or high-level rules in the hierarchical structure, as was the case with the existing hierarchical structures. Furthermore, the system is modular where new low-level FLSs could be easily inserted or deleted to suit the problem at hand.

We have presented the hierarchical fuzzy based tactical resource planner, which significantly outperformed the original crisp based system and resulted in significant uplift in the operations.

For the future work, we intend to extend the proposed system to be based on type-2 fuzzy logic system to be able to handle higher uncertainty levels.

# References

[1] Kern, M., Shakya, S., Owusu, G.: Integrated Resource Planning for Diverse Workforces. In: Computers & Industrial Engineering (CIE 2009), pp. 1169–1173 (July 2009)

[2] Voudouris, C., Owusu, G., Dorne, R., Lesaint, D.: Service Chain Management. Springer (2008)

[3] Wang, S., Gong, L., Yan, S.: The allocation optimization of project human resource based on particle swarm optimization algorithm. In: IITA International Conference on Services Science, Management and Engineering (2009)

[4] Weng, W., Su, J., Chen, G., Wang, Z.: An approach for all allocation optimization of multi-project human resource based on DEA. In: International Conference on Management and Service Science (2010)

[5] Miller, S., Gongora, M., Popova, V.: Optimising resource plans using an interval type-2 fuzzy model. In: Fourth International Workshop on Genetic and Evolutionary Fuzzy Systems, Mieres, Spain (March 2010)

[6] Daojin, F.: Research on the comprehensive evaluation of the human resource allocation based on analytic hierarchy process and fuzzy mathematics. In: Second International Conference on Industrial and Information Systems (2010)

[7] Feili, H., Khoshdoon, M.: A Fuzzy Optimization Model for Supply Chain Production Planning with Total Aspect of Decision Making. The Journal of Mathematics and Computer Science 2(1), 65–80 (2011)

[8] Saffiotti, A.: Fuzzy logic in autonomous robotics: behaviour coordination. In: Proceedings of the 6th IEEE International Conference on Fuzzy Systems, Barcelona, Spain, pp. 573–578 (1997)

[9] Wang, D., Zeng, X., Keane, J.: A survey of hierarchical fuzzy systems. International Journal of Computational Cognition 4(1) (March 2006), http://www.ijcc.us

# An Interval Type-2 Fuzzy Logic System
# for the Modeling and Prediction of Financial Applications

Dario Bernardo, Hani Hagras, and Edward Tsang

The Computational Intelligence Centre,
School of Computer Science and Electronic Engineering,
University of Essex, Wivenhoe Park, Colchester, CO4 3SQ, UK

**Abstract.** In the recent years, there has been growing interest in developing tools for the modeling and prediction of financial applications. The problem of financial applications is that there are huge data sets available which are sometimes incomplete, and almost always affected by noise and uncertainty. Some techniques used in financial applications employ black box models which do not allow the user to understand the behavior and dynamics of the given application. In this paper, we present a type-2 Fuzzy Logic System (FLS) for the modeling and prediction of financial applications. The proposed system avoids the drawbacks of the existing type-2 fuzzy classification systems where the proposed system is able to carry prediction based on a pre-specified rule base size even if the incoming input vector does not match any rules from the given rule base. We have performed several experiments based on the London Stock Exchange data which was successfully used to spot ahead of time arbitrage opportunities. The proposed type-2 FLS has outperformed the existing type-2 fuzzy logic based classification systems and the type-1 FLSs counterparts when using pre-specified rule bases.

**Keywords:** Type 2 fuzzy logic systems, financial applications.

## 1 Introduction

With the current financial applications, there is a pressing need for new comprehensive and accurate approaches to capitalize on economic opportunity without incurring high levels of unexpected risk [1]. The majority of the commercial financial tools employ statistical regression techniques which capture only that information which can be refined into mathematical models to generate two outputs. Moreover, the regression techniques are essentially black box models which cannot be easily understood and analyzed by the normal user. Advanced machine learning and artificial intelligence techniques like Neural Networks suffer from the same problem where they can give good prediction accuracies, however they provide black box models which are very difficult to understand and analyze by a financial manager.

Fuzzy Logic Systems (FLSs) provide white box models which could be easily analyzed and understood by the layman user. However FLSs suffer from the curse of dimensionality problem which causes the FLS-based system to generate a big number

M. Kamel, F. Karray, and H. Hagras (Eds.): AIS 2012, LNCS 7326, pp. 95–105, 2012.

of rules in order to give good model accuracy. Most recently type-2 FLSs that are capable of handling high uncertainty levels have been employed for the generation of classification models [2], [3]. However, the existing type-2 fuzzy classification systems are not suited for the financial domain where such type-2 FLSs generate big rule bases; besides, they make the assumption that all the possible rules are represented in the existing models which is impossible for the huge financial data sets where the generated model will only cover a small subset of the search space. In this paper, we will present a type-2 FLS for the modeling and prediction of financial applications. The proposed system avoids the drawbacks of the existing type-2 fuzzy classification systems because it is able to carry prediction based on a pre-specified rule base size even if the incoming data vector does not match any rules in the FLS rule base. We have performed several experiments based on the London Stock Exchange data which was successfully used to spot ahead of time arbitrage opportunities. The proposed type-2 FLS has outperformed the existing type-2 fuzzy logic based classification systems and also the type-1 FLSs counterparts when using pre-specified rule bases.

In Section 2, we will present a brief overview on type-2 FLSs. Section 3 will present an overview on the fuzzy classification systems. Section 4 will present the proposed type-2 fuzzy based modeling and prediction system for financial applications. Section 5 will presents the experiments on the arbitrage data and the achieved results. Finally Section 6 will present the conclusions and future work.

## 2     A Brief Overview on Type-2 Fuzzy Logic Systems

In the recent years type-2 FLSs have grown in popularity due to their ability to handle high levels of uncertainties. Type-2 FLSs employ type-2 fuzzy sets as shown in Fig. 1 where a type-2 fuzzy set is characterized by a fuzzy membership function, i.e. the membership value (or membership grade) for each element of this set is a fuzzy set in [0,1], unlike a type-1 fuzzy set where the membership grade is a crisp number in [0,1] [4]. The membership functions of type-2 fuzzy sets are three dimensional and include a footprint of uncertainty (shaded in grey in Fig. 1, it is the new third-dimension of type-2 fuzzy sets and the footprint of uncertainty that provide additional degrees of freedom that make it possible to directly model and handle uncertainties [4], [5]. The interval type-2 FLSs use interval type-2 fuzzy sets (such as the type-2 fuzzy set shown in Fig. 1 to represent the inputs and/or outputs of the FLS). In the interval type-2 fuzzy sets all the third dimension values are equal to one. The use of interval type-2 FLS helps to simplify the computation (as opposed to the general type-2 FLS).

The proposed system in the paper is a type-2 fuzzy classification system and hence it does not follow the structure of the type-2 FLSs reported in [4], [5] where the classification system process is summarized in the following section.

An interval type-2 fuzzy set, denoted $\tilde{A}$ is written as follows:

$$\mu_{\tilde{A}}(x) = \int_{x \in X} \int_{u \in \left[\overline{\mu}_{\tilde{A}}(x),\ \underline{\mu}_{\tilde{A}}(x)\right]} 1/u \tag{1}$$

$\bar{\mu}_{\tilde{A}}(x)$, $\underline{\mu}_{\tilde{A}}(x)$ represent the upper and lower membership functions respectively of the interval type-2 fuzzy set $\tilde{A}$. The upper membership function is associated with the upper bound of the footprint of uncertainty $FOU(\tilde{A})$ of a type-2 membership function. The lower membership function is associated with the lower bound of $FOU(\tilde{A})$ [4].

In our system, the generation process of the employed interval type-2 fuzzy sets starts by generating type-1 fuzzy sets which equally partition the input universe of discourse into a given number of partitions. We then blur the type-1 fuzzy sets to the left and the right equally by a given uncertainty factor as shown in Fig. 1a to generate the type-2 fuzzy sets. In the application shown in this paper, we have employed 4 fuzzy sets as shown in Fig. 1b to represent each input variable.

## 3    A Brief Overview on Fuzzy Logic Classification Systems

In fuzzy logic classification systems, for a given $c$-class pattern classification problem with $n$ attributes (or features), a given rule in the FLS rule base could be written as follows:

$$Rule\ R^j\colon If\ x_1\ is\ A_1^j\ and\ \ldots\ and\ x_n\ is\ A_n^j\ then\ Class\ C_j\ with\ CFj,\ j = 1,2,\ldots,N \qquad (2)$$

Where $x_1,\ldots,x_n$ represent the n-dimensional pattern vector, $A_i^j$ is the fuzzy set representing the linguistic label for the antecedent pattern $i$, $C_j$ is a consequent class (which could be one of the possible $c$ classes), $N$ is the number of fuzzy if-then rules in the FLS rule base. $CF_j$ is a certainty grade of rule $j$ (i.e., rule weight). In case each input pattern is represented by $K$ fuzzy sets and given that we have $n$ input patterns, the possible number of rules that will cover the whole search space is $K^n$. In the arbitrage application presented in this paper, we have 7 inputs where each input is represented by 4 fuzzy sets, hence the needed number of rules to cover the whole search space for this given application is $4^7 = 16384$ rules. In our given application (which applies to the vast majority of financial applications), we do not have enough data to generate this huge number of rules. Hence, there will be various cases where the incoming input vector will not fire any rule in the FLS rule base.

Several type-1 fuzzy classification systems have been reported in the literature such as [6], [7], [8], [9], [10], [11], [12], [13], [14], [15] and [16]. However, in the vast majority of these papers, the data was relatively pretty easy to partition, and if an input pattern does not match any of the decision areas previously labeled, the input is discharged. In financial applications this cannot be done where if a new pattern that have never been seen before is proposed, a decision need to be made anyway, and unfortunately discharging it a priory cannot be the solution. A technique to resolve this problem was proposed in [17], [18], this technique keeps in a rule repository all the rules for the minority class in unbalanced data sets. All the inputs that do not match any rule in the repository are considered belonging to the majority class. This technique can work in unbalanced data set but will might not work in all cases.

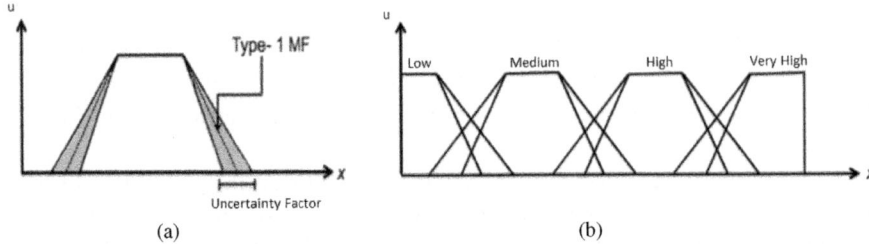

**Fig. 1.** a) The process followed to generate a type-2 fuzzy set from a type-1 fuzzy set. b) The employed interval type-2 fuzzy sets in the application reported in this paper.

## 4   The Proposed Type-2 Fuzzy Modeling and Prediction System for Financial Applications

The proposed system has two phases, a modeling phase and a prediction phase. In the modeling phase the rule base of the type-2 fuzzy classification system is constructed from the existing training dataset. In the prediction phase, the generated rule base is used to predict the incoming input vectors.

### 4.1   The Modeling Phase

The modeling phase operates according to the following steps (as shown in Fig. 2):

**Step 1- Raw Rule Extraction:** For a fixed input-output pair $(x^{(t)}, C^{(t)})$ in the dataset $,t=1,...T$ ($T$ is the total number of data training instances available for the modeling phase) compute the upper and lower membership values $\bar{\mu}_{A_s^q}$, $\underline{\mu}_{A_s^q}$ for each antecedent fuzzy set $q=1,...K$ ($K$ is the total number of fuzzy sets representing the input pattern $s$ where $s=1...n$. ). Generate all rules combining the matched fuzzy sets $A_s^q$ (i.e. either $\bar{\mu}_{A_s^q}>0$ or $\underline{\mu}_{A_s^q}>0$) for all $s=1...n$. Thus the rules generated by $(x^{(t)}, C^{(t)})$ will have different antecedents and the same consequent class $C^{(t)}$. Thus each of the extracted raw rules by $(x^{(t)}, C^{(t)})$ could be written as follows:

$$R^j : \quad If\ x_1\ is\ \tilde{A}_1^{qjt}\ and\ ...\ and\ x_n\ is\ \tilde{A}_n^{qjt}\ then\ Class\ C,\ t = 1,2,...,T \quad (3)$$

For each generated rule, we calculate the firing strength $F^t$. This firing strength measures the strength of the points $x^{(t)}$ belonging to the fuzzy region covered by the rule. $F^t$ is defined in terms of the lower and upper bounds of the firing strength $\underline{f}^{(t)}, \overline{f}^{(t)}$ of this rule which are calculated as follows:

$$\overline{f}^{jt}(x^{(t)}) = \overline{\mu}_{A_1^{qjt}}(x_1) * \cdots * \overline{\mu}_{A_n^{qjt}}(x_n) \quad (4)$$

$$\underline{f}^{jt}(x^{(t)}) = \underline{\mu}_{A_1^{qjt}}(x_1) * \cdots * \underline{\mu}_{A_n^{qjt}}(x_n) \quad (5)$$

The * denotes the minimum or product t-norm. Step 1 is repeated for all the $t$ data points from 1 to $T$ to obtain generated rules in the form of Equation (3). If there are two or more rules generated which have the same antecedents and consequent classes, we will aggregate these rules in one rule having the same antecedents and the same consequent class with the associated $\overline{f^{jt}}$ and $\underline{f^{jt}}$ which result in the maximum average $(\overline{f^{jt}} + \underline{f^{jt}})/2$ amongst these rules.

The financial data is usually highly imbalanced (for example in loan approval applications, we find that the vast majority of the customers are good customers and a minority are bad customers who might never pay their loan back). Hence, we will present a new approach called **"scaled dominance"** which tries to handle imbalanced data by trying to increase the confidence and support for the minority class. In order to compute the scaled dominance for a given rule having a consequent Class $C_j$, we divide the firing strength of this rule by the summation of the firing strengths of all the rules which had $C_j$ as the consequent class. This allows handling the imbalance of data towards a given class. We scale the firing strength by scaling the upper and lower bounds of the firing strengths as follows:

$$\overline{fs^{jt}} = \frac{\overline{f^{jt}}}{\sum_{j \in Class_j} \overline{f^j}} \tag{6}$$

$$\underline{fs^{jt}} = \frac{\underline{fs^{jt}}}{\sum_{j \in Class_j} \underline{fs^{jt}}} \tag{7}$$

**Step 2- Scaled Support and Scaled Confidence Calculation:** Many of the generated rules will share the same antecedents but different consequents. To resolve this conflict, we will calculate the scaled confidence and scaled support which are calculated by grouping the rules that have the same antecedents and conflicting classes. For given $m$ rules having the same antecedents and conflicting classes. The scaled confidence $(\tilde{A}_q \Rightarrow C_q)$ (defined by its upper bound $\overline{c}$ and lower bound $\underline{c}$, it is scaled as it involves the scaled firing strengths mentioned in the step above) that class $C_q$ is the consequent class for the antecedents $\tilde{A}_q$ (where there are $m$ conflicting rules with the same antecedents and conflicting consequents) could be written as follows [2]:

$$\overline{c}\left(\tilde{A}_q \Rightarrow C_q\right) = \frac{\sum_{x_s \in Class\ C_q} \overline{fs^{jt}(x_s)}}{\sum_{j=1}^{m} \overline{fs^{jt}(x_s)}} \tag{8}$$

$$\underline{c}\left(\tilde{A}_q \Rightarrow C_q\right) = \frac{\sum_{x_s \in Class\ c_q} \underline{fs^{jt}(x_s)}}{\sum_{j=1}^{m} \underline{fs^{jt}(x_s)}} \tag{9}$$

The scaled confidence can be viewed as measuring the validity of $A_q \Rightarrow C_q$. The confidence can be viewed as a numerical approximation of the conditional probability [8]. The scaled support (defined by its upper bound $\overline{s}$ and lower bound $\underline{s}$, it is scaled as it involves the scaled firing strengths mentioned in the step above) of $A_q \Rightarrow C_q$ is written as follows:

$$\overline{s}\left(\tilde{A}_q \Rightarrow C_q\right) = \frac{\sum_{x_s \in \, Class \, C_q} \overline{fs^{jt}(x_s)}}{m} \tag{10}$$

$$\underline{s}\left(\tilde{A}_q \Rightarrow C_q\right) = \frac{\sum_{x_s \in \, Class \, c_q} \underline{fs^{jt}(x_s)}}{m} \tag{11}$$

The support can be viewed as measuring the coverage of training patterns by $A_q \Rightarrow C_q$. In this paper we introduce the concept of scaled *dominance*, (defined by its upper bound $\overline{d}$ and lower bound $\underline{d}$) which is calculated by multiplying the scaled support and scaled confidence of the rule as follows:

$$\overline{d}\left(\tilde{A}_q \Rightarrow C_q\right) = \overline{c}\left(\tilde{A}_q \Rightarrow C_q\right) \cdot \overline{s}\left(\tilde{A}_q \Rightarrow C_q\right) \tag{12}$$

$$\underline{d}\left(\tilde{A}_q \Rightarrow C_q\right) = \underline{c}\left(\tilde{A}_q \Rightarrow C_q\right) \cdot \underline{s}\left(\tilde{A}_q \Rightarrow C_q\right) \tag{13}$$

**Step 3- Rule Cleaning:** For rules that share the same antecedents and have different consequent classes, we will replace these rules by one rule having the same antecedents and the consequent class, will be corresponding to the rule that gives the highest scaled dominance value. In [19], the rule generation system generates only the rule with the highest firing strength, however in our method, we generate all rules that are generated by the given input patterns, and this allows covering a bigger area in the decision space.

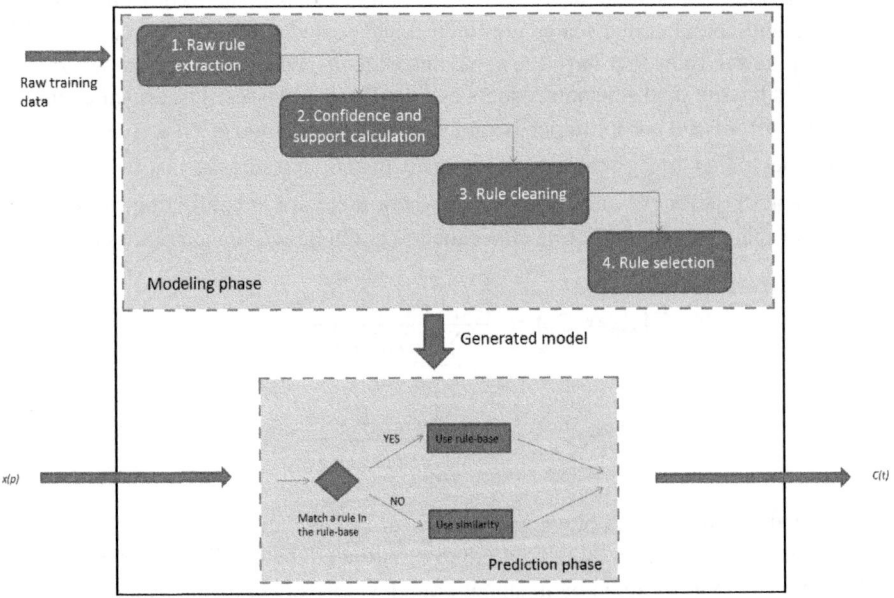

**Fig. 2.** An overview of the proposed modeling and prediction system

**Step 4- Rule Selection:** As fuzzy based classification methods generate big number of rules, hence this can cause major problems for financial applications where it is needed to understand the market or user behavior and dynamics. Hence, in our method, we will reduce the rule base to a pre-specified size of rules that could be easily read, understood and analyzed by the user. In this step, we select only the top Y rules per class (Y is pre-specified by the given financial application) which has the rules with the highest scaled dominance values. This selection is useful because rules with low dominance may actually be not relevant and actually introduce errors. This helps to keep the classification system more balanced between the majority and minority classes. By the end of this step, the modeling phase is finished where we have $X = nC \cdot Y$ rules (with $nC$ the number of classes) ready to classify and predict incoming patterns as discussed below in the prediction phase.

## 4.2    Prediction Phase

When an input pattern is introduced to the generated model, two cases will happen, the first case will happen when the input $x^{(p)}$ matches any of the X rules in the generated model, in this case we will follow the process explained in subsection (4.2.1). If $x^{(p)}$ does not match any of the existing X rules, we will follow the process explained in subsection (4.2.2).

### 4.2.1    Case 1: The Input Matches One of the Existing Rules

In case the incoming input $x^{(p)}$ matches any of the existing X rules, we will calculate the firing strength of the matched rules according to Equations (4) and (5), this will result in $\overline{f^j}(x^{(p)})$, $\underline{f^j}(x^{(p)})$. In this case, the predicted class will be determined by calculating a vote for each class as follows:

$$\overline{Z}Class_h(x^{(p)}) = \frac{\sum_{j\in h} \overline{f^j}(x^{(p)}) * \overline{d}(A_q \rightarrow C_q)}{\max_{j\in h}(\overline{f^j}(x^{(p)}) * \overline{d}(A_q \rightarrow C_q))} \tag{14}$$

$$\underline{Z}Class_h(x^{(p)}) = \frac{\sum_{j\in h} \underline{f^j}(x^{(p)}) * \underline{d}(A_q \rightarrow C_q)}{\max_{j\in h}(\underline{f^j}(x^{(p)}) * \underline{d}(A_q \rightarrow C_q))} \tag{15}$$

The total vote strength is then calculated as:

$$ZClass_h = \frac{\overline{Z}Class_h(x^{(p)}) + \underline{Z}Class_h(x^{(p)})}{2} \tag{16}$$

The class with the highest $ZClass_h$ will be the class predicted for the incoming input vector $x^{(p)}$.

### 4.2.1    Case 2: The Input Does Match Any of the Existing Rules

In case the incoming input vector $x^{(p)}$ does not match any of the existing X rules, we need to find the closest rule in the rule base that matches $x^{(p)}$. In order to do this, we need to calculate the similarity (or distance) between each of the fuzzy rule generated by $x^{(p)}$ and each of the X rules stored in the rule base. The rules generated by $x^{(p)}$ are found by taking each element in $x^{(p)}$ and taking all matching fuzzy sets with either $\overline{\mu}_{A_1^{qj}}(x_i)$ or $\underline{\mu}_{A_1^{qj}}(x_i)$ greater than 0. At this point there will be $k$ rules generated from the input $x^{(p)}$. Let the linguistic labels that fit $x^{(p)}$ be written as $v_{inputr} = (v_{input1r}, v_{input2r},...,v_{inputnr})$ where $r$ is the index of the $r$-th rule generated from the input. . Let the linguistic labels corresponding to a given rule in the rule base be $v_j = (v_{j1},v_{j2},...,v_{jn})$. Each of these linguistic labels (*Low, Medium, etc*) could be decoded into an integer. Hence the similarity between the rule generated by $x^{(p)}$ and a given rule in the rule base could be calculated by finding the distance between the two vectors as follows:

$$Similarity_{input\ r\leftrightarrow j} =(\ (1-\left|\frac{vinput1r-vj1}{v1}\right|)*(1-\left|\frac{vinput2r-vj2}{v2}\right|)*....*(1-\left|\frac{vinputnr-vjn}{vn}\right|)  \quad (17)$$

In the equation $V_s$ represents the number of linguistic labels representing each variable $s$. Each rule in the rule-base will have at this point a similarity associated with the $r$-th rule generated form the input. In this case, the predicted class will be determined by firstly selecting the rules with the highest similarity with the $r$-th generated rule. There will be more than one rule with the same similarity. Considering the rules that will have the same similarity with the $r$-th rule, the winning class for the $r$-th generated rule is calculated as a vote for each class as follows:

$$\overline{Z}Class_{h\ r}\left(x^{(p)}\right) = \sum_{j\in h}\overline{d}\left(A_q \rightarrow C_q\right) \quad (18)$$

$$\underline{Z}Class_{h\ r}\left(x^{(p)}\right) = \sum_{j\in h}\underline{d}\left(A_q \rightarrow C_q\right) \quad (19)$$

The total vote strength is then calculated as:

$$ZClass_{h\ r} = \frac{ZClass_{h\ r}(x^{(p)})+\underline{Z}Class_{h\ r}(x^{(p)})}{2} \quad (20)$$

The class with the highest $ZClass_{h\ r}$ will be the class associated with the $r$-th rule generated from the input. From all the $k$ rules generated from $x^{(p)}$, the final output class is calculated by applying Equations (14), (15) and (16).

## 5      Experiments and Results

We have tested the proposed type-2 fuzzy logic system mentioned in this paper to model and predict arbitrage opportunities. Computers today are able to spot stock misalignment in the market and in milliseconds. This would allow them to make almost risk-free profits. There are two main challenges in this type of operation. Firstly, arbitrage situation do not occur very often. Secondly, the operator must act ahead of others, so the competition is reduced to how fast a computer is, and how fast its connection to the stock exchange is. . Tsang et al [17] showed that arbitrage opportunities do not appear instantaneously. There are patterns in the market which can be recognized 10 minutes ahead.

The proposed system is trained to identify ahead of time arbitrage opportunities, as it is done in [17]. The data reported in this paper was further developed in [18], [21], in order to identify arbitrage situations by analyzing option and futures prices in the

London International Financial Futures Exchange (LIFFE) market. The pre-processed data comprised 1641 instances of which only 401 representing arbitrage opportunities and the rest representing non-arbitrage opportunities. The data was split into 2/3 for modeling and 1/3 for testing.

The type-2 fuzzy modeling and prediction system employed a 20 % uncertainty factor to generate the type-2 fuzzy sets as shown in Fig. 1a. The system was trained with the training data to generate a model of 100 rules per class.

For performance evaluation, the achieved results are compared based on average RECALL on minority and majority class. In pattern recognition and information retrieval, RECALL is calculated as the fraction between the numbers of correctly recognized inputs belonging to the minority (majority) class over the total number of inputs actually belonging to minority (majority) class. In our case, high recall means that an algorithm returned most of the arbitrage opportunities [20]. Recall is important in this application because arbitrage opportunities are rare, hence we cannot afford to lose too many of them.

**Table 1.** The achieved results

|  | RECALL-minority | RECALL-majority | AVG-RECALL |
| --- | --- | --- | --- |
| HFS method | 99.79% | 63.52% | **81.66%** |
| AMRG method / scaled dominance | 94.97% | 84.03% | **89.50%** |
| HFS method / scaled dominance on Type1 | 81.76% | 91.81% | **86.78%** |

AVG-RECALL is the average recall of the $RECALL_{minority}$ and $RECALL_{majority}$. Table 1 shows the results obtained on the testing data when the proposed type-2 fuzzy based system employ the approach which generate in the modeling phase only the rule with Highest Firing Strength (HFS method) where we can see how our approach which involves generating at the modeling phase All the Matched Rule Generated (AMRG method) and scaled dominance have a superior performance which is better by *7.84%* than the HFS method. Table 1 also shows when using our proposed approach how the type-2 fuzzy logic based system is able to produce *2.72 %* better average RECALL than the type-1 based system using the HFS method and scaled dominance. In addition, it is obvious that the proposed system was able to provide prediction for all the incoming input vectors thus avoiding the drawbacks of the fuzzy classification systems counterpart.

# 6    Conclusions and Future Work

In this paper, we have presented a type-2 FLS for the modeling and prediction of financial applications. The proposed system included new techniques such as scaled dominance and similarity matching which enable us to produce predictions when the input vector does not match the rules in the rule base.

We have performed several experiments based on the arbitrage data which was used successfully to spot ahead of time arbitrage opportunities [17]. We have shown that the proposed type-2 FLS outperformed the existing type-2 fuzzy logic based classification systems and the type-1 FLSs counterparts when using pre-specified rule bases.

For our future work, we aim to employ genetic algorithms to tune the type-2 fuzzy sets in order to get better results. We also intend to build a more flexible tool to move along the Receiver Operating Characteristic (ROC) curve with more detail [18].

# References

1. BBC: Meltdown Losses of $4 trillion. BBC News (April 22, 2009),
   http://news.bbc.co.uk
2. Sanz, J., Fernandez, A., Bustince, H., Herrera, F.: Improving the performance of fuzzy rule-based classification systems with interval-valued fuzzy sets and genetic amplitude tuning. Information Sciences 180, 3674–3685 (2010)
3. Sanz, J., Fernandez, A., Bustince, H., Herrera, F.: A genetic tuning to improve the performance of fuzzy Rule-Based Classification Systems with Interval-Valued fuzzy sets: Degree of ignorance and lateral position. International Journal of Approximate Reasoning 52, 751–766 (2011)
4. Hagras, H.: A hierarchical Type-2 Fuzzy Logic Control Architecture for Autonomous Mobile Robots. IEEE Transactions on Fuzzy Systems 12(4), 524–539 (2004)
5. Mendel, J.: Uncertain Rule-Based Fuzzy Logic Systems: Introduction and New Directions. Prentice-Hall (2001)
6. Ishibuchi, H.: Comparison of Heuristic criteria for fuzzy rule selection in classification problems. Fuzzy Optimization and Decision making 3(2), 119–139 (2004)
7. Ishibuchi, H.: Effect of rule weights in fuzzy rule-based classification system. IEEE Transactions on Fuzzy Systems 1, 59–64 (2000)
8. Ishibuchi, H.: Rule Weight specification in fuzzy rule-based classification systems. IEEE Transactions on Fuzzy Systems 13(4), 428–435 (2005)
9. Ishibuchi, H.: Three-objective genetic-based machine learning for linguistic rule extraction. Information Sciences 136(1-4), 109–133 (2001)
10. Ishibuchi, H., Yamamoto, T.: Fuzzy rule selection by multy-objective genetic local search algorithms and rule evaluation measures in data mining. Information Sciences 141(1), 59–88 (2004)
11. Ishibuchi, H., Yamamoto, T., Nakashima, T.: An approach to Fuzzy Default Reasoning for function approximation. Soft Computing 10(9), 850–864 (2006)
12. Shigeo, A., Lan, M.: A method for fuzzy rules extraction directly from numerical data and its application in pattern classification. IEEE Transactions on Fuzzy Systems 3(1), 18–28 (1995)
13. Ahmad, S., Jahormi, M.: Construction accurate fuzzy classification systems: A new approach using weighted fuzzy rules. Computer Graphics, Imaging and Visualisation, 408–413 (2007)
14. Wang, L.: The WM Method Completed: A flexible fuzzy system approach to data mining. IEEE Transactions on Fuzzy Systems 11(6), 768–782 (2003)
15. Mansoori, E., ZolGhadri, M., Katebi, D.: Using distribution of data to enhance performance of fuzzy classification system. Iranian Journal of Fuzzy Systems 4(1) (2006)
16. Fernández, A., del Jesus, M.J., Herrera, F.: Improving the Performance of Fuzzy Rule Based Classification Systems for Highly Imbalanced Data-Sets Using an Evolutionary Adaptive Inference System. In: Cabestany, J., Sandoval, F., Prieto, A., Corchado, J.M. (eds.) IWANN 2009, Part I. LNCS, vol. 5517, pp. 294–301. Springer, Heidelberg (2009)

17. Tsang, E., Markose, S., Er, H.: Chance discovery in stock index option and future arbitrage. New Mathematics and Natural Computation. World Scientific 1(3), 435–447 (2005)
18. Garcia-Almanza, A.: New classification methods for gathering patterns in the context of genetic programming. PhD Thesis, Department of Computing and Electronic Systems, University of Essex (2008)
19. Chi, Z., Yan, H., Pham, T.: Fuzzy Algorithms: With Applications to Image Processing and Pattern Recognition. World Scientific Pub. Co. Inc. (1996) ISBN/ASIN: 9810226977
20. Olson, D., Delen, D.: Advanced Data Mining Techniques, 1st edn., pp. 138–139. Springer (2008) ISBN 3540769161
21. Garcia-Almanza, A.L., Tsang, E.P.K.: Evolutionary Applications for Financial Prediction: Classification Methods to Gather Patterns Using Genetic Programming. VDM Verlag (2011)

# Adaptive Fuzzy Logic Control
# for Time-Delayed Bilateral Teleoperation

Jiayi Zhu and Wail Gueaieb

Machine Intelligence, Robotics and Mechatronics (MIRaM) Laboratory
School of Electrical Engineering and Computer Science
University of Ottawa, Ottawa ON K1N 6N5, Canada
jzhu051@uottawa.ca, wgueaieb@eecs.uottawa.ca

**Abstract.** In recent years, teleoperation has shown great potentials in different fields such as spatial, mining, under-water, etc. When this technology is required to be bilateral, the time delay induced by a potentially large physical distance prevents a good performance of the controller, especially in the case of contact. When bilateral teleoperation is introduced to the field of medicine, a new challenge arises: the controller must perform well in both hard and soft environments. For example, in the context of telesurgery, the robot can enter in contact with both bone (hard) and organ (soft). In an attempt to enrich existing controller designs to better suit the medical needs, an adaptive fuzzy logic controller (AFLC) is designed in this paper. It simulates human intelligence and adapts to environments of different stiffness coefficients. The simulation results prove that this controller demonstrates very interesting potential.

**Keywords:** bilateral teleoperation, time-delay systems, adaptive fuzzy logic.

## 1   Introduction

Teleoperation is an action performed over distance where the human operator at the local site can control a robot at the remote site. When such an operation is qualified as bilateral, the operator can perceive the corresponding force feedback when the remote robot enters in contact with its environment. Such a system is depicted in figure 1.

This technology has a large potential with possible applications in nuclear, mining, spatial and many other fields. While being applied in those areas, teleoperation primarily helped to eliminate or decrease the negative effects of environment's hostility on human operator. However when it comes to the medical field, especially in surgeries, telerobots are mainly praised for their incomparable precision as well as their great potential of expanding medical services to patients far away.

Unilateral teleoperation where the force feedback is non-existent is already in service in many fields. But the incorporation of force feedback under the presence of time delay is still an active area of research. The vast majority

M. Kamel, F. Karray, and H. Hagras (Eds.): AIS 2012, LNCS 7326, pp. 106–115, 2012.

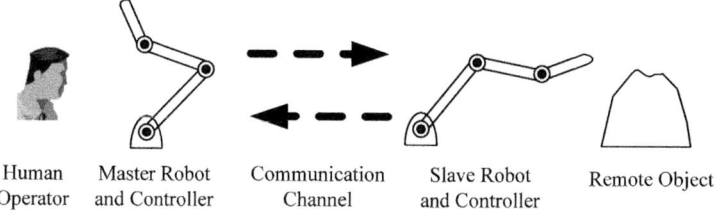

Human          Master Robot          Communication          Slave Robot          Remote Object
Operator       and Controller        Channel                and Controller

**Fig. 1.** An illustration of the bilateral teleoperation system

of the suggested solutions can be categorized into passivity-based, prediction-based or model-based solutions. The passivty-based solutions achieve system passivity independent of the time delay based on system inputs and outputs only. It was originally proposed in [1] and [11] and was considered as a major breakthrough. It was further developed with the use of wave integrals, use of additional transfer and energy methods [3,12]. The prediction-based solutions are based on the concept of Smith predictors which attempts to predict the reaction of the remote site based on a model of the remote environment [16,17]. A third important category consists of the model-based solutions and is heavily built on the robot model dynamics [6,9]. Many traditional control theories are used to cope with the time delayed terms in the dynamic equations. Some important ones include the impedance control, robust control and $H_\infty$ optimal control. There is also much work done in using a combination of solutions. For example, combining the passivity and prediction based solutions [4,10] or combining the model and prediction based solutions [13,15]. A few papers have done surveys of teleoperation control schemes [2,5,7,14,20].

In this paper, a new model-based solution is proposed. It uses the concept of adaptive fuzzy logic control and attempts to explore this method in both hard and soft environments, which is crucial in medical settings.

This paper is organized as follows: Section 2 gives a background of adaptive fuzzy logic. The controller design and stability analysis is covered in section 3. Section 4 shows simulation results of the proposed controller, while section 5 ends with conclusions.

## 2   Adaptive Fuzzy Logic Control

The AFLC is a nonlinear controller which controls a system heuristically using a knowledge base and adaptation laws specified by the designer, hence imitating human logic in order to perform control actions. It is mainly praised for its ability to deal with systems that are uncertain (fuzzy) due to complexity, incompleteness, disturbance, etc. which are difficult to model using conventional controllers. It contains a fuzzifier, a rule base, a fuzzy inference engine and a defuzzifier. In this section, the indirect AFLC is presented. Readers unfamiliar with this controller may refer to [18] for further information and backgrounds. Based on this concept, let $M$ be the total number of fuzzy rules in the fuzzy knowledge base and $l = 1, 2, ...M$ be a rule index. Let the input vector be $x = [x_1, x_2, ...x_N]$ where $N$ is the total number of elements in the input vector. Let $p_i$ be the maximum number of fuzzy sets for the input $x_i$ and $l_i = 1, 2, ..., p_i$ be its counter. A common combination uses the product inference engine, singleton fuzzifier and centre average defuzzifier to give the following:

$$\hat{f}(x|\theta) = \frac{\sum_{l_1=1}^{p_1} \cdots \sum_{l_N=1}^{p_N} \bar{y}^{l_1 \ldots l_N} (\prod_{i=1}^{N} \mu_{A_i^{l_i}}(x_i))}{\sum_{l_1=1}^{p_1} \cdots \sum_{l_N=1}^{p_N} (\prod_{i=1}^{N} \mu_{A_i^{l_i}}(x_i))} = \theta^T \xi(x) \qquad (1)$$

where $\bar{y}^{li}$ denotes the center value for each outputted fuzzy set $B^l$. In order to well illustrate the concept of adaptive fuzzy control, $\hat{f}$ is further broken down into the product of two elements $\theta$ and $\xi$, where $\xi$ contains all the fixed values and $\theta$ contains all the values to be adapted in real-time as follow:

$$\xi_{l_1 \ldots l_N}(x) = \frac{\prod_{i=1}^{N} \mu_{A_i^{l_i}}(x_i)}{\sum_{l_1=1}^{p_1} \cdots \sum_{l_N=1}^{p_N} (\prod_{i=1}^{N} \mu_{A_i^{l_i}}(x_i))} \qquad \theta = [\bar{y}^{l_1}, \bar{y}^{l_2}, ...\bar{y}^{l_N}]$$

With the AFLC, all the elements in $\theta$ are prone to be changed in order to ameliorate the controller output. The adaptation law of $\theta$ is designed in the next section to ensure the AFLC's performance and stability.

## 3    Controller Design

There are two controllers to be designed: one on the master side and one on the slave side. They are both designed as AFLCs using (1). The update law of $\theta$ is derived in the following subsections using Lyapunov's stability theorem [8].

### 3.1    Slave-Side Controller Design

The dynamics of the slave robot can be written as follow:

$$D_s\ddot{q} + C_s\dot{q} + G_s = \tau_{sc} + \tau_e$$

where $D_s$ (short for $D_s(\ddot{q})$) is the inertia matrix, $C_s$ (short for $C_s(q, \dot{q})$) is the centripetal and Coriolis torques, $G_s$ (short for $G_s(q)$) is the gravitational torques, $\tau_{sc}$ is the control torque to be generated by the slave side controller, $\tau_e$ is the external torque produced upon contact and $q$ is the joint position vector of dimension $n$. Note the following robot dynamics property:

$$q^T(\dot{D}_s - 2C_s)q = 0, \quad \forall q \in \mathbb{R}^n. \qquad (2)$$

Let $q_d$ be the desired position and $\Lambda$ be a positive definite matrix whose eigenvalues are strictly in the right-half complex plane. The following terms are defined:

$e = q - q_d$, $\dot{e} = \dot{q} - \dot{q}_d$, $\ddot{e} = \ddot{q} - \ddot{q}_d$, $s = \dot{e} + \Lambda e$, $\dot{s} = \ddot{e} + \Lambda \dot{e}$, $\dot{q}_r = \dot{q}_d - \Lambda e$, $\ddot{q}_r = \ddot{q}_d - \Lambda \dot{e}$. Then, let Lyapunov's function candidate be

$$V = \frac{1}{2}(s^T D_s s + \sum_{i=1}^{n} \tilde{\theta}_i^{T} \Gamma_i \tilde{\theta}_i)$$

where $\Gamma_i$ is a positive constant and $\tilde{\theta}_i$ is the error vector representing the difference between the adaptive parameters of the AFLC, $\theta_i$, and its optimal value $\theta_i^*$. That is,

$$\tilde{\theta}_i = \theta_i^* - \theta_i. \tag{3}$$

Deriving $V$ by substituting the dynamic equation while using model property gives:

$$\dot{V} = -s^T (D_s \ddot{q}_r + C_s \dot{q}_r + G_s - \tau_e - \tau_{sc}) + \sum_{i=1}^{n} \dot{\tilde{\theta}}_i^T \Gamma_i \tilde{\theta}_i$$

Taking into consideration that $D_s$, $C_s$ and $G_s$ are not exactly known due to friction, disturbance, etc., consider the following controller:

$$\tau_{sc} = \hat{D}_s \ddot{q}_r + \hat{C}_s \dot{q}_r + \hat{G}_s - \tau_e - K_D s + \hat{F}(\ddot{q}_r, \dot{q}_r, q, \dot{q}, \ddot{q}|\theta) \tag{4}$$

where $\hat{D}_s$, $\hat{C}_s$ and $\hat{G}_s$ are estimates of $D_s$, $C_s$ and $G_s$, respectively, and $K_D$ is a positive definite matrix. $\hat{F}$ is the adaptive fuzzy control that compensates for the uncertainties in robot dynamics.

Then, rewriting $\dot{V}$ and substituting $\tau_{sc}$ in (4) gives

$$\dot{V} = -s^T \left( \tilde{D}_s \ddot{q}_r + \tilde{C}_s \dot{q}_r + \tilde{G}_s - \hat{F}(\ddot{q}_r, \dot{q}_r, q, \dot{q}, \ddot{q}|\theta) \right) - s^T K_D s + \sum_{i=1}^{n} \dot{\tilde{\theta}}_i^T \Gamma_i \tilde{\theta}_i$$

where $\tilde{D}_s = D_s - \hat{D}_s$, $\tilde{C}_s = C_s - \hat{C}_s$ and $\tilde{G}_s = G_s - \hat{G}_s$ are the estimation errors of the robot dynamics. $\hat{F}(\ddot{q}_r, \dot{q}_r, q, \dot{q}, \ddot{q}|\theta)$ is the estimate of the function $F(\ddot{q}_r, \dot{q}_r, q, \dot{q}, \ddot{q})$, where $F(\ddot{q}_r, \dot{q}_r, q, \dot{q}, \ddot{q}) = \tilde{D}_s \ddot{q}_r + \tilde{C} \dot{q}_r + \tilde{G}$. Consider $\theta$ to represent the array of adaptive parameters in an AFLC as discussed in section 2. Then a minimum approximation error vector is defined as $w = F(\ddot{q}_r, \dot{q}_r, q, \dot{q}, \ddot{q}) - \hat{F}(\ddot{q}_r, \dot{q}_r, q, \dot{q}, \ddot{q}|\theta^*)$.

Reformulating $\dot{V}$ by subbing in $F(\ddot{q}_r, \dot{q}_r, q, \dot{q}, \ddot{q})$ and $w$ gives

$$\dot{V} = -s^T \left( w + \hat{F}(\ddot{q}_r, \dot{q}_r, q, \dot{q}, \ddot{q}|\theta^*) - \hat{F}(\ddot{q}_r, \dot{q}_r, q, \dot{q}, \ddot{q}|\theta) \right) - s^T K_D s + \sum_{i=1}^{n} \dot{\tilde{\theta}}_i^T \Gamma_i \tilde{\theta}_i$$

The two $\hat{F}$ terms can be combined to $\hat{F}(\ddot{q}_r, \dot{q}_r, q, \dot{q}, \ddot{q}|\theta)$ by equation 3. This term then represents the AFLC $\tilde{\theta}^T \xi(\ddot{q}_r, \dot{q}_r, q, \dot{q}, \ddot{q})$. Further expanding the equation gives

$$\dot{V} = -s^T w - s^T K_D s + \sum_{i=1}^{n} (\dot{\tilde{\theta}}_i^T \Gamma_i \tilde{\theta}_i - s_i^T \tilde{\theta}_i^T (\ddot{q}_r, \dot{q}_r, q, \dot{q}, \ddot{q}))).$$

If the fuzzy logic controller is rich enough to capture the system's nonlinearities, then it can be concluded that $w$ is practically negligible [19]. $K_D$ is a positive definite matrix, hence $-s^T K_D s < 0$. Thus, to ensure that $\dot{V} \leq 0$, it is sufficient to impose that the summation terms add up to zero. This yields

$$\dot{\theta}_i = -\Gamma_i^{-1} s_i^T \xi_i(\ddot{q}_r, \dot{q}_r, q, \dot{q}, \ddot{q})) \tag{5}$$

## 3.2   Master-Side Controller Design

The same design procedure is applied to the master-side AFLC with slight modifications. In this subsection, $D_m$, $C_m$ and, $G_m$ are defined as $D_s$, $C_s$ and, $G_s$ except they are used for the master robot. All other parameters are as defined in the previous subsection but in the context of the master side control unless otherwise specified, hence the subscript "m".

The dynamics of the master robot is written as

$$D_m \ddot{q} + C_m \dot{q} + G_m = \tau_h + \tau_{mc}$$

where, $\tau_h$ is the control torque coming from the human operator and $\tau_{mc}$ is the control torque to be generated by the controller on the master side.

Consider the closed-loop impedance error equation:

$$\bar{D}_m \ddot{q} + \bar{C}_m \dot{q} + \bar{G}_m = \tau_h + \tau_e^{dy}$$

where the superscript "$dy$" indicates a time delayed term. $\bar{D}_m, \bar{C}_m, \bar{G}_m > 0$ are the desired D, M, G matrices which are are pre-set by the user.

Design wise, the master controller is different than the slave controller. The slave AFLC is solely responsible for estimating and compensating the errors caused by uncertainties in the robot dynamics parameters. However the master AFLC is responsible for the uncertainties in both robot dynamics parameters as well as the desired impedance parameters pre-fixed by the user. Therefore the AFLC designed for the master robot contains two parts. Whenever applies, the different parts will be identified by subscripts 1 and 2.

Let Lyapunov's function candidate for the master-side AFLC be

$$V = \frac{1}{2}(s^T D_m s + \sum_{i=1}^{n} \tilde{\theta}_{1i}^T \Gamma_{1i} \tilde{\theta}_{1i} + \tilde{\theta}_{2i}^T \Gamma_{2i} \tilde{\theta}_{2i})$$

where $\Gamma_{1i}$ and $\Gamma_{2i}$ are positive constants, $\tilde{\theta}_1$ and $\tilde{\theta}_2$ are the adaptive parameter error vectors for both parts of this AFLC. Deriving $V$ and substitute robot dynamic equation while using the model property gives:

$$\dot{V} = -s^T(D_m \ddot{q}_r + C_m \dot{q}_r + G_m + \tau_e^{dy} - \bar{D}_m \ddot{q} - \bar{C}_m \dot{q} - \bar{G}_m - \tau_{mc})$$
$$+ \sum_{i=1}^{n}(\dot{\tilde{\theta}}_{1i}^T \Gamma_{1i} \tilde{\theta}_{1i} + \dot{\tilde{\theta}}_{2i}^T \Gamma_{2i} \tilde{\theta}_{2i})$$

The dynamical terms $D_m$, $C_m$, $G_m$ are considered as unknown while the dynamical terms $\bar{D}_m$, $\bar{C}_m$, $\bar{G}$ are pre-fixed by the user. Consider the controller

$$\tau_{mc} = \hat{D}_m \ddot{q}_r + \hat{C}_m \dot{q}_r + \hat{G}_m + \tau_e^{dy} - \hat{\bar{D}}_m \ddot{q} - \hat{\bar{C}}_m \dot{q} - \hat{\bar{G}}$$
$$-K_D s + \hat{F}_1(\ddot{q}_r, \dot{q}_r, q, \dot{q}, \ddot{q}|\theta) + \hat{F}_2(\ddot{q}, \dot{q}, q|\theta)$$

where $\hat{F}_1$ is the AFLC that compensates for the uncertainties in robot dynamics and $\hat{F}_2$ adjusts for the errors in impedance parameters approximations. $\hat{\bar{D}}_m$, $\hat{\bar{C}}_m$ and $\hat{\bar{G}}_m$ are estimates of the ideal $\bar{D}_m$, $\bar{C}_m$, $\bar{G}_m$ respectively. Those approximations are used to cope with environments of different stiffness coefficients while $\bar{D}_m$, $\bar{C}_m$, $\bar{G}_m$ are pre-set by the user. Then,

$$\dot{V} = -s^T[\tilde{D}_m \ddot{q}_r + \tilde{C}_m \dot{q}_r + \tilde{G}_m - \tilde{\bar{D}}_m \ddot{q} - \tilde{\bar{C}}_m \dot{q} - \tilde{\bar{G}}_m + K_D s$$
$$-\hat{F}_1(\ddot{q}_r, \dot{q}_r, q, \dot{q}, \ddot{q}|\theta) - \hat{F}_2(\ddot{q}, \dot{q}, q|\theta)] + \sum_{i=1}^{n}(\tilde{\theta}_{1i}^T \Gamma_{1i} \dot{\tilde{\theta}}_{1i} + \tilde{\theta}_{2i}^T \Gamma_{2i} \dot{\tilde{\theta}}_{2i})$$

where $\tilde{\bar{D}}_m = \bar{D}_m - \hat{\bar{D}}_m$, $\tilde{\bar{C}}_m = \bar{C}_m - \hat{\bar{C}}_m$, $\tilde{\bar{G}}_m = \bar{G}_m - \hat{\bar{G}}_m$ are the approximation errors of desired master robot impedance. $\hat{F}_1(\ddot{q}_r, \dot{q}_r, q, \dot{q}, \ddot{q}|\theta)$ is the estimate of the function $F_1(\ddot{q}_r, \dot{q}_r, q, \dot{q}, \ddot{q})$ and $\hat{F}_2(q, \dot{q}, \ddot{q}|\theta)$ is that of the function $F_2(q, \dot{q}, \ddot{q})$, where $F_1(\ddot{q}_r, \dot{q}_r, q, \dot{q}, \ddot{q}) = \tilde{D}_m \ddot{q}_r + \tilde{C}_m \dot{q}_r + \tilde{G}_m$ and $F_2(q, \dot{q}, \ddot{q}) = -\tilde{D}_m \ddot{q} - \tilde{B}_m \dot{q} - \tilde{G}_m$.

Two minimum approximation error vectors are defined as $w_1 = F_1(\ddot{q}_r, \dot{q}_r, q, \dot{q}, \ddot{q}) - \hat{F}_1(\ddot{q}_r, \dot{q}_r, q, \dot{q}, \ddot{q}|\theta_1^*)$ and $w_2 = F_2(q, \dot{q}, \ddot{q}) - \hat{F}_2(q, \dot{q}, \ddot{q}|\theta_2^*)$ where $\theta_1^*$ and $\theta_2^*$ are the vector of optimal parameters. Sub in $F_1$, $F_2$ and $w_1$, $w_2$ into $\dot{V}$, then use $\theta_1^* - \theta_1 = \tilde{\theta}_1$ and $\theta_2^* - \theta_2 = \tilde{\theta}_2$ gives

$$\dot{V} = -s^T w_1 - s^T w_2 - s^T K_D s + \sum_{i=1}^{n}[\tilde{\theta}_{1i}^T \Gamma_{1i} \dot{\tilde{\theta}}_{1i} + \tilde{\theta}_{2i}^T \Gamma_{2i} \dot{\tilde{\theta}}_{2i}$$
$$-s_i^T \tilde{\theta}_{1i} \xi_1(\ddot{q}_r, \dot{q}_r, q, \dot{q}, \ddot{q}) - s_i^T \tilde{\theta}_{2i} \xi_2(q, \dot{q}, \ddot{q})]$$

As in the case for the slave robot, the $w_1$ and $w_2$ are extremely small and $K_D$ being a positive definite matrix gives $-s^T K_D s < 0$. Therefore to ensure that $\dot{V} \leq 0$, it is sufficient to impose that the sum of all terms within the summation sign should be zero. This yields

$$\dot{\theta}_{1i} = -\Gamma_{1i}^{-1} s_i^T \xi_1(\ddot{q}_r, \dot{q}_r, q, \dot{q}, \ddot{q}) \qquad \dot{\theta}_{2i} = -\Gamma_{2i}^{-1} s_i^T \xi_2(q, \dot{q}, \ddot{q})$$

## 4   Simulation and Results

With the proposed controllers, simulation is performed for both hard and soft environments with stiffness coefficients of $450\,\text{N·m}^{-1}$ and $45\,\text{N·m}^{-1}$ respectively. Both environments are modelled as pure spring. In the simulation, the master and slave robots are identical and modelled using Newton-Euler dynamics algorithm. They consist of two one-link robots rotating around a fixed base each

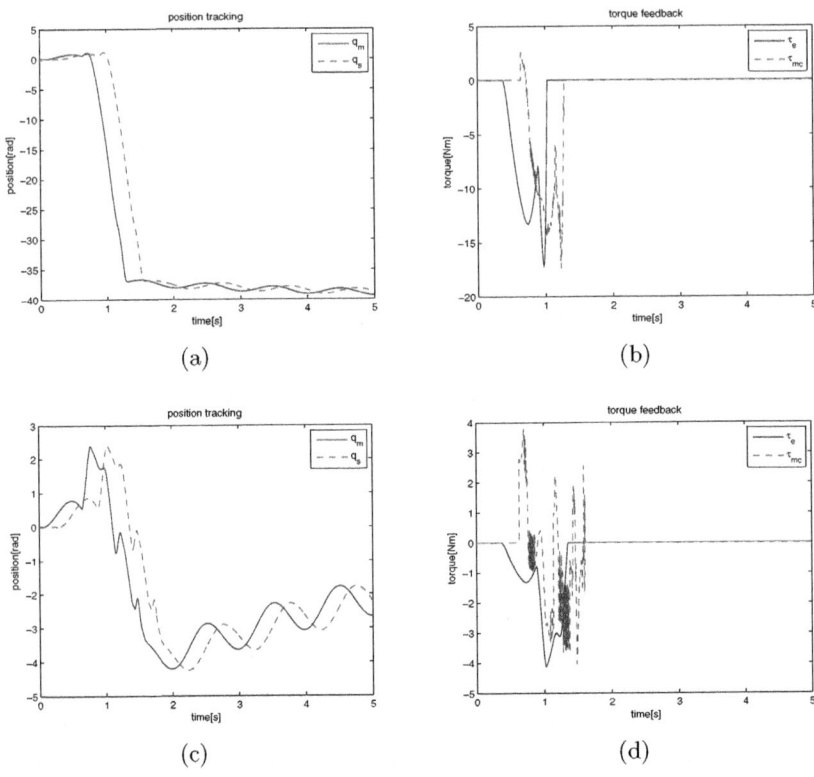

**Fig. 2.** Simulations with constant time delay. (a)–(b) hard contact; and (c)–(d) soft contact.

with an arm measuring 0.2 m. Both robots are starting from the horizontal position at the beginning of each simulation test. The gravity force is considered to act downward and the human input torque is defined as $f(t) = 0.5\sin(2\pi t)$, measured in Newton-meter (N·m), where $t$ is time in second. It is assumed that the force applied by the end-effector is always acting perpendicularly to the robot arm.

The simulations run under a constant time delay of 0.25 s are shown in figure 2. In all the simulations, the contact point is located at 0.1 rad and only occurs at the beginning of the graphs. It can be seen that the position tracking is fairly precise in both environments. The graphs show some slight position drift at the very beginning. This is natural as it corresponds to the AFLCs' (short) adaptation phase. In the hard contact environment, the robot was pushed quite far by the contact force, but it eventually reached a stable point at the position $-12.5\pi$ rad. Therefore, the controller is still stable. The force tracking performance is fairly satisfactory in the hard contact environment simulations as $\tau_e$ was faithfully reproduced by the AFLC. The constantly changing parameters within the controller are to be praised for being able to adapt to environments

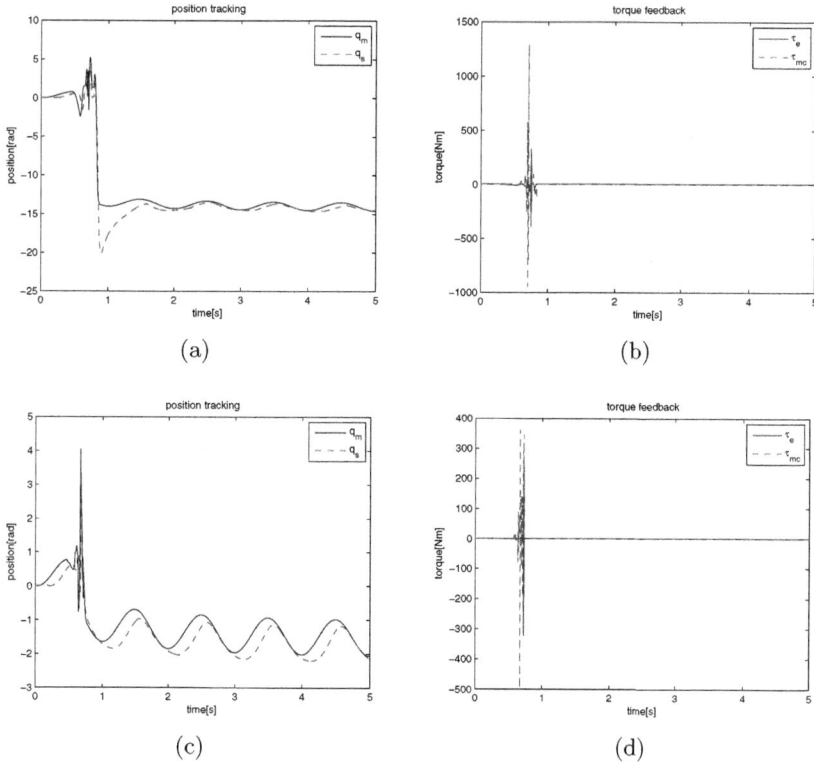

**Fig. 3.** Simulations with variable time delay. (a)–(b) hard contact; and (c)–(d) soft contact.

with different stiffness coefficients. In the soft contact environment, the differences between $\tau_e$ and $\tau_{mc}$ curves are expected to be ameliorated with further tuning of certain parameters specific to robot dynamics. However, it should be noted that if there was no collision detected between slave robot and environment, any output of the master side AFLC is reduced to zero. This step greatly helped for the seemingly good performance graphs.

The system has also been tested under a variable time delay of $0.1\sin(2\pi t) + 0.1$, where $t$ denotes time in second. The simulation results are shown in figure 3. Both figures show that the controller is stable in the face of variable time delay and variable contact point's stiffness. They also show the controller's satisfactory position and force transparencies under such conditions.

An important weakness of this controller is its computational complexity which affects its real-time execution. It would be desirable to find a way to simplify it for example by pruning the number of fuzzy membership functions or fuzzy rules. Another disadvantage is the dependency on typically noisy signals, such as acceleration and force/torque. One way to alleviate this shortcoming is by using observers, for example.

## 5  Conclusion

In this text, the design of an AFLC was attempted for bilateral teleoperation for a better integration of this technology in the field of telemedicine. The controller was rigorously proven stable using Lyapunov's stability theorem. With the simulation results, it can be seen that this controller showed a fairly satisfactory position and force tracking performance for both hard and soft environments. Hence, it is reasonable to conclude that it has motivating potentials and is worth further investigation.

## References

1. Anderson, R., Spong, M.: Bilateral control of teleoperators with time delay. IEEE Transactions on Automatic Control 34(5), 494–501 (1989)
2. Arcara, P., Melchiorri, C.: Control schemes for teleoperation with time delay: a comparative study. Robotics and Autonomous Systems 38(1), 49–64 (2002)
3. Aziminejad, A., Tavakoli, M., Patel, R., Moallem, M.: Wave-based time delay compensation in bilateral teleoperation: two-channel versus four-channel architectures. In: Proceedings of the 2007 American Control Conference, pp. 1449–1454 (July 2007)
4. Ching, H., Book, W.: Internet-based bilateral teleoperation based on wave variable with adaptive predictor and direct drift control. Transactions of the ASME, Journal of Dynamic Systems, Measurement and Control 128, 86–93 (2006)
5. Ferre, M., Buss, M., Aracil, R., Melchiorri, C., Balaguer, C.: Advances in Telerobotics. Springer Tracts in Advanced Robotics, vol. 31. Springer, Herdelberg (2007)
6. Garcia-Valdovinos, L., Parra-Vega, V., Arteaga, M.: Observer-based sliding mode impedance control of bilateral teleoperation under constant unknown time delay. Robotics and Autonomous Systems 55(8), 609–617 (2007)
7. Hokayem, P., Spong, M.: Bilateral teleoperation: An historical survey. Automatica 42, 2035–2057 (2006)
8. Khalil, H.: Nonlinear Systems, 3rd edn. Prentice Hall (2002)
9. Lee, D., Spong, M.: Passive bilateral teleoperation with constant time delay. In: Proceeding of the 2006 IEEE International Conference on Robotics and Automation, vol. 22(2), pp. 2902–2907 (2006)
10. Liu, X., Wilson, W., Fan, X.: Pose reflecting teleoperation using wave variables with wave prediction. In: Proceedings of the IEEE International Conference on Mechatronics & Automation, pp. 1642–1647 (July 2005)
11. Niemeyer, G., Slotine, J.: Stable adaptive teleoperation. IEEE Journal of Oceanic Engineering 16(1), 152–162 (1991)
12. Niemeyer, G., Slotine, J.: Telemanipulation with time delays. International Journal of Robotics Research 23(9), 873–890 (2004)
13. Pan, Y., Gu, J., Meng, M., Jayachandran, J.: Bilateral teleoperation of robotic systems with predictive control. In: 2007 IEEE International Conference on Robotics and Automation, pp. 1651–1656 (April 2007)
14. Rodriguez-Seda, E., Lee, D., Spong, M.: Experimental comparison study of control architectures for bilateral teleoperators. IEEE Transactions on robotics 25(6), 1304–1318 (2009)

15. Sadeghi, M.S., Momeni, H.: A new impedance and robust adaptive inverse control approach for a teleoperation system with varying time delay. Science in China Series E: Technological Sciences 52(9), 2629–2643 (2009)
16. Shahdi, A., Sirouspour, S.: Improved transparency in bilateral teleoperation with variable time delay. In: The 2009 IEEE/RSJ International Conference on Intelligent Robots and Systems, pp. 4616–4621 (2009)
17. Smith, A., Hashtrudi-Zaad, K.: Smith predictor type control architectures for time delayed teleoperation. The International Journal of Robotics Research 25(8), 797–818 (2006)
18. Wang, L.X.: A Course in Fuzzy Systems and Control. Prentice Hall (1997)
19. Yoo, B., Ham, W.: Adaptive control of robot manipulator using fuzzy compensator. IEEE Transactions on Fuzzy Systems 8(2), 186–199 (2000)
20. Zhu, J., He, X., Gueaieb, W.: Trends in the Control Schemes for Bilateral Teleoperation with Time Delay. In: Kamel, M., Karray, F., Gueaieb, W., Khamis, A. (eds.) AIS 2011. LNCS, vol. 6752, pp. 146–155. Springer, Heidelberg (2011)

# Bearing Only SLAM: A New Particle Filter Based Approach

Mohammad Hossein Mirabdollah and Baerbel Mertsching

GET Lab, University of Paderborn,
33098 Paderborn, Germany
{mirabdollah,mertsching}@get.upb.de
http://getwww.upb.de

**Abstract.** In this paper a new method to address bearing-only SLAM using particle filters is proposed. We use a set of line pieces to model the uncertainties of landmarks and derive a proper formulation to modify the joint robot and landmark assumptions in the context of a particle filter approach.

**Keywords:** SLAM, particle filter, line piece.

## 1   Introduction

The Simultaneous Localization And Mapping problem has been addressed in many works in the recent years; however, using a monocular camera is relatively new and more challenging. Nevertheless, since a monocular camera provides only bearing data, it is essential to come up with depth estimation algorithms which utilize different shots of the environment at different robot positions. Referring to geometry, we know that the precise depth can be calculated if the camera's (robot's) positions are precisely known, while in SLAM problem we assume uncertainty for the robot's position. Hence, the bearing only SLAM problem will be much more complex than the usual SLAM in which both distances and bearings to the landmarks are available.

In recent years, several papers have been published about bearing SLAM with different approaches. However, most of them have focused more on feature extraction and used simple Kalman filter based methods to implement their algorithms [11] [9] [2] [7]. Obviously, these methods should have some assumptions about the uncertainties of the landmarks' depths, otherwise they would diverge (since the Gaussian assumption of the uncertainties in a Kalman filter would be violated). Therefore, it is important to come up with some methods which can deal with larger depth uncertainties. Previous works to handle the fusion problem of bearing-only SLAM can be divided into two different categories: The first and important one includes delayed or un-delayed methods. Delayed methods postpone the modification of the hypothesized robot's position until they can have a reliable depth measurement of landmarks [1], whereas, un-delayed methods which are more recent, try to decrease the uncertainty of the robot's position

M. Kamel, F. Karray, and H. Hagras (Eds.): AIS 2012, LNCS 7326, pp. 116–125, 2012.
© Springer-Verlag Berlin Heidelberg 2012

from the beginning [8] [3]. The second category is based on filtering methods. The majority applies Extended Kalman filters with some variations. They differ in the way they handle the uncertainties of the landmarks taking into account that the distances to them are not known initially. The main idea behind EKF based methods is the use of multiple Gaussian distributions which are placed along the ray between the robot and the landmark. It is known that using EKF requires the representation of each variable's uncertainty in a Gaussian form. To address this problem two main methods have been proposed so far: The first method assumes different hypotheses for each landmark as distinctive landmarks and include them in the estimated vector separately [8] [3]. The second one uses Gaussian Sum Filter (GSF) which are the extensions of EKF when the uncertainties can be expressed as a weighted sum of some individual Gaussians [10]. It is known that the GSF based method has the best performance among the EKF based approaches, however its complexity increases exponentially as the number of landmarks increase linearly. Additionally, the GSF based method can easily diverge when the landmarks are located relatively near to the robot or there is an average amount of odometry noise. Since EKF based approaches use linear approximations for the motion and the measurement models of the robot, the models cannot be accurate enough in the mentioned situations as to not result in divergence. That is the reason why even the normal SLAM problem has been addressed in the recent works by Rao-Blackwellized particle filters [4]. The other shortcoming of EKF is the slow convergence rate which can cause serious problems when a consumer camera with a limited field of view is used since most of the features would be available only for a few frames.

As mentioned before, a proper substitution for an EKF is the Rao-Blackwellized particle filter which is also known in literature as Fast-SLAM in which the robot's position and the landmarks' positions are decomposed. The uncertainties related to the robot's position Are Expressed By Particles And The Uncertainties For The Measurement Are Handled By Ekfs. The Extension Of Fast-Slam For Vslam has been done in [5][6]. However, they used just a single Gaussian to model landmarks' uncertainties which results in slow convergence or even divergence of their algorithms. As a result, they cannot properly limit the uncertainties when a camera with limited field of view is applied. In this paper, we propose a full particle filter approach which has a very simple landmark initialization and unlike the usual particle filter methods, the complexity of the algorithm grows linearly with respect to the landmark numbers. Besides, the algorithm converges very fast (just in a few steps).

The paper is organized as follows: in section 2 the outline of bearing-only SLAM for a robot in 2D space is described, while in section 3 the formulation of the new recursive algorithm is derived. Section 4 gives the implementation of the proposed algorithm using a particle filter. The complexity of the algorithm is discussed in section 5. The performance of the proposed algorithm is evaluated in section 6 and section 7 concludes this paper.

## 2    The Statement of Bearing Only SLAM Problem

Assuming a mobile robot in a 2D space in which some desired landmarks are available, the robot position can be stated by three parameters $x^R, y^R, \theta^R$, where $x^R, y^R$ show the absolute location of the robot on the $x - y$ plane and $\theta^R$ gives the orientation of the robot with respect to the $x$ axis. The robot can also move on the plane, based on linear velocity $u_1(m/s)$ and angular velocity $u_2(rad/s)$. Considering these parameters, the motion model of the robot in a discrete time case will be as follows:

$$
\begin{aligned}
x^R_{t+1} &= x^R_t + (u_{1,t} + w_{1,t})\cos(\theta^R_t) \\
y^R_{t+1} &= y^R_t + (u_{1,t} + w_{1,t})\sin(\theta^R_t) \quad ; t = 0, 1, 2, ... \\
\theta^R_{t+1} &= \theta^R_t + u_{2,t} + w_{2,t}
\end{aligned}
\tag{1}
$$

where $w_{1,t}$ and $w_{2,t}$ are normal processes which represent the uncertainties of the linear and angular speeds. As we know from real robots, $u_{1,t}$ and $u_{2,t}$ can usually be measured by odometry system on which the basic navigation systems of mobile robots are established. However, due to some environmental disturbances such as slippage there are always uncertainties related to these values which can be modeled as normal processes. The presence of these uncertainties gives rise to accumulating errors which may cause the robot have a complete wrong assumption of its position. To address this problem, relative measurements between the robot and some special features (landmarks) in the environment are employed. In the case of bearing only measurements (for example using a monocular camera), if a set of landmarks such as $\{L_1, ..., L_M\}$ which are uniquely identifiable can be extracted, the available measurements will be $\{\phi_{1,t}, ..., \phi_{M,t}\}$. Then, the mathematical model of the measurement is obtained as follows:

$$
\phi_{j,t} = \tan^{-1}\frac{y^L_j - y^R_t}{x^L_j - x^R_t} - \theta^R_t + v_t \quad ; j = 1, ..., M
\tag{2}
$$

where $(x^L_j, y^L_j)$s represent the absolute positions of the landmarks; however, initially they are unknown. $v_t$ is a measurement noise which can be modeled as a normal process.

Now, the problem is the estimation of the robot's and the landmarks' positions, given the following measurement sequence:

$$
\{\Phi_{j,t}\} = [\Phi_{1,t}, ..., \Phi_{M,t}]
$$

where $\Phi_{j,t} = [\phi_{j,t}, \phi_{j,t-1}, ..., \phi_{j,0}]$.

Speaking in probabilistic terms, if the vectors $\mathbf{x}^R_t = [x^R_t \ y^R_t \ \theta^R_t]^T$ and $\mathbf{x}^L_j = [x^L_j \ y^L_j]^T$ are defined, the following PDF (Probability Density Function) should be estimated:

$$
p(\mathbf{x}^R_t, \{\mathbf{x}^L_{j,t}\}|\{\Phi_{j,t}\})
\tag{3}
$$

such that the norm of the covariance matrix of the estimated joint random vector $[\mathbf{x}^R_t; \{\mathbf{x}^L_{j,t}\}]$ becomes minimized.

As we know for linear systems, the optimal solution of this estimation is a Kalman filter. This filter can also be applied in nonlinear cases by the linearizing the equations around the operating point of the system. However, in the case of bearing only measurements, since the range is not initially available, the uncertainties of the landmarks cannot be modeled by a simple Gaussian distribution. As already mentioned, one approach is to consider a set of Gaussians which are placed along the ray between the robot and the landmark and then estimate the joint vector using a modified EKF filter.

## 3    Derivation of a New Recursive Algorithm

As illustrated in section II, the main problem of bearing only SLAM is modeling a landmark's uncertainty. To deal with this problem when considering the particle filter approach, we attempt to model the landmark uncertainty using a set of line pieces.

As soon as the robot observes the landmark $L_j$ at the angle $\phi_{j,t}$ for the first time, the following PDF can be defined:

$$p(\mathbf{x}_j^L | \mathbf{x}_t^R, \phi_{j,t}) = \begin{cases} 1/(r_{max} - r_{min}), & \text{if condition} \\ 0, & \text{else} \end{cases} \tag{4}$$

where the condition is:

$$y_j^L = (m_{j,t}\, x_j^L + b_{j,t}) : (r_{min} \cos \psi_{j,t} + x_t^R < x_j^L < r_{max} \cos \psi_{j,t} + x_t^R)$$

$\psi_{j,t} = \phi_{j,t} + \theta_t^R$, $m_{j,t} = \tan \psi_{j,t}$, $b_{j,t} = y_t^R - \tan(\psi_{j,t}) x_t^R$. $r_{min}, r_{max}$ are constants which state the minimum and maximum ranges in which we expect the landmark. It means that we assume a uniform distribution of a landmark in a predefined range given the position of the robot. The above probability can be described using a general function such as:

$$\Gamma(\mathbf{x}_{j,c}^L, \lambda_j, \alpha_j) \tag{5}$$

where $\mathbf{x}_{j,c}^L = (x_{j,c}^L, y_{j,c}^L)$ represents the center of the line piece with the length $\lambda_j$ and angle $\alpha_j$ with respect to the $x$ axis.

After the initialization of a landmark, the goal is to minimize the uncertainties of the robot and the landmark while the robot moves. The main idea is that the bounded uncertainty of the landmark results in the bounded uncertainty of the robot's position and vice-versa. Such a mutual relation in EKF based methods is implemented using the covariance matrix. In the following, considering the proposed model of landmarks' uncertainties, such a relation in the context of a probability density function will be derived.

First, we start with the depth estimation of a feature when at least two different bearing observations of the feature are available. Referring to analytical geometry, we know that if the landmark $L_j$ is observed at two different angles

$\phi_{j,1}$ and $\phi_{j,2}$ at two different robot positions $\mathbf{x}_1^R$ and $\mathbf{x}_2^R$, then the landmark is located on the intersection of the lines along which the landmark is observed, meaning:

$$y_j^L = m_{j,1}x_j^L + b_{j,1} \quad \text{and} \quad y_j^L = m_{j,2}x_j^L + b_{j,2}$$

Solving the above equation system results in:

$$x_j^L = (b_{j,2} - b_{j,1})/(m_{j,1} - m_{j,2}); \quad y_j^L = m_{j,1}x_j^L + b_{j,1}$$

Nevertheless, this calculation is valid if and only if the exact position of the robot is available and also the measurement is noiseless. Otherwise, minor changes in theses parameters can cause major errors in the calculation of the landmark's position especially when it is much further from the robot. However, this fact can be used in a recursive algorithm including two parts, prediction and update, to minimize the uncertainties.

To derive the necessary formulations, we consider the following decomposition of (3) which can be easily verified using Bayesian rules:

$$p(\mathbf{x}_t^R, \{\mathbf{x}_{j,t}^L\}|\{\Phi_{j,t}\}) = p(\{\mathbf{x}_{j,t}^L\}|\mathbf{x}_t^R, \{\Phi_{j,t}\}) \times p(\mathbf{x}_t^R|\{\Phi_{j,t}\}) \tag{6}$$

Assuming (6) the recursive algorithm can be formed as follows:

**Prediction**

$$p(\mathbf{x}_{t+1}^R|\{\Phi_{j,t}\}) = \int p(\mathbf{x}_{t+1}^R|\mathbf{x}_t^R, \mathbf{u}_t, \mathbf{w}_t) \times p(\mathbf{x}_t^R|\{\Phi_{j,t}\})p(\mathbf{w}_t)d\mathbf{x}_t^R \, d\mathbf{w}_t \tag{7}$$

**Update**

$$p(\mathbf{x}_{t+1}^R|\{\Phi_{j,t+1}\}) = \frac{p(\{\phi_{j,t+1}\}|\mathbf{x}_{t+1}^R, \{\Phi_{j,t}\})}{p(\{\phi_{j,t+1}\}|\{\Phi_{j,t}\})} \times p(\mathbf{x}_{t+1}^R|\{\Phi_{j,t}\}) \tag{8}$$

Since the positions of the landmarks are independent of each other, (8) becomes:

$$p(\mathbf{x}_{t+1}^R|\{\Phi_{j,t+1}\}) = \frac{\displaystyle\prod_{j=1}^{M} p(\phi_{j,t+1}|\mathbf{x}_{t+1}^R, \{\Phi_{j,t}\})}{\displaystyle\prod_{j=1}^{M} p(\phi_{j,t+1}|\{\Phi_{j,t}\})} \times p(\mathbf{x}_{t+1}^R|\{\Phi_{j,t}\}) \tag{9}$$

where:

$$p(\phi_{j,t+1}|\{\Phi_{j,t}\}) = \int p(\phi_{j,t+1}|\mathbf{x}_{t+1}^R, \{\Phi_{j,t}\}) \times p(\mathbf{x}_{t+1}^R|\{\Phi_{j,t}\})d\mathbf{x}_{t+1}^R \tag{10}$$

and

$$p(\phi_{j,t+1}|\mathbf{x}_{t+1}^R, \{\Phi_{j,t}\}) = \int p(\phi_{j,t+1}|\mathbf{x}_{t+1}^R, \mathbf{x}_{j,t}^L) \times p(\mathbf{x}_{j,t}^L|\mathbf{x}_t^R, \{\Phi_{j,t}\}) \times$$
$$p(\mathbf{x}_t^R|\{\Phi_{j,t}\}) \, d\mathbf{x}_{j,t}^L \, d\mathbf{x}_t^R \tag{11}$$

The term $p(\mathbf{x}_{j,t}^L|\mathbf{x}_t^R, \{\Phi_{j,t}\})$ in (11) has a uniform distribution over a piece of line which goes out from the robot position along the $\theta_t^R + \phi_{j,t}$ angle and is bounded by the border of the last estimation of $\mathbf{x}_{j,t}^L$, (Fig.1).

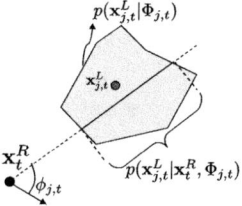

**Fig. 1.** The possible landmark position given the robot's position and all the previous measurements

Now the next step is to update the PDF of $\mathbf{x}_j^L$, meaning:

$$p(\mathbf{x}_{j,t+1}^L|\{\Phi_{j,t+1}\}) = \int p(\mathbf{x}_{j,t+1}^L|\mathbf{x}_{t+1}^R, \{\Phi_{j,t+1}\})p(\mathbf{x}_{t+1}^R|\{\Phi_{j,t+1}\})d\mathbf{x}_{t+1}^R \quad (12)$$

In (12), $p(\mathbf{x}_{j,t+1}^L|\mathbf{x}_{t+1}^R, \{\Phi_{j,t+1}\})$ again represents a uniform distribution over a line piece which is limited by the border of $p(\mathbf{x}_{j,t}^L|\{\Phi_{j,t}\})$. This results in the required rule to trim the length of the line piece. However, the implementation of such an algorithm needs to store the landmarks' regions as polygons and then find the intersecting points, which gives rise to a very time consuming algorithm. To ease this problem, we attempt to generalize the idea of using a line piece attached to each robot position which belongs to $p(\mathbf{x}_{j,t}^L|\{\Phi_{j,t}\})$ in order to represent part of the landmark uncertainty, but not necessarily along the ray between the robot and the landmark. Obviously, the constraint is not to lose any piece of information concerning landmarks' positions which can be guaranteed by a pessimistic trimming method discussed in the following. For the sake of simplicity, the initialized angles of the line pieces, $\alpha_{j,t}$s , are kept constant, and then the center of the line pieces and their length will be modified. For this purpose,(12) is decomposed again as follows:

$$p(\mathbf{x}_{j,t+1}^L|\{\Phi_{j,t+1}\}) = \int p(\mathbf{x}_{j,t+1}^L|\mathbf{x}_{t+1}^R, \bar{\mathbf{x}}_{j,t}^L, \mathbf{x}_t^R, \{\Phi_{j,t+1}\}) \times$$
$$p(\mathbf{x}_{t+1}^R|\mathbf{x}_t^R, \{\Phi_{j,t+1}\}) p(\bar{\mathbf{x}}_{j,t}^L|\mathbf{x}_t^R) p(\mathbf{x}_t^R|\{\Phi_{j,t}\}) d\mathbf{x}_{t+1}^R d\mathbf{x}_t^R d\bar{\mathbf{x}}_{j,t}^L \quad (13)$$

where $\bar{\mathbf{x}}_{j,t}^L$ denotes a line piece related to the landmark $L_j$. Considering the fact that each line piece can be uniquely defined when $\mathbf{x}_t^R$ is given, then $p(\bar{\mathbf{x}}_{j,t}^L|\mathbf{x}_t^R) = 1$. Besides, in the first probability term, the conditions $(\mathbf{x}_t^R, \{\Phi_{j,t}\})$ do not give any more information when $\bar{\mathbf{x}}_{j,t}^L$ is available. Therefore, (13) becomes:

$$p(\mathbf{x}_{j,t+1}^L|\{\Phi_{j,t+1}\}) = \int p(\mathbf{x}_{j,t+1}^L|\mathbf{x}_{t+1}^R, \bar{\mathbf{x}}_{j,t}^L, \phi_{j,t+1}) \times$$
$$p(\mathbf{x}_{t+1}^R|\mathbf{x}_t^R, \{\Phi_{j,t+1}\})p(\mathbf{x}_t^R|\{\Phi_{j,t}\}) d\mathbf{x}_{t+1}^R d\bar{\mathbf{x}}_{j,t}^L d\mathbf{x}_t^R \quad (14)$$

This decomposition gives us the possible location of $\mathbf{x}_{j,t+1}^L$ over the line piece $\bar{\mathbf{x}}_{j,t}^L$. Obviously, based on the distribution of this location over the line piece the new centers and the lengths of the line pieces can be defined. In (14), the terms $p(\mathbf{x}_{j,t+1}^L|\mathbf{x}_{t+1}^R, \bar{\mathbf{x}}_{j,t}^L, \phi_{j,t+1})p(\mathbf{x}_{t+1}^R|\mathbf{x}_t^R, \{\Phi_{j,t+1}\})$ can be interpreted the way that assuming a robot's position $\mathbf{x}_t^R$, a set of $\mathbf{x}_{t+1}^R$s denoted as $\bar{\mathbf{x}}_{t+1}^R|\mathbf{x}_t^R$ can be achieved. Then for each $\mathbf{x}_{t+1}^R \in \bar{\mathbf{x}}_{t+1}^R$ and a given $\phi_{j,t+1}$ a line can be assumed which intersects $\bar{\mathbf{x}}_{j,t}^L$. These intersection points are selected as the centers of the new line pieces $\mathbf{x}_{j,c,t+1}^L|\mathbf{x}_{t+1}^R$. Now, to select the lengths of the line pieces, we use a pessimistic way to guarantee the algorithm does not get stuck in local minima. For this purpose, the new lengths are selected such that:

$$\text{support of} \int p(\mathbf{x}_{j,t+1}^L|\mathbf{x}_{t+1}^R, \bar{\mathbf{x}}_{j,t}^L, \phi_{j,t+1})\, p(\mathbf{x}_{t+1}^R|\mathbf{x}_t^R, \{\Phi_{j,t+1}\})d\mathbf{x}_{t+1}^R$$

$$\in$$

$$\text{support of} \int \Gamma(\mathbf{x}_{j,c,t+1}^L|\mathbf{x}_{t+1}^R, \lambda_{j,t+1}, \alpha_j)d\mathbf{x}_{t+1}^R \qquad (15)$$

## 4    Particle Filter Implementation

The recursive algorithm mentioned in the previous section can be implemented using a modified particle filter approach. In such an approach the possible positions of the robot are assumed as a finite set of points in $R^3$ space, while the uncertainty of landmarks is modeled by a set of line pieces each of which is bound to a robot position. Thus, the estimated vector will be: $[\mathbf{x}_t^R, \{\bar{\mathbf{x}}_{j,t}^L\}]^T$ where $p(\bar{\mathbf{x}}_{j,t}^L|\mathbf{x}_t^R) = \Gamma(\mathbf{x}_{j,c,t}^L, \lambda_{j,t}, \alpha_{j,t})$

Based on the derived formulation in the previous section, the following algorithm can be generated:

**Prediction**
Generate $N$ samples (particles) from the PDF $p(\mathbf{x}_t^R|\{\Phi_{j,t}\})$ such as $X_t^R = \{\mathbf{x}_{t,1}^R, ..., \mathbf{x}_{t,N}^R\}$, and initialize the $N$ attached line pieces for each landmark such as $\bar{X}_{j,t}^L = \{\bar{\mathbf{x}}_{j,t,1}^L, ..., \bar{\mathbf{x}}_{j,t,N}^L\}$, where $j = 1, .., M$. Generate also $N$ samples from $p(\mathbf{w}_t)$ such as $W_t = \{\mathbf{w}_{t,1}, ..., \mathbf{w}_{t,N}\}$. By plugging $X_t^R$ and $W_t$ into (1) the next possible robot positions $\hat{X}_{t+1}^R = \{\hat{\mathbf{x}}_{t+1,1}^R, ..., \hat{\mathbf{x}}_{t+1,N}^R\}$ are obtained.

Considering $\{\bar{X}_{j,t}^L\}$ and $\hat{X}_{t+1}^R$ , a predicted measurement range for each pair of robot and landmark positions $(\hat{\mathbf{x}}_{t+1,i}^R, \bar{\mathbf{x}}_{j,t}^L)$ like $\bar{\phi}_{j,t+1,i}$ based on the end points of a line piece is obtained: $\bar{\phi}_{j,t+1,i} = [\phi_{j,t+1,i,min}, \phi_{j,t+1,i,max}]$

**Update**
As soon as the new measurements $\{\phi_{j,t+1}\}$ are obtained, the following weights can be calculated:
$$\omega_i = p^r \times q^{M-r}$$
where $r$ is the number of $\phi_{j,t+1,i}$s which belong to $\bar{\phi}_{j,t+1,i}$, $p$ and $q$ are desired positive values such that $p > q > 0$. Then the updated PDF can be built as follows:

$$p(\mathbf{x}_{t+1,i}^R | \{\Phi_{j,t+1,i}\}) = \frac{\omega_i}{c}$$

where $c = \sum \omega_i$ is a normalization term. Consequently, the line pieces are modified as follows:

$$p(\bar{\mathbf{x}}_{j,t+1,i}^L | \{\Phi_{j,t+1,i}\}, \mathbf{x}_{t+1,i}^R) = \Gamma(\mathbf{x}_{j,c,t+1,i}^L, \alpha_{j,i}, \lambda_{j,t+1,i})$$

where $\mathbf{x}_{j,c,t+1,i}^L$ is obtained from the intersection of the two line pieces: $\bar{\mathbf{x}}_{j,t,i}^L$ and $\bar{\mathbf{x}}_{j,t+1,i}^L | \hat{\mathbf{x}}_{t+1}^R$. The lengths of the line pieces are modified as follows:

$$\lambda_{j,t+1,i} = 2\sigma_{w2}$$

## 5    Complexity Analysis

The main achievement of such an algorithm is reducing the complexity of a particle based bearing-only SLAM algorithm. To handle this problem using a usual particle filter method, it is required to add a new dimension to the estimated vector for each new landmark, because of which the particles' number should be increased exponentially. Whereas, using line pieces to model the landmarks' uncertainties, given $N$ particles, $N$ new line pieces are added to the model for a new landmark. This results in the addition of a complexity $O(N)$, which is related to the calculation of the intersection points during each iteration. Therefore, assuming $M$ available landmarks the complexity of the problem is $O(MN)$, i.e. it increases linearly with respect to the number of landmarks. In addition, unlike the EKF based approaches, the initialization of a new landmark is a very simple task which can be done using (4) for any particle at any time step.

## 6    Simulation

To show the performance of the proposed algorithm, an environment including one robot and 5 landmarks was simulated. We ran our method, GSF based and fast SLAM methods in this environment for 35 time steps to evaluate their abilities to track the real path of the robot and generating a proper map. The linear and angular speeds of the robot were set as $0.6\,m/s$ and $0.2\,rad/s$; the input noises were normal processes with the variances $\sigma_{w1}^2 = \sigma_{w2}^2 = 0.05$. In our method, we used 150 particles and the line pieces were initialized at the range[1 8] meters. In the GSF based case, each landmark was initialized by 2 Gaussians at the range 4 and 8 meters, meaning that $2^5$ parallel EKF filters were used. For the fast SLAM algorithm, also 150 particles were used and for each pair of particle-landmark one Gaussain was placed at the depth of about 4 meters. The results of the position estimation of the robot and the landmarks can been seen in Fig. 2a, Fig. 2b, Fig. 2c and Fig. 2d. As can be observed, our algorithm managed to estimate the landmark positions properly in few steps and also tracked the real path of the robot closely, while the GSF based method lost the position of one landmark ($L_2$) and the estimated path was far from the real path at many points. Fast SLAM, as expected, showed also unacceptable performance (since the depth uncertainties is modeled just by one Gaussian for each landmark). We tested our algorithm also for many more landmarks and as expected the running time increased linearly, whereas the GFS based method converged tediously more and more for more landmarks. We ran the simulation for a wide range of input noises $\sigma_{w1} = \sigma_{w2} = 0.02...0.15$ and observed that the GSF based approach started to diverge in some runs (paths) for the noise variances of more than 0.04, but our algorithm managed to converge robustly in almost all of the cases. To compare their performances in this aspect, the percentage of the times that the GSF based method and our algorithm diverged in different runs with respect to the noise variances is shown in Fig. 2e.

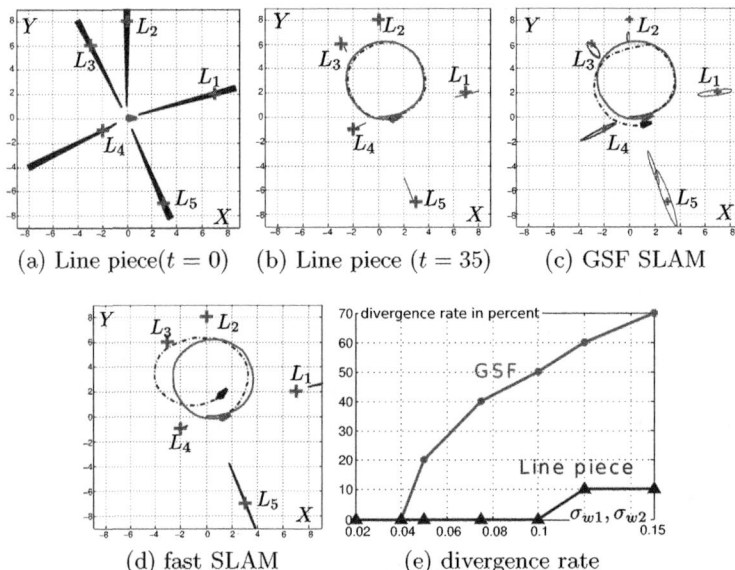

(a) Line piece($t = 0$)    (b) Line piece ($t = 35$)    (c) GSF SLAM

(d) fast SLAM    (e) divergence rate

**Fig. 2.** Localization and mapping using line pieces at the time steps 0 (a) and 35 (b), GSF method (c) and fast SLAM (d) for $\sigma_{w1}, \sigma_{w2} = 0.05$. The continuous paths show the real paths and the dashed paths show the estimated paths. The crosses show the real landmarks' positions. The line pieces or the ellipses show the landmarks' uncertainties. The divergence rate of the GSF based method and our method in different runs w.r.t the noise variances $\sigma_{w1}, \sigma_{w2}$ (e).

## 7   Conclusion

A new algorithm based on a particle filter approach was proposed to address the bearing-only SLAM problem. The novelty of this method is the use of line pieces to model the landmark uncertainties finding an approach to trim the joint assumptions over robot's and landmarks' positions. It was also shown that unlike the normal particle filter approach the complexity of the algorithm grows linearly. In the end, the performance of the algorithm was verified using simulation.

We have implemented the proposed algorithm on a real robot equipped with a simple web-cam. The measurement is obtained by using the Lucas-Kanade method to track some stable features within the frames. In this case wrong feature association or correspondence problems should be augmented in the formulations. These variations and also the practical aspects of the implementation of algorithm will be presented in future work.

# References

1. Bailey, T.: Constrained initialisation for bearing-only SLAM. In: Proceedings of ICRA 2003. International Conference on Robotics and Automation, pp. 1966–1971. IEEE (2003)
2. Strasdat, H., et al.: Visual bearing-only simultaneous localization and mapping with improved feature matching. Autonome Mobile Systeme 15–21 (2007)
3. Sola, J., et al.: Undelayed initialization in bearing only SLAM. In: Proceedings of IROS 2005. International Conference on Intelligent Robots and Systems, pp. 2499–2504. IEEE (2005)
4. Montemerlo, M., et al.: FastSLAM: A Factored Solution to the Simultaneous Localization and Mapping Problem. In: Proceedings of the AAAI National Conference on Artificial Intelligence, pp. 593–598 (2002)
5. Sim, R., et al.: Vision-based SLAM using the Rao-Blackwellised particle filter. In: IJCAI Workshop on Reasoning with Uncertainty in Robotics, pp. 9–16 (2005)
6. Farrokhsiar, M., Najjaran, H.: A higher order Rao-Blackwellized particle filter for monocular vSLAM. In: American Control Conference (ACC), pp. 6987–6992. IEEE (2010)
7. Jeong, W.Y., Lee, K.M.: Visual SLAM with line and corner features. In: Proceedings of IROS 2007. International Conference on Intelligent Robots and Systems, pp. 2570–2575. IEEE (2007)
8. Kwok, N.M., Dissanayake, G.: An efficient multiple hypothesis filter for bearing-only SLAM. In: Proceedings of IROS 2005. International Conference on Intelligent Robots and Systems, pp. 736–741. IEEE (2005)
9. Lee, Y.J., Song, J.B.: Visual SLAM in Indoor Environments Using Autonomous Detection and Registration of Objects. In: International Conference on Multisensor Fusion and Integration for Intelligent Systems. IEEE (2008)
10. Kwok, N.M., Dissanayake, G., Ha, Q.P.: Bearing-only slam using a SPRT based Gaussian sum filter. In: Proceedings of ICRA 2006. International Conference on Robotics and Automation, pp. 1109–1114. IEEE (2006)
11. Castle, R.O., Klein, G., Murray, D.W.: Combining monoSLAM with Object Recognition for Scene Augmentation using a Wearable Camera. Image and Vision Computing 28(11), 1548–1556 (2010)

# An Interval Type-2 Fuzzy Logic System for Human Silhouette Extraction in Dynamic Environments

Bo Yao[1], Hani Hagras[1], Danniyal Al Ghazzawi[2], and Mohammed J. Alhaddad[2]

[1] School of Computer Science and Electronic Engineering, University of Essex,
Wivenhoe Park, Colchester, CO4 3SQ, UK
[2] King Abdulaziz University, Jeddah, Saudi Arabia

**Abstract.** In this paper, we present a type-2 fuzzy logic based system for robustly extracting the human silhouette which is a fundamental and important procedure for advanced video processing applications, such as pedestrian tracking, human activity analysis and event detection. The presented interval type-2 fuzzy logic system is able to detach moving objects from extracted human silhouette in dynamic environments. Our real-world experimental results demonstrate that the proposed interval type-2 fuzzy logic system works effectively and efficiently for moving objects detachment where the type-2 approach outperforms the type-1 fuzzy system while significantly reducing the misclassification when compared to the type-1 fuzzy system.

**Keywords:** Interval type 2 fuzzy logic, Silhouette extraction, Human tracking.

## 1 Introduction

Accurate human silhouette (or outline) segmentation from a video sequence is important and fundamental for advanced video applications such as pedestrian tracking and recognition, human activity analysis and event detection. Advanced human detection and identification approaches like [1], [2] can be utilized for silhouette extraction. However, such methods are commonly of high computational complexity and hence not suitable for dynamic and complex environments. Hence, there is a need for silhouette extraction methods which are computationally efficient and that are able to operate in dynamic and complex environments.

The first step for silhouette extraction is background modeling and subtraction to detect moving targets as foreground objects. In [3], an approach based on a single Gaussian modal was developed which employed a simple robust method to handle moving objects and slow illumination changes. However, there are several limitations in this method such as learning stage necessity for background distribution, and robustness deficiency for situations like sudden illumination changes, slow moving objects, etc. To address these problems, the Gaussian Mixture Model (GMM) was proposed [4]. In this model, each pixel is modeled using $n$ Gaussian distributions. GMM is effective to overcome the shortcomings of single Gaussian model and hence GMM is extensively recognized as a robust approach for background modeling and subtraction. Therefore, in this paper, GMM is utilized for foreground detection. However, it is unreasonable to simply consider GMM foreground as human silhouette in

M. Kamel, F. Karray, and H. Hagras (Eds.): AIS 2012, LNCS 7326, pp. 126–134, 2012.

real-life environments because there are numerous noise factors and uncertainties to handle which include:

- Varying light condition
- Reflections and   shadows
- Moving objects attached to human silhouette (a book, a chair, etc.).

To handle these problems and detach the moving objects from the human silhouette, a type-1 Fuzzy Logic System (T1FLS) was proposed [5]. This T1FLS is capable of handling to an extent the uncertainties mentioned above, however, the extracted silhouette will be degraded due to misclassification of the proposed T1FLS. Hence, in this paper, we will present an Interval Type-2 Fuzzy Logic System (IT2FLS) which will be able to handle the high uncertainty levels present in real-world dynamic environments while also reducing the misclassification of extracted silhouette. The IT2FLS used similar type-1 membership function as the ones presented in [5] as principal membership functions which are then blurred to produce the type-2 fuzzy sets used in this paper. We have also used the same rule base as [5] to allow for a fair comparison with the results reported in [5].

In this proposed system, GMM is adopted for original foreground detection, then a IT2FLS is performed to detach the moving objects from the human silhouette. We have performed several real-world experiments where it was shown that the proposed IT2FLS is effective to reduce the misclassification and the quality of the extracted human silhouette is much improved when compared to the T1FLS.

The rest of this paper is organized as follows. In section 2, we provide a brief overview of type-2 FLSs. Section 3 presents the proposed IT2FLS. Section 4 presents the experiments and results and finally the conclusions and future work are presented in section 5.

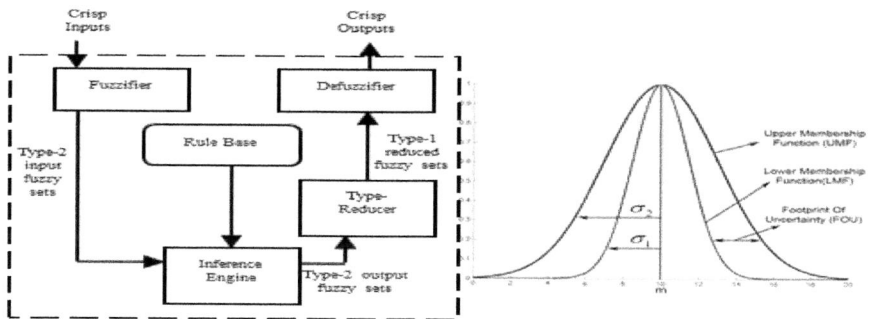

**Fig. 1.** (a) Structure of the type-2 FLS. (b) An interval type-2 fuzzy set.

## 2    A Brief Overview of the IT2FLS

The IT2FLS depicted in Fig. 1a) [6] uses interval type-2 fuzzy sets (such as the type-2 fuzzy set shown in Fig. 1b) [6] to represent the inputs and/or outputs of the FLS. In the interval type-2 fuzzy sets all the third dimension values are equal to one [6], [7].

The use of interval type-2 FLS helps to simplify the computation (as opposed to the general type-2 FLS) [8].

The interval type-2 FLS works as follows [6], [7], [8]: the crisp inputs from the input sensors are first fuzzified into input type-2 fuzzy sets; singleton fuzzification is usually used in interval type-2 FLS applications due to its simplicity and suitability for embedded processors and real time applications. The input type-2 fuzzy sets then activate the inference engine and the rule base to produce output type-2 fuzzy sets. The type-2 FLS rule base remains the same as for the type-1 FLS but its Membership Functions (MFs) are represented by interval type-2 fuzzy sets instead of type-1 fuzzy sets. The inference engine combines the fired rules and gives a mapping from input type-2 fuzzy sets to output type-2 fuzzy sets. The type-2 fuzzy output sets of the inference engine are then processed by the type-reducer which combines the output sets and performs a centroid calculation which leads to type-1 fuzzy sets called the type-reduced sets. There are different types of type-reduction methods. In this paper we will be using the Centre of Sets type-reduction as it has reasonable computational complexity that lies between the computationally expensive centroid type-reduction and the simple height and modified height type-reductions which have problems when only one rule fires [6], [7]. After the type-reduction process, the type-reduced sets are defuzzified (by taking the average of the type-reduced set) to obtain crisp outputs that are sent to the actuators. More information about the interval type-2 FLS and its benefits can be found in [6], [7], [8].

## 3    The Proposed IT2FLS for Human Silhouette Extraction

Fig. 2 provides an overview of proposed IT2FLS approach for human silhouette extraction. In the video capturing stage, source images are captured using a stationary camera. The images are then analyzed using GMM to detect foreground. After that, the foreground detected by GMM is partitioned into small $n \times n$ blocks, for example $2 \times 2$ blocks. Then Human tracking is performed by global nearest neighbor (GNN) [9] to obtain the human centroids. Based on the partitioned foreground and the obtained human centroids, human silhouette extraction is carried out.

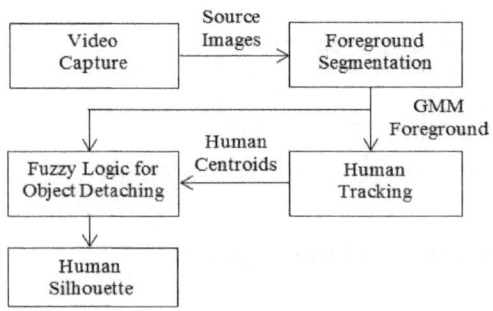

**Fig. 2.** An overview of the proposed system

Indeed, advanced human detection and identification approaches can be utilized for silhouette extraction. However, those methods are commonly of high computational complexity and not robust for dynamic and complex environments. Hence in [5] a T1FLS was presented and in this paper we will present an IT2FLS for silhouette extraction.

Suppose that we are working on frame $i$, and the foreground image of frame $i$ obtained by GMM is denoted by $O_i$, and the silhouettes in frame $i$-$1$ that have been properly segmented are denoted by $O_{i-1}$. As described above, the foreground in $O_i$ may contain the human body and moving non-human objects attached to silhouette. To detach the moving objects and refine the human silhouette, the fuzzy logic system is developed based on the following observations:

1. If an image block in $O_i$ belongs to the human body, it is of high probability to match an image block in $O_{i-1}$ in a good match degree. SAD (sum of absolute difference) between their corresponding blocks in frame $i$ and frame $i$-$1$, is used to measure the matching degree between the image block in $O_i$ and its best match block in $O_{i-1}$.
2. If the distance between this block and human centroid is far, the probability that this block belongs to the human body is low.
3. If the amount of its neighbor blocks with high probability belonging to the human body is huge, for example having good matches in $O_{i-1}$, or having low distance to human centroids etc., then, the probability of this block also belongs to the human body is high.

Based on observations above, the following variables of each block are calculated.

— SAD of motion estimation. For every image block in $O_i$, its best match image block in frame $i$-$1$ is searched. And the matching degree is used to describe its SAD variable.
— The distance between this block and the human centroid.
— The amount of its neighborhood with high probability of belonging to human body, for example, have a good match block in human body, low centroid distance, etc.

The rules of the type-2 fuzzy system should remain the same as the T1FLS in [5] and we will use the similar type-1 fuzzy sets in [5] as the principal membership functions which are then blurred by 10% (the 10 % was determined empirically to balance between robustness to noise and system performance) to produce the type-2 fuzzy sets in our IT2FLS. The membership functions for the inputs and output of the IT2FLS are shown in Fig 3. The rule base of the IT2FLS is the same as [5] and it is as follows:

1. If SAD is very low AND Neighborhood is huge AND Distance is close,
   THEN Silhouette is high.
2. If SAD is large AND Neighborhood is small AND Distance is very far,
   THEN Silhouette is low.
3. If SAD is low AND Neighborhood is large AND Distance is medium,
   THEN Silhouette is high.
4. If SAD is medium AND Neighborhood is medium AND Distance is medium,
   THEN Silhouette is medium.

5. If SAD is large AND Neighborhood is medium AND Distance is medium,
   THEN Silhouette is medium.
6. If SAD is large AND Neighborhood is large AND Distance is close,
   THEN Silhouette is high.
7. If SAD is medium AND Neighborhood is large AND Distance is medium,
   THEN Silhouette is high.
8. If SAD is medium AND Neighborhood is large AND Distance is close,
   THEN Silhouette is high.
9. If SAD is very large AND Neighborhood is small AND Distance is far,
   THEN Silhouette is low.
10. If SAD is medium AND Neighborhood is small AND Distance is very far,
    THEN Silhouette is low.
11. If SAD is low AND Neighborhood is huge AND Distance is medium,
    THEN Silhouette is high.
12. If SAD is low AND Neighborhood is medium AND Distance is medium,
    THEN Silhouette is high.

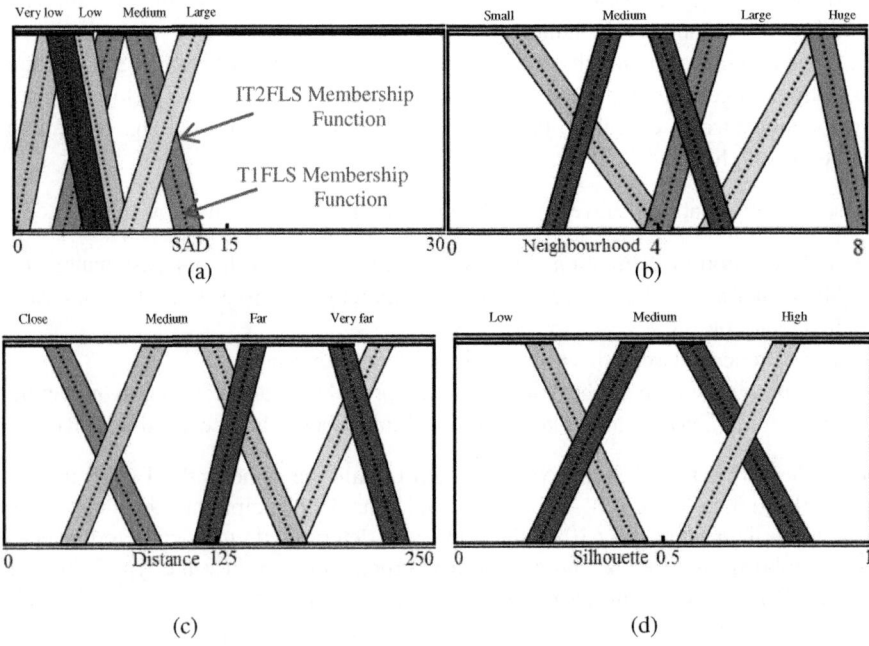

**Fig. 3.** The interval type-2 fuzzy sets employed in the inputs of our IT2FLS for (a) SAD, (b) Neighborhood, (c) Distance, (d) The Silhouette output of the IT2FLS

## 4    Experimental Results

We have performed several real-world experiments to validate our proposed approach and to compare the performance of the IT2FLS and the T1FLS presented in [5]. The ground truth data is captured from cameras deployed around our laboratory of smart living room to analyze people's regular activity. The aim of the experiments was to validate that the proposed IT2 fuzzy system is effective to detach the moving objects from human silhouette with much fewer misclassifications than T1 fuzzy system.

As shown in Fig. 4 to 8, column (a) shows the source images; column (b) shows the original foreground detected by GMM; column (c) illustrates the results after the T1FLS; column (d) exhibits the results after using IT2FLS. In our case, human silhouette is represented by pixels having a higher degree than the 0.5 degree of the fuzzy silhouette.

In Fig. 4, the results of "Raising a book" demonstrate that the book attached to human is eliminated after using the fuzzy based systems. The experiment shows that two fuzzy systems extract a proper silhouette and that they are able to detach the book from human body. However, as mentioned above, the silhouette extracted by T1FLS is degraded due to misclassification (as confirmed by [5]) while IT2FLS can address this problem and reduce the misclassification.

In Fig. 5, enlarged silhouette images of "Raising a book" are provided. As can be seen in Fig. 5(b), the edge the silhouette of T1FLS, (highlighted with red rectangles) is degraded to coarser edge due to misclassification when compared to the original foreground. However, as displayed in Fig. 5(c), the IT2FLS achieves a same human silhouette as the original foreground while the book is detached.

To demonstrate the robustness of the proposed system, more experiments have been done in various environments. In Fig. 6, results in an outdoor environment of single pedestrian are shown, via this experiment, we can see that the proposed system is working effectively in an outdoor environment. In Fig. 7, it can be seen in the images that the reflection of human body which is a noise factor is detected as foreground by GMM while the reflection is eliminated by utilizing fuzzy logic. In Fig. 8, results in an outdoor environment crowded with people are shown. In this complex outdoor environment with more noises and uncertainties, the proposed system demonstrates the robustness by extracting human silhouette with a promising result. In these experiments, for the purpose of comparison, Table 1 provides the average and standard deviations for the misclassification and accuracy for T1 and IT2 fuzzy systems. In Table 1, it can be seen clearly that the misclassification of the proposed IT2FLS is reduced significantly compared to T1FLS while the IT2FLS results also in higher accuracy than the T1FLS.

**Table 1.** Comparison of misclassification and accuracy

| Experiment Name | T1 Average Misclassify | T1 STDEV Misclassify | T1 Average Accuracy | IT2 Average Misclassify | IT2 STDEV Misclassify | IT2 Average Accuracy | Frame Used |
|---|---|---|---|---|---|---|---|
| Single person | 125.74 pixels | 41.19 pixels | 93.23% | 0.61 pixels | 1.37 pixels | 99.9791% | 383 |
| Multi-person with reflection | 221.38 pixels | 74.73 pixels | 95.23% | 0.54 pixels | 1.39 pixels | 99.9951% | 246 |
| Crowded environment | 637.62 pixels | 180.59 pixels | 90.21% | 12.84 pixels | 16.46 pixels | 99.9004% | 255 |

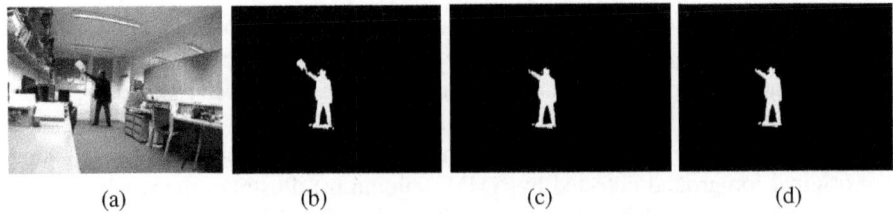

(a)                (b)                (c)                (d)

**Fig. 4.** The experiment of "Raising a book"; (a) source images, (b) foreground detected by GMM, (c) extracted silhouette after using T1FLS, (d) extracted silhouette after using IT2FLS

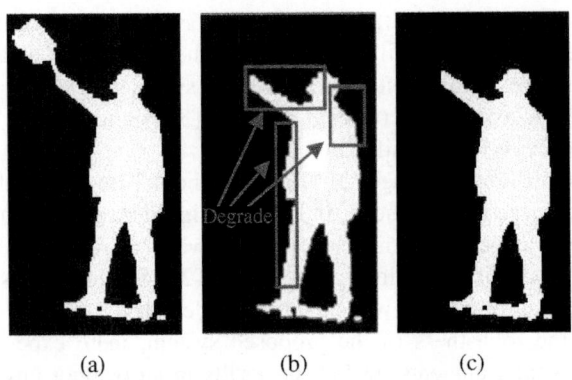

(a)                (b)                (c)

**Fig. 5.** Enlarged images of "Raising a book"; (a) foreground detected by GMM, (b) extracted silhouette after using T1FLS, (c) extracted silhouette after using IT2FLS

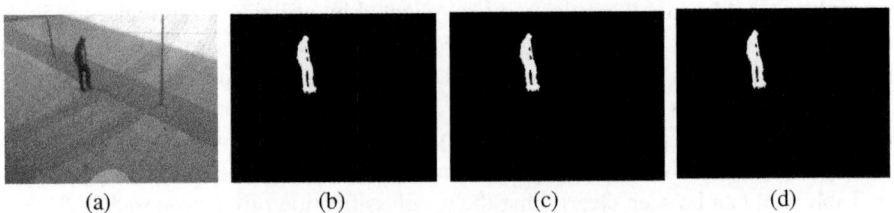

(a)                (b)                (c)                (d)

**Fig. 6.** The experiment of "Single person"

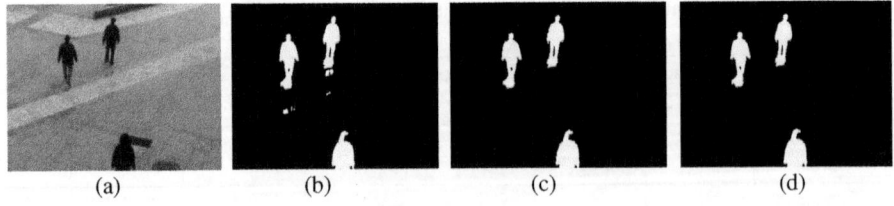

(a)                (b)                (c)                (d)

**Fig. 7.** The experiment of "Multi-person with reflection"

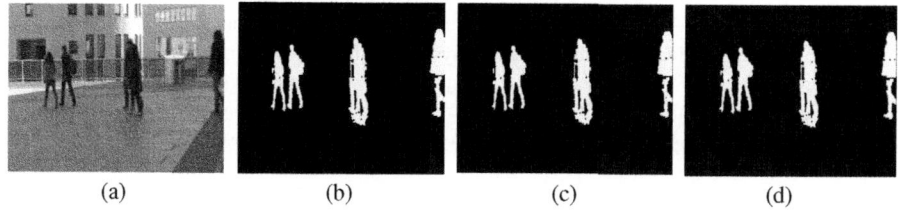

(a)                (b)                (c)                (d)

**Fig. 8.** The experiment of "Crowded environment"

## 5    Conclusions and Future Work

In this paper, we have presented an IT2FLS for improved silhouette extraction in dynamic real-world environments. Due to the huge complexity of the dynamic and real-life environment, the problem of detaching moving objects from human silhouette gets quite complicated. To address this problem without high computational complexity, we firstly use GMM to detect foreground, then an IT2FLS is employed for moving objects detachment. We have conducted several real-world experiments which have shown that the proposed IT2FLS is effective to detach objects and the misclassification is greatly reduced compared to a similar T1FLS while the IT2FLS results also in high accuracy for silhouette extraction compared to the T1FLS.Hence, by utilizing IT2FLS, the proposed system obtain silhouette extraction with good robustness to noise factors and uncertainties such as light condition changes, reflection of human body, and moving objects attached to the human silhouette, etc., in dynamic indoor/outdoor environments.

For our future ongoing research, we intend to extend the proposed algorithm for automatic learning which will enable the system to be more robust in dynamic environments. We also aim to apply the proposed system in high-level vision applications such as event detection, human activity recognition, etc.

## References

[1] Dalal, N., Triggs, B.: Histograms of oriented gradients for human detection. In: Conference on Computer Vision and Pattern Recognition, CVPR (2005)

[2] Wu, Y., Yu, T.: A field model for human detection and tracking. IEEE Transactions on Pattern Analysis and Machine Intelligence 28(5), 753–765 (2006)

[3] Aubert, G., Kornprobst, P.: Mathematical Problems in Image Processing. Partial Differential Equations and the Calculus of Variations. Applied Mathematical Sciences, vol. 147. Springer, New York (2001)

[4] Katz, B., Lin, J., Stauffer, C., Grimson, E.: Answering questions about moving objects in surveillance videos. In: AAAI Spring Symposium on New Directions in Question Answer (2003)

[5] Chen, X., He, Z., Keller, J., Anderson, D., Skubic, M.: Adaptive Silhouette Extraction in Dynamic Environments Using Fuzzy Logic. In: Proc. WCCI, Vancouver, BC (July 2006)

[6] Mendel, J.: Uncertain Rule-Based Fuzzy Logic Systems: Introduction and New Directions. Prentice-Hall, Upper Saddle River (2001)

[7] Hagras, H.: A Type-2 fuzzy controller for autonomous mobile robots. In: Proc. of the 2004 IEEE International Conference on Fuzzy Systems, Budapest Hu, pp. 965–970 (2004)

[8] Liang, Q., Mendel, J.: Interval type-2 fuzzy logic systems: theory and design. IEEE Trans. on Fuzzy Systems 8, 535–550 (2000)

[9] Blackman, S., Popoli, R.: Design and Analysis of Modern Tracking Systems. Artech (1999)

# Sparse VSLAM with Camera-Equipped Quadrocopter

Matevž Bošnak and Sašo Blažič

Faculty of Electrical Engineering, University of Ljubljana, Slovenia
matevz.bosnak@fe.uni-lj.si

**Abstract.** A successful Video-based Simultaneous Localization And Mapping (VSLAM) implementation usually requires a vast amount of feature points to be detected in the environment, which makes the VSLAM problem s computationally demanding operation in mobile robot navigation. This paper presents a VSLAM implementation that is based on a sparse distribution of high-informative artificial landmark features. Additionally, our approach combines the video system analysis results and the inertial measurement unit (IMU) measurements that define the orientation of the video camera. Successful implementation of the VSLAM system can enable autonomous quadrocopter navigation in the structured environment without the presence of the additional external positioning systems.

## 1 Introduction

A quadrotor helicopter or quadrocopter is an aerial vehicle propelled by four fixed-pitch rotors, which makes it mechanically simpler than an ordinary helicopter and by using multiple smaller rotors arranged around the central base, the rotors can also be protected against hitting obstacles. This makes the quadrocopter appropriate for indoor use. Small quadrotors with on-board stabilization which can be bought off-the-shelf, like the Parrot AR.Drone, make it possible to shift the research from basic control of the platform towards applications that make use of their versatile scouting capabilities. Still, the limited sensor suite, fast response rates along with the low inertia make it a challenge to fully automate the navigation of such platform. One of the prerequisites for autonomous navigation is the capability to make and use the map of the environment[10].

The dominant approach to the SLAM problem was introduced in a seminal paper by Smith, Self, and Cheeseman[8]. This paper proposed the use of the extended Kalman filter (EKF) for incremental estimation of the posterior distribution over camera pose along with the positions of the landmarks. In the last decade, this approach has found widespread acceptance in the field of the mobile robotics and many variants of the SLAM algorithm have been produced from the original idea (the OpenSLAM.org project lists more than 20 different implementations of this algorithm).

Our study is focused on implementation of the FastSLAM algorithm[9,5] using the video camera observations of the high-informative artificial landmark features, distributed across the environment. Because the AR.Drone quadrocopter already contains two video cameras, inertial measurement unit and telemetry via WiFi network[1], it was selected as a test platform for our research.

## 2    Computer-Vision System

Video frames, captured by the AR.Drone vertical camera (observing the floor under the vehicle) is wirelessly transferred to a personal computer, running our custom video-recognition software. The software, developed in Visual C# using the combination of ARDroneControl library, AForge.NET and Emgu CV frameworks for image processing and glyph Recognition library GRATF for glyph extraction, estimates the optical flow vectors and extracts the glyph's position, which is used to calculate the position of the origin and the orientation of the glyphs in regards to the quadrocopter current position.

The goal of the optical flow estimation is to compute an approximation of the motion field from the time-varying position of the image-features. By measuring the angular velocity of the camera and the distance to the object by other means, the feature velocity with respect to the camera can be accurately estimated. Sparse optical flow vector field was produced using the combination of FAST corner detection algorithm [7] and the Lucas-Kanade differential method[4]. The resulting motion field vectors were then averaged by the lengths and the orientations, which were then used to identify all significantly different vectors and label them as outliers. The remaining vectors were used to recalculate the average($\Delta x^{OF}, \Delta y^{OF}$), which was then used as the estimation of the optical flow due to the camera's combined lateral and angular motion. The apparent lateral motion of the camera due to the camera's rotation ($\Delta x_r^{OF}, \Delta y_r^{OF}$) can then be safely approximated by the circular motion of the features at the radius $z$. By subtracting these approximations from the total observed motion, the lateral motion of the camera can be expressed as

$$\Delta x_m^{OF} = z K_1^{OF} \left[ \Delta x^{OF} - K_2^{OF} \omega_\phi \right], \qquad \Delta y_m^{OF} = z K_1^{OF} \left[ \Delta y^{OF} - K_2^{OF} \omega_\theta \right] \quad (1)$$

where $z$ is the distance from the camera's origin to the ground plane, $K_1^{OF}$ and $K_2^{OF}$ are constants that are obtained with the camera's calibration and $\omega_\theta$, $\omega_\phi$ are the radial velocities about the $x$ and $y$-axes of the quadrocopter.

### 2.1    Glyph Recognition and the Position Determination

Graphical glyphs, represented by a black border and a square central grid with the glyph code, were used as the artificial landmark features. Glyphs with their bounding boxes were extracted from each video frame and used for determining the position and orientation of the camera. Instead of determining the homography matrix between the glyph plane and the camera plane, we followed a more specialized approach, taking advantage of the sensory data already present in the system. In the first step we assume that the camera frame is levelled horizontally, with the camera's central axis directed straight down and that the projective geometry of the pinhole camera is modelled by a perspective projection [2].

A point on the glyph plane $(T_x, T_y, T_z)$, whose coordinates are expressed with respect to the quadrocopter, will project onto the image plane at a point $C'_x, C'_y$, given by

$$\begin{bmatrix} C'_x \\ C'_y \end{bmatrix} = \frac{\lambda \nu}{T_z} \begin{bmatrix} T_x \\ T_y \end{bmatrix}, \qquad T_z = \lambda \nu \sqrt{\frac{\Delta T_x^2 + \Delta T_y^2}{\Delta C_x^2 + \Delta C_y^2}} = \lambda \nu \frac{a_T}{a_C} \qquad (2)$$

where $\lambda$ denotes the distance of the viewpoint origin behind the image plane [3], $\nu$ is the pixel density (in pixels per millimeter) and $T_z$ is the distance between the viewpoint origin and the ground plane, perpendicular to the image plane, calculated from the size of the recognized glyph $(\Delta T_x, \Delta T_y)$ and the size of the glyph in the image $(\Delta C_x, \Delta C_y)$ ($a_T$ is the size of the target in millimeters and $a_C$ is the size of the image of the target in pixels). The coordinates $T_x$ and $T_y$ (3, left) can be obtained by combining equations (2). Coordinates of the glyph's top-right $(T_{x,0}, T_{y,0})$ and the bottom-left $(T_{x,2}, T_{y,2})$ corners can then be used to define the camera z-axis rotation angle $\psi$ (3, right).

$$\begin{bmatrix} T_x \\ T_y \end{bmatrix} = \frac{a_T}{a_C} \begin{bmatrix} C'_x \\ C'_y \end{bmatrix}, \qquad \psi = \arctan 2 \frac{T_{y,0} - T_{y,2}}{T_{x,0} - T_{x,2}} \qquad (3)$$

To determine the true position of the quadrocopter with respect to the observed target, the effect of the video-camera tilt and orientation is compensated using the angles measured by IMU unit. The point $\mathbf{T}$ is rotated about the $x$-axis by the angle $\phi$, about the $y$-axis by the angle $\theta$ and about the $z$-axis by the angle $\psi$.

$$(T_x^f, T_y^f, T_z^f) = \mathbf{R}_z(\psi) \mathbf{R}_y(\theta) \mathbf{R}_x(\psi) \begin{bmatrix} T_x \\ T_y \\ T_z \end{bmatrix} \qquad (4)$$

where $\mathbf{R}_x(\psi)$, $\mathbf{R}_y(\theta)$ and $\mathbf{R}_z(\psi)$ are the standard rotation matrices about the axes of rotation, $x$, $y$ and $z$, respectively. The coordinates $(-T_x^f, -T_y^f, -T_z^f)$ finally define the position of the quadrocopter with respect to the observed target.

## 2.2   Estimation of the Position Covariance Matrix

In order to implement robust localization and mapping, the position measurement must be supplemented with the measurement error covariance matrix. Let's assume that both the position of the camera $p_C = (p_{C,x}, p_{C,y}, p_{C,z})$ and the position of the glyph target $p_T = (p_{T,x}, p_{T,y}, p_{T,z})$ were successfully estimated in the global coordinate system (the one that the map of the environment is referenced to).

Let the $\vec{v_1'} = p_T - p_C = (p_{T,x} - p_{C,x}, p_{T,y} - p_{C,y}, p_{T,z} - p_{C,z})$. Then the vertical covariance matrix eigenvectors $\vec{v_1'}, \vec{v_2'}, \vec{v_3'}$ are defined as

$$\vec{v_1'} = \frac{\vec{v_1'}}{|\vec{v_1'}|} \qquad \vec{v_2'} = \vec{v_1'} \times ([0\ 0\ 1] \times \vec{v_1'}) \qquad \vec{v_3'} = \vec{v_1'} \times \vec{v_2'} \qquad (5)$$

and the eigenvalues matrix as

$$\mathbf{L} = \begin{bmatrix} \sigma_x d & 0 & 0 \\ 0 & \sigma_y d & 0 \\ 0 & 0 & \sigma_z d \end{bmatrix} \tag{6}$$

where $d = |\vec{v_1}|$ and $\sigma_x$, $\sigma_y$, $\sigma_z$ constants that describe the shape of the 3-dimensional ellipsis that is described by the error covariance matrix $\mathbf{C}$ defined as

$$\mathbf{C} = \begin{bmatrix} \vec{v_1} & \vec{v_2} & \vec{v_3} \end{bmatrix} (2\mathbf{L}^2) \begin{bmatrix} \vec{v_1} & \vec{v_2} & \vec{v_3} \end{bmatrix}^T \tag{7}$$

## 3   FastSLAM Implementation

In the paper [5] authors approached the SLAM problem from a Bayesian point of view and observed that the SLAM problem exhibits important conditional independences. In particular, knowledge of the camera's path renders the individual landmark feature measurements independent. Based on this observation, the paper [5] describes an efficient SLAM algorithm called FastSLAM, which decomposes the SLAM problem into a camera localization problem, and a collection of landmark feature estimation problems that are conditioned on the camera pose estimate. It uses a modified particle filter for estimating the posterior over camera paths. Each of these particles estimates a camera pose and possesses Kalman filters that estimate the landmark feature locations conditioned on the camera pose estimate. Unlike the traditional SLAM techniques where sensors return the observed feature's bearing and range or only the feature's bearing (image-based features like SURF, SIFT, FAST or others), our glyph-based features together with the measured tilt angles define both the 3-dimensional position and 3-dimensional orientation of the camera. This enables us to use far less artificial markers for successful localization and mapping.

**Particle Filter Initialization.** Without any information on position of the camera and the features in the environment, particle filter is initialized with randomly distributed particles over the environment, defined only by the initial borders.

**Particle Filter Propagation Step.** In the first step of particle filter evaluation, all particles are propagated using the optical flow-based prediction and random noise. If the $i$-th particle position in the step $k$ is denoted as $p_i(k) = (p_{i,x}(k), p_{i,y}(k), p_{i,z}(k))$, the camera yaw orientation as $\psi_i(k)$ and the random noise for each of the component as $n_{p,x}$, $n_{p,y}$ and $n_{p,z}$, then the propagation step can be expressed as

$$p_{i,x}(k+1) = p_{i,x}(k) - p_{i,z}(k)(\cos \psi_i(k)\Delta x_m^{OF} - \sin \psi_i(k)\Delta y_m^{OF}) \cdot 0.01 + n_{p,x} \tag{8}$$
$$p_{i,y}(k+1) = p_{i,y}(k) + p_{i,z}(k)(\sin \psi_i(k)\Delta x_m^{OF} + \cos \psi_i(k)\Delta y_m^{OF}) \cdot 0.01 + n_{p,y} \tag{9}$$
$$p_{i,z}(k+1) = p_{i,z}(k) + n_{p,z} \tag{10}$$

**The measurement update** step was divided into the following steps.

1. Identifying the correspondences between features in the map and newly observed features is done by employing matching by the closest Bhatacharyya distance, which is a generalization of the Mahalanobis distance.
2. Camera position is re-estimated based on the correspondences from the previous step and the matching is executed again to produce more accurate sum of all distances in matched pairs. This sum is later treated as the particle weight.
3. The correction step of the map features' filters in each particle is executed along with the correction step of the camera position filter.

**Importance Weights.** In order to evaluate each particle, a weight is assigned to each of them with the rule that all particle weights sum up to 1. In the process of resampling, these weights are used to determine, which particles will "survive" and which particles will be replaced by newly generated random particles. In our case, the sum of all distances between the observed features positions and the positions of the same features in the map defines the particle weight.

**Adding of the New Map Features.** For each observed feature that does not have a match in the map, a new map feature entry is created. This comes at a price of a greater particle weight, which in the end creates a particle filter with a tendency for particles that do not add additional map features to the map. This way, a map is generated with the least number of false features.

**Resampling.** One of the problems that appear with the use of the particle filters is the depletion of the population after a few iterations. To solve this, the particles are resampled once the measure named $ESS$ (effective particle size) goes below a predefined threshold (in our case, this threshold was set to $0, 5$). $ESS$ is calculated from sample weights $w_i$ with the following equation

$$ESS = \frac{M}{1 + \frac{1}{M} \sum_{i=1}^{M} (Mw_i - 1)^2} \qquad (11)$$

where $M$ is the number of particles. For our implementation, we selected *Select with replacement* resampling algorithm, described in the technical report by Rekleitis[6].

## 4   Results and Future Development

The FastSLAM implementation was tested on a series of video frames (Figure 1a) that were captured along a path around our laboratory. The result is a map of the features in our laboratory and the camera path estimate that illustrates the explored positions in the map (Figure 1b). The camera path estimate is based on the particle with the highest weight, which corresponds to a particle with the highest probability.

Although the SLAM in the field of the mobile robotics is a widely adopted approach, our team just recently started with the work in this field. This paper

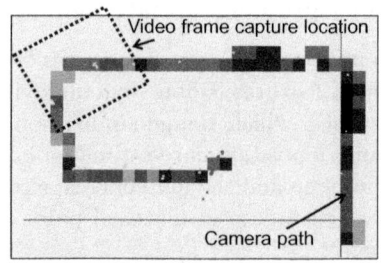

(a) Captured video frame.     (b) Feature map with highlighted most probable camera path.

**Fig. 1.** One of camera frames (a) used to build a map of the environment (b)

presents the results of our approach that still lacks some important algorithms. The system will be extended with the loop-closure algorithm that will allow us to build bigger and more accurate maps, thus allowing us to autonomously navigate a quadrocopter through the structured environment using the artificial landmark features.

# References

1. Bristeau, P.J., Callou, F., Vissiere, D., Petit, N.: The Navigation and Control technology inside the AR. Drone micro UAV. In: Proceedings of the 18th IFAC World Congress, pp. 1477–1484 (2011)
2. Forsyth, D.A., Ponce, J.: Image formation: Cameras. In: Computer Vision: A Modern Approach, pp. 3–27 (2003)
3. Hutchinson, S., Hager, G., Corke, P.: A tutorial on visual servo control. IEEE Transactions on Robotics and Automation 12, 651–670 (1996)
4. Lucas, B.D.: Generalized Image Matching by the Method of Differences. Phd thesis, Carnegie Mellon University (1984)
5. Montemerlo, M., Thrun, S., Koller, D., Wegbreit, B.: FastSLAM: A Factored Solution to the Simultaneous Localization and Mapping Problem. In: Proceedings of the AAAI National Conference on Artificial Intelligence, pp. 593–598. AAAI (2002)
6. Rekleitis, I.M.: A Particle Filter Tutorial for Mobile Robot Localization, Technical Report TR-CIM-04-02. Tech. rep., Centre for Intelligent Machines, McGill University, Montreal, Quebec, Canada (2004)
7. Rosten, E., Drummond, T.: Machine Learning for High-Speed Corner Detection. In: Leonardis, A., Bischof, H., Pinz, A. (eds.) ECCV 2006. LNCS, vol. 3951, pp. 430–443. Springer, Heidelberg (2006)
8. Smith, R., Self, M., Cheeseman, P., Park, M.: The Stochastic Map. In: Proceedings 1987 IEEE International Conference on Robotics and Automation, pp. 850–850 (1987)
9. Thrun, S., Burgard, W., Fox, D.: Probabilistic Robotics. Massachusetts Institute of Technology (2005)
10. Visser, A., Dijkshoorn, N., Veen, M.V.D., Jurriaans, R.: Closing the gap between simulation and reality in the sensor and motion models of an autonomous AR. Drone. In: Proceedings of the International Micro Air Vehicle Conference and Flight Competition, IMAV 2011 (2011)

# Search-and-Rescue-Operation
# with an Autonomously Acting Rescue Boat

Martin Kurowski[1], Holger Korte[2], and Bernhard P. Lampe[1]

[1] University of Rostock, Center for Marine Information Systems, Rostock, Germany
{martin.kurowski,bernhard.lampe}@uni-rostock.de
[2] Jade University of Applied Sciences, Department of Shipping, Elsfleth, Germany
holger.korte@jade-hs.de

**Abstract.** A novel, satellite-guided rescue system is under development, utilizing an autonomously acting rescue boat to salvage a person overboard to increase significantly the chance of survival. This advanced technology requires new approaches for naval architecture and integration of computer-aided tools to develop and operate such devices. A substantial challenge is the design and the automation of the self-acting autonomous rescue boat which navigates to the person overboard automatically. The design of this free fall rescue vessel guarantees that it will be self-righting. The developed cascaded control concepts were designed to ensure the fastest possible approach to the casualty without endangering the person. A specifically integrated monitoring system supports the entire rescue operation.

In this paper, a complex Search-and-Rescue-System for a satellite-supported rescue operation at sea will be presented which couples efficiently independent engineering tasks like naval architecture, control system design as well as information processing and monitoring.

**Keywords:** Autonomous vehicles, control systems, Search-and-Rescue-Operation, maritime systems, intelligent systems.

## 1 Introduction

Going overboard is one of the worst accidents at sea with a substantial endangerment for the life of the person overboard (POB) and the seamen taken part in the rescue operation. Therefore the ship operators are obligated in the context of the international convention to the protection of the human life at sea to initiate all conceivable activities to find and rescue the casualty. At sea, the crew of the ship with reduced number is often relying on itself alone. Furthermore the accident is determined mostly too late. Further time is required for the return maneuver of the ship, the disembarkation of a rescue vessel and the real rescue operation according to conventional methods. The POB drowns very often within a few minutes because of exhaustion or hypothermia. A faster rescue becomes an essential significance especially with regard to the conquering of the Northwest and the Northeast Passage through the Arctic Ocean for the commercial shipping

M. Kamel, F. Karray, and H. Hagras (Eds.): AIS 2012, LNCS 7326, pp. 141–148, 2012.
© Springer-Verlag Berlin Heidelberg 2012

In order to solve this problem, different research groups of the University of Rostock started together with different scientific and industrial partners the interdisciplinary research project AGaPaS. This project is supported by the German Federal Ministry of Economics and Technology (BMWi) under the registration number FKZ 03SX259.

Main task of the project was to reduce substantially the time interval between the accident and the rescue of the POB by the development of an automatic alarming system combined with an autonomous and remote-controlled rescue vehicle. So this system covers the entire Search-and-Rescue-Process.

## 2     Search-and-Rescue-Scenario

Motivated by the common Search-and-Rescue-Process, a Search-and-Rescue-Scenario was developed which is shown in Fig. 1.

An alarm will be released automatically if a person goes overboard. The POB carries a specially modified rescue vest which determines sequentially the position of the POB by global navigation satellite systems such as (D)GPS, GLONASS or GALILEO. The vest system sends the position message via the Automatic Identification System (AIS) to the Search-and-Rescue-Control-Station located onboard of the parent ship as well as to the autonomous rescue vehicle. After receiving the alarm, the Search-and-Rescue-Control-Station activates the autonomous rescue vessel.

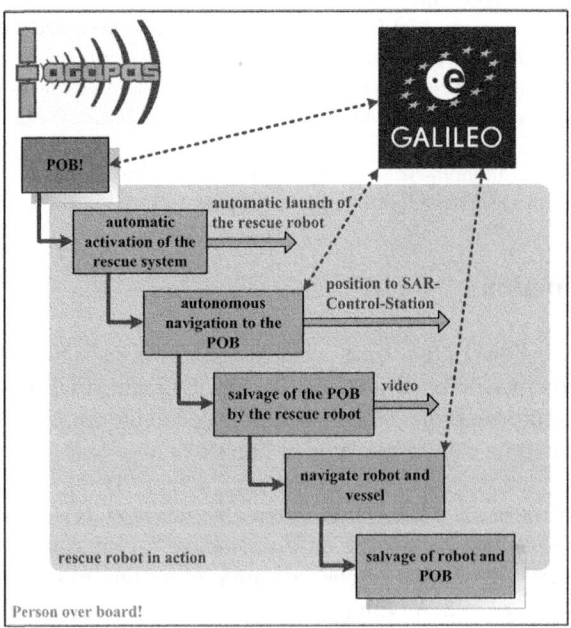

**Fig. 1.** Search-and-Rescue-Scenario

The automatically lowered rescue vessel is supplied constantly with position data of the POB and moves autonomously as fast as possible to the POB. In order to realize this task, different cascaded track and attitude control systems were developed, which involves additionally the environmental situation into the control task. After reaching a minimal distance to the POB, the rescue vehicle will be remote-controlled on the basis of radio communication and video transmission. It will be brought manually into a salvage position which allows the operator to take up the POB. Then, the POB will be saved with a particularly developed pick-up-device. After the successful salvage of the POB, the rescue vessel and the parent ship maneuver to each other. A constant radio link helps the seamen taken part in the Search-and-Rescue-Operation.

To cover the defined Search-and-Rescue-Scenario, different individual systems had to be developed and integrated in a complex Search-and-Rescue-System.

## 3    Rescue Vessel

### 3.1    Design

After analyzing different mono and twin hull concepts, the rescue vessel was designed as a catamaran (twin hull). This concept offers a better agility and the possibility to save the person with a basket between the hulls. The final draft combines an optimal longitudinal and lateral stability, a high payload, sufficient storage space, and good free fall characteristics. The salvage of the POB takes place via a hydraulically operated net basket, which is fixed at its rear edge. The hull speed is about 5.5kn and the free fall accelerations from 7.5m height could be reduced to approx. 6.5g [5].

The full-scale vehicle is 4.5m long and almost 2.3m wide. It is equipped with two electrical podded drives, which are supplied by 48V-Lithium-Accumulators. For the supply of the on-board electronics a 24V voltage rail is used. To the further equipment belong navigation equipment, sensors, communication technology as well as the camera equipment with infrared and daylight cameras.

Fig. 2 shows the developed autonomous rescue catamaran.

**Fig. 2.** Rescue catamaran *AGaPaS*

## 3.2   Automation

One of the most challenging tasks within the Search-and-Rescue-Process is the autonomous approach of the POB by the rescue vehicle.

The approach of the POB at the open sea with cruising speed is realized by subordinated heading (direction), and superior track (distance) controllers which use the model background of a slim rigid body, explained in [7]. The novelty is the on-line adjustment of the final point which changes temporally by drifting of the POB. After reaching a certain distance to the POB, the rescue vessel stops and maintains this distance without endangering him until the manual rescue operation is initiated. These operations cannot be realized with conventional cascaded track control systems, since special algorithms are necessary for these. Such algorithms are used by so called Dynamic Positioning Systems. In this case drive powers of the forward direction also have to be used for the direction control. This is possible with special propulsion systems like the podded drives used at the rescue vehicle. In addition, more complex models and control strategies become necessary for realizing such operations [1], [2].

In order to fill the gap between these both systems, a subordinated speed cascade was integrated into the control system. The used dynamic model describes a 3DoF[1] motion of the vessel. Within this cascade the controller design is decomposed into two steps: a controller and a dynamical allocation to the device configuration of the actual operating point of the vehicle. The allocation as part of the controller distributes force and moment demands $\eta$ on the individual aggregates $u$ according to their efficiency and other boundary conditions. The control system itself consists of a feedforward and a feedback control structure to decouple the reference input $x_r$ from the disturbed states $x$, like it is shown in Fig 3. This 2DoF MIMO (Multiple Input Multiple Output) speed control loop itself is a substructure of superior control algorithms like heading and bearing control, Line-of-Sight and waypoint track control as well as distance control structures. From the viewpoint of the superior cascades, the speed control loop is one part of the process and the reference input of the speed cascade is generated by these superior algorithms or the operator.

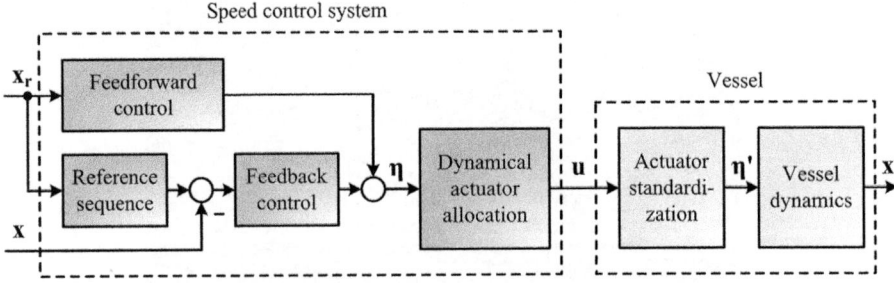

**Fig. 3.** Open speed control structure

---

[1] Three degrees of freedom (longitudinal speed, transversal speed, and yaw rate).

## 4     Integration to a Complex Search-and-Rescue-System

After developing the individual components the project partners integrated the parts to a complex Search-and-Rescue-System. It consists of the main elements rescue vest (POB), rescue vehicle, and parent ship like it is shown in Fig. 4.

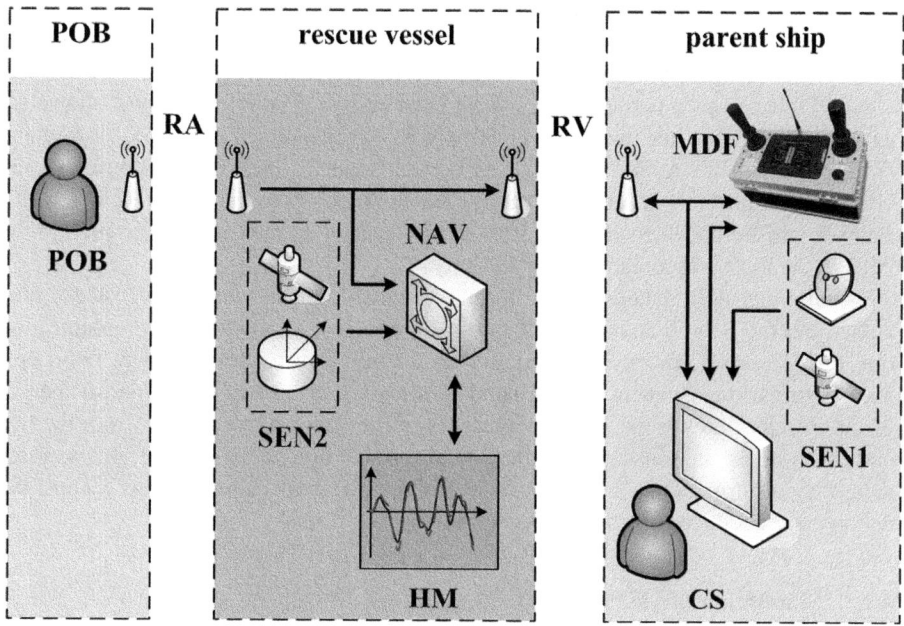

**Fig. 4.** Search-and-Rescue-System

The central element is the Search-and-Rescue-Control-Station (CS), located on the parent ship. It represents the main monitoring and control system of the Search-and-Rescue-Process onboard of the parent ship and is provided with an electronic chart display (ECDIS). The Control-Station is supplied with data of the onboard navigation sensors (SEN1) to monitor the whole traffic situation including the AIS[2] data send by the POB rescue vest. Furthermore it processes the video data and controls the rescue vessel in manual mode. The whole Search-and-Rescue-Process will be recorded automatically by the Control-Station. This enables the possibility of simulated training based on real data.

The portable remote control unit (MDF) is used for manual and partially automated remote control of the rescue vehicle as well as the rescue basket. The Search-and-Rescue-Operator has the possibility to control the vessel with different control modes like manual mode controlling the two podded drives separately, single joystick mode controlling heading and speed of the vessel, and course-to-steer track mode regarding

---

[2] Automatic Identification System.

the influence of drift. Furthermore it shows conditions, warnings and malfunctions of the rescue equipment.

The midsection of Fig. 4 describes the different elements concerning the rescue vehicle. The communication and control system (NAV), build as embedded PC/104, has mainly two tasks. Firstly, it controls the radio and video data transmissions and collects the data from the infrared camera and from the daylight camera onboard of the rescue vessel. Secondly, another PC/104 system navigates and controls the rescue vessel itself. The control system is fed with motion and navigation data of different sensors (SEN2) like gyroscope, Multi-(D)GNSS[3], echo sounder and log.

The POB carries a particularly modified rescue vest. The integrated AIS transmitter activates itself after the contact with saline water and sends the position of the POB continuously. Additionally the associated medical breast belt determines the vital data of the POB and transfers it via Bluetooth (RA) to the rescue vehicle. This channel will be activated if the rescue vessel reaches an appropriate distance to the POB or the POB was already salvaged.

Finally, the radio data communication and video transmission between the rescue vessel and the Search-and-Rescue-Control-Station is realized by the communication link RV. To ensure the required quality and stability, different redundant radio systems with various frequencies and bandwidths are used to exchange control data of the approach of the rescue vessel, vital signs of the POB and video data from the two camera systems. For instance to increase the reliability of the control and actuator data, they are transferred redundantly with three different radio modems[4]. Thus, the delays can be reduced during the transmission.

### 4.1   Simulation Mode

The whole Search-and-Rescue-System is modularly structured and can be used as a Hardware-In-the-Loop Simulator. This offers the possibility to design and test control algorithms and information processing systems, to investigate communication techniques, and develop human-machine-interfaces. The training of the crew of the ship is another aspect of the system. It allows the training of extreme situations during a Search-and-Rescue-Operation for example the handling of the rescue vessel and the manual salvage of a simulated POB [4]. During this simulation mode, the developed hydrodynamic model module (HM) provides the motion data of the rescue vessel including the influence of disturbances like waves. Different methods were used to realize the hydrodynamic model. On the one hand, empirical methods were verified with existing model test data and optimized with free maneuvering trails. And on the other hand CFD[5] computations for quasi-stationary forces and moments were validated successfully with bounded model tests [3].

---

[3] Provides data of different Global Navigation Satellite Systems GPS, GLONASS, and GALILEO (ready).

[4] Short-Range-Device- (SRD) radio modems (433,7 and 869,4MHz) and Worldwide-Interoperability-for-Microwave-Access- (WiMAX-) radio modem (5,47GHz).

[5] Computational fluid dynamics.

## 5     System Tests

In the first test phase, the project partners accomplished maneuvering, swell and free fall tests with a 1:4,5 model of the rescue vessel. For these tests was used the 100m long and 8m wide swell basin of the Technical University of Berlin. A special attention was paid to the free fall trails because this release represents a substantial requirement for the small response time of the rescue system and the structural loads of the free fall are particularly high for a catamaran. In order to determine the optimal release conditions, systematic CFD analyses have been accomplished and were validated by model tests [6]. With these tests, the basic concept of the rescue system could be confirmed. Additionally, the salvage of the POB was tested in presence of waves with a maximum height of two meters in relation to the full-scale catamaran. The tests have shown that the relative motion between the POB and the rescue vehicle impacted by the waves is small and the person can be salvaged even in rough seas.

In a second step, the entire Search-and-Rescue-System was proved. The focus was on the autonomous approach of the POB by the rescue vessel, maintaining a safety distance to the POB, and keeping the bearing to the POB automatically. Furthermore, the salvage of the POB was processed with single joystick control mode heading and speed control in combination with the infrared video stream and the monitoring displays of the Search-and-Rescue-Control-Station. These tests using the full-scale vehicle in presence of small till medium disturbances have shown promising results. The procedure of saving a POB succeeded without problems and could be repeated any number of times. Nevertheless, exercise is required to achieve the necessary skills to handle the autonomous rescue vehicle and the subsidiary control systems.

In addition, it is essential to validate the entire Search-and-Rescue-System in presence of more bad weather conditions and heavy seas.

 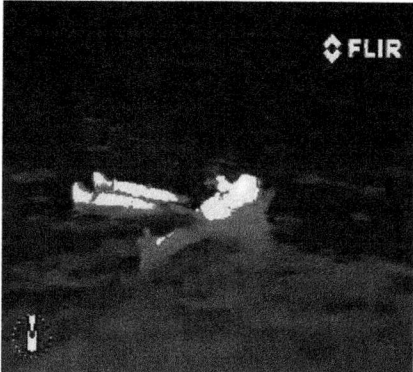

**Fig. 5.** Model tests in the swell basin (left), miscolored image of the infrared camera (right)

# 6     Conclusions

The paper presents a complex Search-and-Rescue-System which covers the entire Search-and-Rescue-Process. It combines efficiently interdisciplinary research sectors like naval architecture, automation engineering, control and feedback control systems, hydrodynamic simulation, medical aspects as well as data processing and data communication technologies. The developed Search-and-Rescue-System makes it possible to find and salvage a person overboard without endangering additional crew members. Thus, an important contribution is made for maritime safety. With these characteristics, the developed unmanned rescue system has the potential to revolutionize the market of offshore rescue systems.

Today, standard AIS technology uses exclusively (D)GPS to calculate the position data. With that GNSS-System it is not possible to salvage a POB completely autonomous, since the update rate of position data is too small, the accuracy is too low, and the availability is reduced because of flushing and shading the receiver. Special image processing methods which can determine the relative position of the POB to the rescue vessel may be able to solve the problem and make a completely autonomous salvage possible.

The additionally established integrated Hardware-In-the-Loop development environment represents a platform for further development of modules for satellite-based rescue with an autonomous rescue catamaran under almost real conditions. It allows to design and test control systems, human machine interfaces or data processing systems efficiently. Furthermore, the crew of a ship or an offshore platform can train Search-and-Rescue-Operations to gain experience in such extreme situations.

# References

1. Korte, H., Kurowski, M., Baldauf, M., Lampe, B.P.: AdaNav - A Modular Control and Prototyping Concept for Vessels with variable Gear Configurations. In: 8th IFAC Conference on Manoeuvring and Control of Marine Craft, Guarujá, Brasilia, pp. 91–96 (2009)
2. Rosenwasser, E., Lampe, B.P.: Multivariable computer controlled systems – a transfer function approach. Springer, London (2006)
3. Haase, M., Bronsart, R., Kornev, N., Nikolakis, D.: Simulation of the Dynamics of an Autonomously Acting Small Catamaran for Search and Rescue Process. In: 8th IFAC Conference on Control Applications for Marine Systems, Rostock, Germany, pp. 207–212 (2010)
4. Bronsart, R., Buch, T., Haase, M., Ihde, E., Kornev, N., Kurowski, M., Lampe, B.P.: Integrated Software-in-the-Loop Simulation of an Autonomously Acting Rescue Boat. In: International Conference on Computer Applications in Shipbuilding, Trieste, Italy (2011)
5. Clauss, G.F., Kauffeldt, A., Otten, N., Stuppe, S.: Hull Optimization of the unmanned AGaPaS Rescue Vessel. In: 29th OMAE - International Conference on Offshore Mechanics and Arctic Engineering, Shanghai, China (2010)
6. Clauss, G.F., Kauffeldt, A., Otten, N., Stuppe, S.: Identification of favourable Free Fall Parameters for the AGaPaS Rescue Catamaran. In: 30th OMAE - International Conference on Ocean, Offshore and Arctic Engineering, Rotterdam, Netherlands (2011)
7. Fossen, T.I.: Marine Control Systems. Marine Cybernetics AS, Trondheim, Norwegen (2002)

# Automatic Planning in a Robotized Cell

Federico Guedea-Elizalde and Yazmin S. Villegas-Hernandez

ITESM, Manufacture Center
{fguedea,a00640118}@itesm.mx

**Abstract.** This research presents the implementation of a language to describe a robot-based assembly and welding domain to generate programs for manufacturing robots. This language is used to describe an automated robotic-workcell. The developed system consists of the following subsystems: a High-level-language Planner, a Generic-level-language Parser, a Wrapper-generic-level language, and Graphic User Interface as well. Within this paper, it is presented an example of a generated program for a pick-and-place robot.

**Keywords:** Domain language, planner, pick-and-place task, automatic assembly.

## 1 Introduction

In automotive manufacturers, the robots are widely used in production lines for flexible assembly automation. Most of them work in "a teach and playback" mode. They must be taught with some work points, work sequences as well as several parameters. Therefore, they can not adapt to any workspace change such as position references, material raw dimensions, assembly type, among others. Because this type of robots are not able to adapt to changes, robot programming sequences must be done for each assembly type. In order to resolve this problem, it is presented an approach to generate a robot program using artificial intelligence tools such as the Graphplan planner. Planning is a branch of artificial intelligence that have a strong interaction with logical-based knowledge representation and reasoning schemas. Autonomous planners for robotic assembly lines are not commonly used in industries, so in this paper it is proposed the usage of a planner to generate robot programs where this program contains the plan for executing the required tasks. In order to use this planner as well as represent the planning problem, it was developed a new domain language for the description of an automated welding domain. The key for the generation of plans is the translation of the original problem to a domain language in order to use a planner to generate high-level plans.

## 2 Related Work

Automatic planning for production lines is a mature field with basic research moving to implementation cases. Car maker industries have many benefits from

M. Kamel, F. Karray, and H. Hagras (Eds.): AIS 2012, LNCS 7326, pp. 149–158, 2012.

the implementation of these planning techniques. Even though, there are previous works related with automatic planning systems as the work of Cho *et al.*, that describes the development of an automated welding operation planning system for block assembly in shipbuilding [1]. Cho's system was divided in four modules that perform the determination of welding postures, welding methods, welding equipment and welding materials. Another work of Cho *et al*, is about the development of an automatic process planning system for block assembly in shipbuilding [2]. In this work, a case-based reasoning approach was adopted for block assembly process planning and rule-based reasoning for process planning of cutting and welding operations.

Another related work regarding of planning systems is the work of Zaho *et al.*, where they present an intelligent computer-aided assembly process planning system (ICAAPP) [3] developed for generating an optimal assembly sequence for mechanical parts. In the generation of assembly sequences for any product, the critical problems to be addressed include determining the base part, selection of subassemblies, defining all necessary constraints, and finally quantifying and solving these constraints.

In the innovative artificial intelligence approach of Rabemanantsoa *et al.* [4] for generating assembly sequences on a consortium of database emulating expert systems, a CAD (Computer-aided design) analyser is used for shape and feature recognition, data structure and modelling, knowledge-based representation, and inference processing throughout a set of heuristics and rules.

In the work of Ming, Peng *et al*, a new approach for planning process is proposed, they developed a collaborative process planning and manufacturing in product life-cycle management[5] which in contrast to the method described in this research both the planning and the task performance processes are done by the planner and the robot respectively.

In the work of Seo,Y. *et al.* it is used a case-based reasoning (CBR) for block assembly in shipbuilding[6]. Also, in their work, it is used an interface system for extracting information out of CAD models to obtain data for the planning method.

In general, all of these systems consist of a planning process, use assembly graphs to generate sequences of assembly, or use another method; but none of these systems generate the code to program the robots online to perform welding and pick-and-place operations like the approach suggested in this paper.

## 3   Planner System

The objective of the system is to produce a logical program based on the provided assumptions and on the established goal. In order to accomplish these objectives, a planner system was developed which is able to generate programs and have ready a pick-and-place robot to execute them.

In order to solve these problems, the system, which is shown in Fig. 1, is divided in the next subsystems: High-level-language Planner, Generic-level-language Parser, Wrapper-generic-level-language, and Graphic User Interface.

This system produces, as output, a robot program that contains generic commands to do the assembly tasks. The program gets helped from a dynamic library that communicates with the Wrapper-generic-level-language functions program for the robot to perform the needed goals.

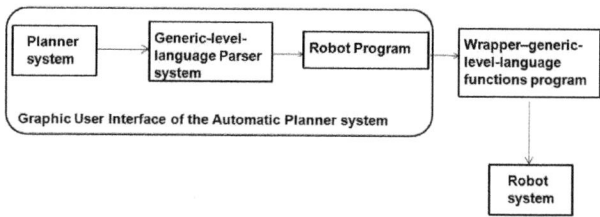

**Fig. 1.** Automatic Planner System

### 3.1  High-Level-Language Planner Subsystem

In this research, it was used the Graphplan planner algorithm for the generation of a high-level-language plan. To work with any planner, it is necessary to develop a language which is the input for any planner firstly. The language consists of logical propositions, also named facts which describe the initial state and operators. In order to create a logical plan that describes the assembly operations, it was developed a language that characterize these actions, the objects that are used as well as the initial state. The language developed in this work is divided in two parts: facts, and operators.

The planning process is defined as a tuple $< P, A, S_o, G >$, where $P$ is a set of propositions in a language $\Sigma$, $A$ is a set of actions, $S_o \subset P$ is the initial state, and $G \subset P$ is the goal state.

For example, an action $A$, meaning to pick and place an object should written as: ($PickAndPlace\ robot\ container1\ fastener1$), where $container1$ is the picking location and $fastener1$ is the placing position.

An example of the language to describe the initial state, $S_o$, is ($have\ container1\ tube1$), which represents that the $container1$ has the $tube1$.

Following the same reasoning, it is possible to describe an assembly goal $G$, as ($assemblyCorner90\ tube1\ tube2$) which represent a Corner joint with an angle of $90\,°$ using the tubes: $tube1$ and $tube2$.

Let $\Sigma =< Pn, L, C >$ be a first-order language composed of a set of predicate names $Pn = \{pn_1, pn_2, pn_3, ..., pn_w\}$, a set of literal constants $L = \{l_1, l_2, l_3, ..., l_n\}$, and a set of classes $C = \{c_1, c_2, c_3, ..., c_m\}$. The function $f_c$:

$$f_c : L \to C$$

associates each literal constant to a class (or object type). Then the set of literals (or propositions) $P = \{p_1, p_2, p_3, ..., p_w\}$, where $p_i = (pn_x, l_{i1}....l_{im})$ is a list involving constants and a predicate name. Also $\hat{P} = (\hat{p}_1, \hat{p}_2, \hat{p}_3, ..., \hat{p}_q)$ is defined as a set of propositions, where $\hat{p}_i = (pn_y, l_{i1}....l_{ik})$ is a list involving constants and a predicate name too. Then we define the function $f_p$ by:

$$f_p : P \to \hat{P}$$
$$\forall p_i = (pn_x, l_{i1}, ..., l_{ik}), f_p(p_i) = \hat{p}_j :$$
$$\hat{p}_j = (pn_x, f_c(l_{i1})...f_c(l_{ik})).$$

This function is overload as:

$$F_p : \{P\} \to \{\hat{P}\} : \forall J \subset P, F_p(J) = f_p(j_i), \forall j_i \in J.$$

An example of a literal is $(have\ container1\ tube1)$, where $have$ is a predicate name while $container1$ and $tube1$ are literal constants of the class $container$ and $tube$.

Given an action $A = \{Pre, Post\}$ as a set of two kind of propositions: pre-conditions and post-conditions of the initial state, $S_o$, where preconditions are the requirements to be accomplished before action, and the post-conditions are the description of the state after action. The function $F_A$ is defined as:

$$F_A : \{P\} \to \{\hat{P}\}$$
$$\forall a = \{Pre, Post\}, F_A(a) = \{F_p(Pre), F_p(Post)\}.$$

Using these definitions, the planning problem can be written in the proposed language with the form $\{F_p(P), F_A(A), F_p, S_o, F_p(G)\}$. The high-level plan is obtained by applying the actions $A$ to the initial state $F_p(S_o)$ to reach the goal $F_p(G)$. This high-level-language plan will be parsed to obtain the robot tasks program.

## 3.2   Generic-Level-Language Parser Subsystem

In the previous section, it was generated a high-level-language plan, $Plan_s \leftarrow Find-Plan(S_o^s, A^s, G^s)$, that will be used as the input to the Generic-Language Parser subsystem. This subsystem has the purpose to be used as an intermediate stage between the High-level Language Planner Subsystem and the Robot Subsystem, because this subsystem generates Generic-Language Commands for the pick-and-place robot.

In order to accomplish this purpose, it was decided to parse each high-level-operation to produce generic-level commands for the pick-and-place robot. This parsing consists in recognizing a string (high-level-operation) by dividing it into a set of symbols (or tokens) and analysing each one of them against the developed grammar in this work.

Algorithm 1 provides a way to parse a set of high-level-language expressions to generic-level-language expressions. Step one retrieves every high-level-language expression. Next, step two parse every expression from high to generic level. Then, step three adds every parsed expression to the robot program. finally, step four compile the robot program with the parsed expressions.

---

**Algorithm 1.** ParseGenerator(E)

---
**Require:** A set of expressions E.
**Ensure:** Change every expression from high to generic level.
1: **for** $e \in E$, $x \in instancesof e$ do
2: parsing: e(x)
3: Add e(x) to the robot program
4: **end for**
5: Compile robot program

---

The output of this parser is a program written in C-Sharp-style for the CRS-F3 arm manipulator to pick-and-place objects using the Generic-level-language-commands of the developed class library. Fig. 2 depicts the sentences (high-level-language commands) that are parsed to generic-level language commands. Each of these sentences are parsed by the ParseProgram class in the same order as they are written in the plan generated by the planner.

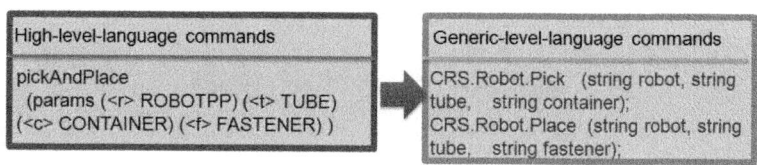

**Fig. 2.** High-level-language commands parsed to Generic-level-language commands

### 3.3   Wrapper-Generic-Level-Language-Functions Program

It was decided to use a DLL (Dynamic-Link-Library) that works as an intermediary between the Generic-language-level commands and the executable functions for the wrapper of the robot.

The Wrapper-generic-level-language-functions program consists in a set of wrapper methods written in C ++ that uses the original functions of the CRS-F3 arm manipulator library, to make a set of generic functions that have a standard form. These wrapper methods are equivalent to the generic-level-language methods in the Dynamic-Library-Link developed in this research. Therefore, using this library it will be able to do generic-robot programs, which can be generated and executed using the developed Graphic-User-Interface.

Algorithm 2 provides a way to execute each requested action. Step one retrieves all actions requested by the planner and step two executes them all.

---

**Algorithm 2.** Wrapper(A)

---
**Require:** A set of action literals A.
**Ensure:** Do all requested actions
1: **for** $a \in A$, $x \in instances\ of\ a$ **do**
2: execute a(x)
3: **end for**

---

## 3.4    Graphic-User-Interface for the Planner System

It was developed a GUI (Graphic-User-Interface),Fig. 3, for the user to run each subsystem separately such as the High-level-language Graphplan Planner, the Generic-level-language Parser and the Wrapper-generic-level-language-functions as well.

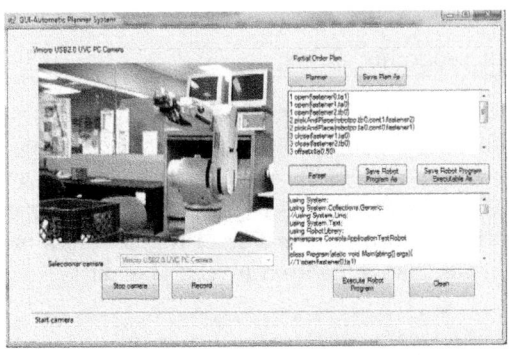

**Fig. 3.** GUI for the Planner system

This GUI is divided in three components: the planner, the parser, and the camera. This GUI was designed specially for the user to get total control of the automatic planner at each part of the process. This GUI contains sections where it is possible to generate the plan and to parse the high-level-language plan to generic-level-language commands for the CRS-F3 arm manipulator. Also, there is the camera window where visual aid is provided to follow the arm manipulator actions.

## 4    Experiments of the System

### 4.1    Robot-Program-Generation Example

An example of a test for the proposed system was done using tubes with different cutting angles for the assembly goals. This example starts describing the initial state of the environment to generate a high-level-language plan using the graphplan algorithm which will be parsed to generate a robot program to do the assembly task.

The utility of the proposed system is demonstrated. First at all, it is assumed that a robotized workcell has one pick-and-place robot and one welding robot, some tube containers, metallic tubes of different sizes, and a table where the tubes should be set prior to the welding process. The implementation of

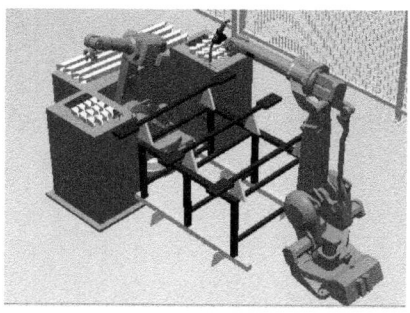

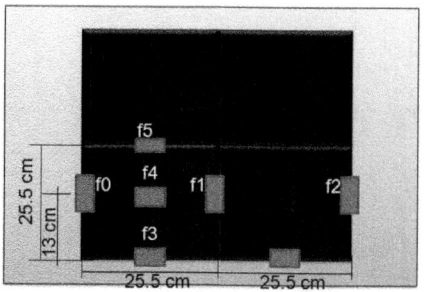

(a) Workcell with two robots: a pick-and-place robot and a welding robot

(b) Table with fasteners

**Fig. 4.** Workcell

the experiment was applied in a real robotized workcell. This workcell has only the pick-and-place robot but the other. Fig. 4(a) shows a model of the robotized workcell. Depicted in Fig. 4(b) is a representation of the welding table with the location of its fasteners used in this experiment.

In Fig. 5(a) is depicted the planning problem written in the proposed language and it is splitted into objects, a description of the initial state, and goal requirement conditions. The initial state can be described as $F_p(G)$ :

$$(assemblyCorner90\ tc0\ tb0)\ (assemblyCorner90\ tc0\ td0)\ (offset\ td0\ 25.5)$$
$$(assemblyT90\ tc0\ td1)\ (offset\ td1\ 13)$$

where $(assemblyCorner90\ tc0\ tb0)$ and $(assemblyCorner90\ tc0\ td0)$ means that the robot should form two corner joints of 90 °: one assembly using $tc0$ and $tb0$ and other assembly using $tc0$ and $td0$. The literal $(assemblyT90\ tc0\ td1)$ represent a T joint using $tc0$ and $td1$. The literals $(offset\ td0\ 25.5)$ and $(offset\ td1\ 13)$ represent that the assembly of $tc0$ and $td0$ has an offset in y-axis of 25.5 cm, and that the assembly of $tc0$ and $td1$ has an offset in y-axis of 13 cm. Fig. 6(b) is the graphical representation of the assembly goal.

In Fig. 5(b), it is shown the generated plan by the planner. This partial-order plan of 10 sets of high-level-commands was generated to accomplish the goal in Fig. 6(b). This plan is divided in several actions, $A$ ,as: *open* (to open fasteners), *close* (to close fasteners), *pickAndPlace* (to pick and place an object), *weldingCorner90* (to do Corner joint), *weldingT90oy* (to do T joint), *offsety* (to establish an offset in y-axis). This plan will be parsed in order to obtain the robot program to perform the needed actions.

In Fig. 6(a) is shown the robot program generated by the parser system, which contains generic-level-language commands to pick and place tubes.

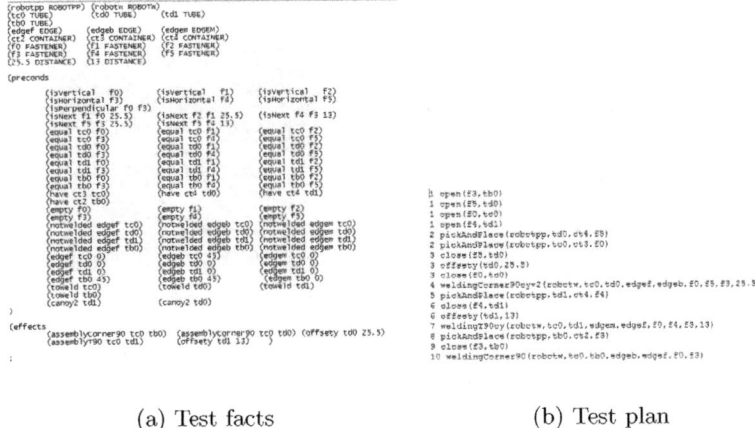

(a) Test facts        (b) Test plan

**Fig. 5.** Input and output data that belongs to the Planner system

## 4.2 Experimental Results

The processing time of the High-level-language Planner subsystem and the Generic-level-language Parser subsystem was tested in order to know the behaviour of this factor. In this test, it was used the same initial state and the number of goals were changed in order to obtain the mathematical model of the processing time. In Fig. 7 it can be seen the processing time is polynomial, i.e. when the number of goals increase the processing time increase in a polynomial way.

## 4.3 Language Experimental Results

This high-level language was developed for both planners: Graphplan and PDDL Graphplan (Planning Domain Definition Language). In both planners it was described the initial state of the environment and the welding table. The description of the welding table includes the orientation, size, and order sequence of the fasteners. For this research work, the released language describes three types of joints: corner, T, and butt. The difference between the developed languages for each planner was that the language for the graphplan planner is an abstract form because it is not possible to apply mathematical operations with it. However, the PDDL planner can handle mathematical computations quite well and also apply logical operators. A significant difference between these two languages is the processing time. In the PDDL planner is not able to generate plans for more than three goals, because the planner makes the computing system run out of memory. In the case of the graphplan planner, it is possible to generate plans for at most ten goals.

In Fig. 8 is depicted the response time of both planners. The response time of the Graphplan planner is lower than the PDDL planner. However, the tendency of both planners is polynomial.

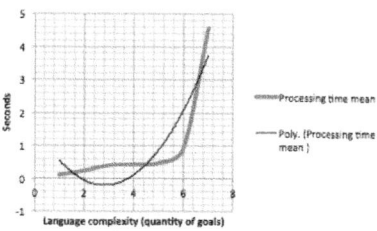

(a) Test program generated by the parser system

(b) Test goal

**Fig. 6.** (a)Robot program for accomplish (b) the graphical representation for the assembly goal

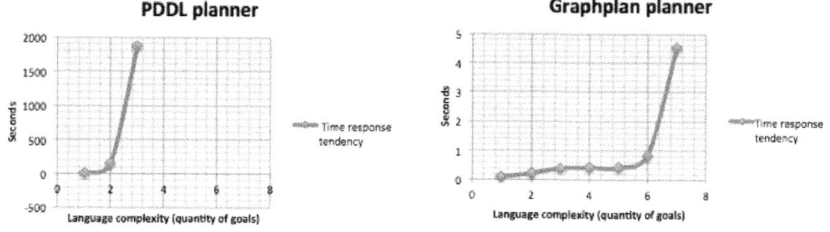

**Fig. 7.** Polynomial Processing Time

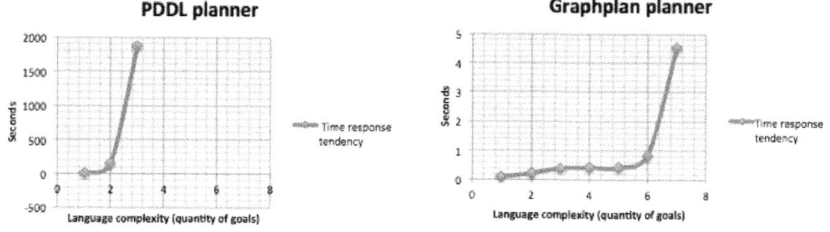

(a) PDDL planner time tendency

(b) Graphplan planner time tendency

**Fig. 8.** Planner response times

# 5   Conclusions

The use of a specific domain language to describe an automated welding domain, which is used in combination with a parser in order to automatically generate a robot program for pick-and-place operations, has been proposed using a Graphical User Interface (*GUI*).

Two types of languages were developed for two different planners: Graphplan and PDDL Graphplan. The response time of each language with their respective planner describe the same polynomial tendency, however the language for the graphplan planner was better than the other one as it can be reviewed on the experiments section. So, it was decided to use the Graphplan algorithm and the developed language for this planner in the Planner system.

The result of the experiments of the system was that the developed language accomplish the goals in seconds, in case of ten or less goals, using the Graphplan algorithm in the Planner system. Also, it is possible to establish three joints: Corner, T and Butt joint as goals. The results of the generated robot program were satisfactory, as the robot do not have problem to perform the assigned tasks for the pick and place operations.

# References

1. Cho, K., Sun, J., Oh, J.: An automated welding operation planning system for block assembly in shipbuilding. Production Economics 60-61, 203–209 (1999)
2. Cho, K., Lee, S., Chung, D.: An Automatic Process-Planning System for Block Assembly in Shipbuilding. CIRP Annals - Manufacturing Technology 45, 41–44 (1996)
3. Zhao, J., Masood, S.: An Intelligent Computer-Aided Assembly Process Planning System. The International Journal of Advanced Manufacturing Technology 15, 332–337 (1999)
4. Rabemanantsoa, M., Pierre, S.: An artificial intelligence approach for generating assembly sequences in CAD/CAM 10, 97–107 (1996)
5. Ming, X., Yan, J., Wang, X., Li, S., Lu, W., Peng, Q., Ma, Y.: Collaborative process planning and manufacturing in product lifecycle management 59, 154–166 (2008)
6. Seo, Y., Sheen, D., Kim, T.: Block assembly planning in shipbuilding using case-based reasoning 32, 245–253 (2007)

# Pattern-Based Trace Correlation Technique
# to Compare Software Versions

Maher Idris[1], Ali Mehrabian[1], Abdelwahab Hamou-Lhadj[1], and Richard Khoury[2]

[1] Software Behaviour Analysis Lab
Electrical and Computer Engineering, Concordia University, Montréal, Canada
{m_idris,al_meh,abdelw}@ece.concordia.ca
[2] Department of Software Engineering, Lakehead University, Thunder Bay, Canada
rkhoury@lakeheadu.ca

**Abstract.** Understanding the behavioural aspects and functional attributes of an existing software system is an important enabler for many software engineering activities including software maintenance and evolution. In this paper, we focus on understanding the differences between subsequent versions of the same system. This allows software engineers to compare the implementation of software features in different versions of the same system so as to estimate the effort required to maintain and test new versions. Our approach consists of exercising the features under study, generate the corresponding execution traces, and compare them to uncover similarities and differences. We propose in this paper to compare feature traces based on their main behavioural patterns instead of a mere event-to-event mapping. Two trace correlation metrics are also proposed and which vary whether the frequency of the patterns is taken into account or not. We show the effectiveness of our approach by applying it to traces generated from an open source object-oriented system.

**Keywords:** Dynamic analysis, Trace correlation, Software evolution, Software maintenance.

## 1    Introduction

One of the main challenges that engineers face when maintaining an existing system is to answer questions like what the system does, how it is built, and why it is built in a certain way [1]. Understanding an existing system has been shown to account for almost 80% of the cost of the software life cycle [2, 3]. Documentation is normally the main source of information where answers to these questions should be found, but in practice documentation is rarely up to date when it exists at all. The problem is further complicated by the fact that the initial designers of the system are often no longer available.

Execution traces have been used in various studies to observe and investigate the behavioural aspects of a software system. In most cases, traces have been found to be difficult to work with due to the large size of typical traces. Although many trace analysis tools and techniques have been proposed (e.g., [4, 5, 6, 7, 8]), most of them do not tackle the problem of correlating trace content. One of the few research studies that focuses on comparing traces is the work of Wilde [9], in which the author introduced

M. Kamel, F. Karray, and H. Hagras (Eds.): AIS 2012, LNCS 7326, pp. 159–166, 2012.

the concept of Software Reconnaissance. The author compared traces based on their distinct components. The objective was to identify the components that implemented a specific feature (also known as solving the problem of feature location). However, the author's approach did not take into account the interaction between the components in the trace, which is needed to understand differences in the execution trace.

In this paper, we focus on the problem of understanding the differences between subsequent versions of the same system, an activity that can help in many software engineering tasks including estimating the time and effort required to maintain new versions of the system, uncovering places in the code where faults have been introduced, understanding the rationale behind some design decisions, and so on. We propose a novel approach that allows software engineers to compare the implementation of software features in different versions of the same software system. Our approach is based on information gathered from two sources. First, we generate execution traces (dynamic analysis) by exercising the target features of the system to identify the differences between implementation. Several studies (e.g. [10, 6]) have showed that trace patterns often characterize the main content of a trace. Consequently, we propose two new metrics to measure the correlation between two traces based on their patterns, and we measure the extent of the differences between them. Once these differences are identified, we refer to the second source of information, the source code (static analysis), to understand the underlying changes. In other words, the result of the dynamic analysis not only shows the variation in the two implementations but is also used to guide software engineers in understanding where these variations appear.

The rest of this paper is structured as follows: In the next section, we briefly define the concept of trace patterns and we present our novel pattern-based approach for correlating traces. Our two proposed correlation measures are also presented in that section. A case study is presented in Section 3. We conclude the paper in Section 4.

## 2     Trace Correlation Approach

The aim of our approach is to compare traces generated from different versions of the same system. Both versions of the system are first instrumented and run using the same usage scenario. The generated traces then go through two main phases. The first phase consists of pre-processing the traces by removing continuous repetitions and noise caused by the presence of low-level utility components. The second phase consists of extracting similar patterns common to both traces resulting from the first phase and using them to compare the traces.

### 2.1     First Phase: Trace Pre-processing

During the pre-processing phase, we begin by eliminating contiguous repetitions due to the existence of loops and recursion in the code. We then filter out utility routines such as accessing methods (sets and gets) from the raw traces. These routines encumber the traces without adding much to their content. We rely on naming conventions to identify such utilities. For example, any routine that starts with 'set' or 'get' is automatically removed. We can also refer to the system folder structure to identify utilities packages. This is aligned with Hamou-Lhadj and Lethbridge study and in which the authors showed that an effective analysis of a trace should include a utility removal stage that cleans up the trace from noise [5].

## 2.2    Second Phase: Trace Correlation

The trace correlation phase is comprised of two main steps: the pattern detection step and the trace correlation measure. The goal is to take the two traces, extract their behavioural patterns, and compare the extracted patterns using different similarity measures. Two traces exhibit the same behaviour if the pattern sets are deemed similar. A threshold needs to be identified beyond which one can consider two traces similar. We anticipate, however, that this threshold is application-dependent and that a tool that supports our technique should provide enough flexibility to modify this threshold.

To reduce the size of the pattern space, several matching criteria have been proposed in the literature to measure the extent to which two sequences of events could be deemed similar without being necessarily identical (see [6, 10] for some examples). Some of these criteria include ignoring contiguous repetitions, ignoring the order of calls, or treating subtrees as a set. For example, the sequence $A_{BBBCCC}$[1] and $A_{BC}$ can be considered similar if the contiguous repetitions are ignored. Similarly, the sequence $A_{BBBCC}$ and $A_{CBBB}$ could be considered instances of the same pattern if the contiguous repetitions and the order of calls are ignored during the correlation process. Although it is still unclear how these matching criteria can be used for different maintenance tasks, the common consensus is that some sort of generalization is needed to reduce the size of the pattern space.

In [10], Hamou-Lhadj et al. presented an algorithm to detect and extract the patterns from a trace using predefined matching criteria. The algorithm uses one criterion at a time. Our method adopts the idea of the algorithm while providing the ability for using and applying more than one matching criteria to extract the similar patterns as desired. We also improved the performance of the algorithm when applied to large traces.

In the rest of this section, we use the sample traces presented in Figure 1 to illustrate the concept of correlating traces based on their behavioural patterns. Tables 1 and 2 show the patterns extracted from these traces. In this example, two matching criteria are used, which are ignoring the order of calls and removing utilities. We assume, in this example, that the utilities are the routines that start with 'u'.

## 2.3    Trace Correlation Metrics

We developed two new metrics to calculate the similarity between the traces of two versions of the same system based on their behavioural patterns. The two trace correlation metrics are the non-weighted trace correlation metric and the weighted trace correlation metric. Both correlation metrics range between 0 and 1, where 0 is complete dissimilarity and 1 is absolutely identical.

The non-weighted trace correlation metric, NW_TCM, is used to compare two execution traces based on the proportion of similar extracted patterns they share in common. More formally, NW_TCM is defined as follows:

$$NW\_TCM(T1, T2) = \frac{\left(\frac{CPtrnN}{T1TotalPtrnN}\right) + \left(\frac{CPtrnN}{T2TotalPtrnN}\right)}{2} \tag{1}$$

---

[1] We use the notation $A_B$ to mean A calls B.

where CPtrnN is the total number of patterns common to both traces, and T1TotalPtrnN and T2TotalPtrnN are the total number of patterns of Trace 1 and Trace 2 respectively.

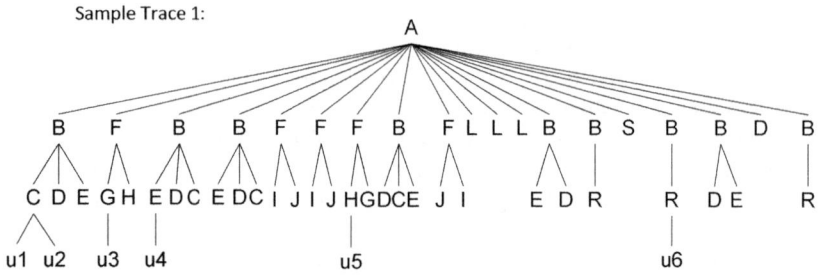

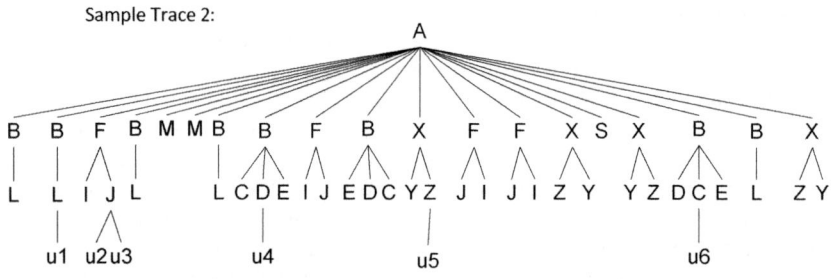

**Fig. 1.** Two sample routine (method) call traces

**Table 1.** Similar Patterns extracted from Sample Trace 1 of Figure 1

| Number | Pattern | Frequency |
|:---:|:---:|:---:|
| 1 | $B_{CDE}$ | 3 |
| 2 | $F_{IJ}$ | 3 |
| 3 | $F_{GH}$ | 2 |
| 4 | $B_{DE}$ | 2 |
| 5 | $B_R$ | 3 |
| Total | | 13 |

**Table 2.** Similar Patterns extracted from Sample Trace 2 of Figure 1

| Number | Pattern | Frequency |
|:---:|:---:|:---:|
| 1 | $B_{DCE}$ | 3 |
| 2 | $F_{IJ}$ | 4 |
| 3 | $B_L$ | 4 |
| 4 | $X_{YZ}$ | 4 |
| Total | | 15 |

The weighted trace correlation metric, W_TCM, modifies the previous metric by taking into account the frequency of the patterns, i.e., the number of times the patterns occur in the traces. More formally, W_TCM can be calculated as follows:

$$W_{TCM(T1,T2)} = \frac{\left(\frac{T1CPtrnFreqN}{T1TotalFreqN} \times \frac{CPtrnN}{T1TotalPtrnN}\right) + \left(\frac{T2CPtrnFreqN}{T2TotalFreqN} \times \frac{CPtrnN}{T2TotalPtrnN}\right)}{2} \quad (2)$$

where T1CPtrnFreqN and T2PtrnFreqN are the number of occurrence of the common similar patterns of both traces in Trace 1 and Trace 2 respectively, T1TotalFreqN and T2TotalFreqN are the total frequency of all patterns found in Trace 1 and Trace 2 respectively, and CPtrnN, T1TotalPtrnN and T2TotalPtrnN have the same meaning as before.

After we performed the pattern detection algorithm on the example traces of Figure 1, we obtained the five frequent patterns of Trace 1 and four frequent patterns of Trace 2 presented in Tables 1 and 2 respectively. The two common patterns between both traces, $B_{CDE}$ and $F_{IJ}$, can be discovered from these frequent patterns. Referring back to Tables 1 and 2, we can see that these patterns account for 6 of the 13 frequent pattern occurrences in Trace 1, and 7 of the 15 occurrences in Trace 2. The results of applying the non-weighted and weighted correlation metrics are as follows:

$$NW\_TCM(T1, T2) = \frac{\left(\frac{2}{5}\right) + \left(\frac{2}{4}\right)}{2} = 0.45 = 45\%$$

$$W\_TCM(T1, T2) = \frac{\left(\frac{6}{13} \times \frac{2}{5}\right) + \left(\frac{7}{15} \times \frac{2}{4}\right)}{2} = 0.21 = 21\%$$

Applying the NW_TCM metric results in a 45% similarity between Trace 1 and Trace 2, reflecting the fact that nearly half the frequent patterns of each trace are common to both traces. The W_TCM metric, on the other hand, gives a similarity of 21% due to the fact that it also takes pattern frequencies into account. We anticipate that the decision of which metric best reflects the similarity or dissimilarity between the traces will depend on the task at hand. For example, if one needs to understand the impact of a particular input on the resulting traces then the weighted metric could be considered since it takes into account the frequency of the patterns (which are often subject to the input data used to trigger the system). Further studies should provide more insight on situations where each of these metrics can be most informative.

# 3    Case Study

## 3.1    Target System

We have applied our proposed trace correlation technique to traces generated from two versions of the Java-based software system called Weka [11], an open-source software which was developed in the University of Waikato, New Zealand. It is a machine learning tool that supports several algorithms such as classification algorithms, regression techniques, clustering, and association rules. We selected this system because its framework and components are well documented [12], and detailed description of its architecture are available online. The versions of Weka that have been selected for this case study are versions 3.4 and 3.7. Weka version 3.4 is comprised of 55 packages, 732 classes, 8,980 methods and 147,335 lines of code (approximately 147 KLOC) while Weka version 3.7 contains 76 packages, 1129 classes, 14111 methods and 224,556 lines of code (approximately 224 KLOC).

## 3.2     Usage Scenario

We have applied our trace correlation technique to a specific software feature supported in both versions of Weka, namely the J48 classification algorithm used to construct efficient decision trees. In order to generate the execution traces for the selected feature, we instrumented Weka using the open source Eclipse Test and Performance Tool Platform Project (TPTP) [13]. Probes were inserted at each entry and exit method of the intended system, including the constructor and all invoked routines, in order to instrument it.

We used a sample input data provided in the documentation and the source code package of the Weka system to exercise the J48 feature. Executing the two instrumented versions of Weka with that data generated the two execution traces for that feature that we used in this study.

## 3.3     Applying the Trace Correlation Algorithm

The first phase of the algorithm is to pre-process the traces by filtering out utilities, contiguous repetitions, and the methods responsible for generating the GUI and initializing the environment. This allows us to focus only on those parts of the traces concerned with the implementation of the J48 algorithm.

In Table 3, we present statistical information regarding the size of the traces before and after the pre-processing stage. We can see that the removal of contiguous repetitions and utilities reduces the size of the raw traces considerably, but the resulting traces are still in the order of thousands of method calls. The size of the initialization routine is also very small compared to the total size of the entire traces. We can also see in Table 3 that the J48 trace in Weka 3.7 is considerably longer than the equivalent trace generated from the older version Weka 3.4. This indicates that significant changes have been made to this algorithm in the newer version. The objective of our research is to evaluate the extent of these changes and to uncover the exact nature and location of these changes in the source code.

**Table 3.** The size of execution traces of Weka system for versions 3.4 and 3.7

| Properties of Execution Traces | Weka 3.4 | Weka 3.7 |
|---|---|---|
| Original (raw) trace size | 35,974 | 103,009 |
| Original trace after removing contiguous repetitions | 6,850 | 26,978 |
| Initialization trace size | 5,919 | 17,534 |
| Initialization trace after removing contiguous repetitions | 682 | 1,288 |
| Original trace after removing initialization trace | 5,510 | 24,700 |

**Quantitative Analysis.** The second phase of our approach begins by applying the pattern detection algorithm. We used the "ignore order" matching criterion during the extraction process. Future work should focus on experimenting with other matching criteria to study their impact on the final result. We discover 162 frequent patterns in Weka 3.4 and 299 frequent patterns in Weka 3.7, almost twice as many. This result shows immediately that the implementation of the J48 algorithm in Weka has undergone several changes.

Next, we apply the correlation metrics to measure the differences between the two pattern sets extracted from the Weka traces. This finds only 64 similar patterns common to both traces. This means that less than half of the total patterns relevant to each version of Weka are shared in both versions. Applying our trace correlation metrics gives a non-weighted correlation (NW_TCM) of 30.45% and a weighted correlation (W_TCM) of only 5%. The results show again that these traces are considerably different from each other. To be able to explain these differences, we perform next a qualitative examination of the patterns that are not common between the two traces by exploring the source code of the two Weka versions.

**Qualitative Analysis.** The dissimilarity between the two versions in terms of the number patterns (without taking into account the frequency) is almost 70%. After exploring the content of both traces, we found that the number of distinct methods of the J48 trace in Weka 3.4 was 656, whereas the number of distinct methods in the trace generated from Weka 3.7 was 1024. This led to the generation of many patterns that were in one trace and not in the other trace, patterns that were triggered by the new methods.

We focused on the patterns that exist in both traces and which derive from the same root nodes. This revealed that refactoring tasks have been used in Weka 3.7 to modify the way these methods were implemented. This includes adding new classes and methods, changing the names of existing methods, and moving classes and routines to other (existing or new) classes and components. For example, the size() method of the FastVector class of Weka 3.4 has been replaced by the utility routine size() implemented in the built-in Collection interface class of the Java package java.util. Consequently, the size() method was included in patterns in Weka 3.4 but did not appear in the extracted patterns of Weka 3.7 since it is an external utility method of that system.

Another difference we have observed when examining the patterns and the source code of both versions is that some invoked methods have been moved to new classes that were introduced in Weka 3.7. For example, the method hasMoreElements() was part of the FastVectorEnumeration class of the old version but moved to the new class WekaEnumeration in the new version. Refactoring the code in this manner leads to the discovery of patterns that begin from the same parent nodes in both version of the trace, but diverge in the lower levels.

The above discussion demonstrates the usefulness of using a pattern-directed approach not only to measure the similarity between two versions of the same system but also to guide the process of investigating the root causes of dissimilarities.

# 4     Conclusions and Future Directions

In this paper, we presented a new approach for comparing the implementation of different features in subsequent versions of the same system. Our method discovers similarities between two traces generated from the different versions of the software. In particular, we focused on calculating the trace correlation based on all the behavioural patterns extracted from the execution traces. Future work includes the need to conduct more experiments with multiple versions of a given software system to further assess the efficiency of our approach. There is also a need to compare our results with other software comparison techniques; this could lead us to expand our method by adapting some of their strengths.

**Acknowledgement.** This work is partly supported by the Natural Sciences and Engineering Research Consortium (NSERC), Canada.

# References

1. Dunsmore, A., Roper, M., Wood, M.: The role of comprehension in software inspection. Journal of Systems and Software 2(3), 121–129 (2000)
2. Martin, J., Mcclure, C.: Software Maintenance: The Problem and its Solutions. Prentice-Hall, Englewood Cliffs (1983)
3. Pigoski, T.M.: Practical Software Maintenance: Best Practices for Managing Your Software Investment, p. 384. John Wiley and Sons, New York (1997)
4. Cornelissen, B., Moonen, L.: On Large Execution Traces and Trace Abstraction Techniques. Software Engineering Research Group, Delft (2008) ISSN 1872-5392
5. Hamou-Lhadj, A., Lethbridge, T.C.: Summarizing the content of Large Traces to Facilitate the Understanding of the Behaviour of a Software System. In: Proceedings of the 14th IEEE International Conference Program Comprehension, pp. 181–190 (2006)
6. De Pauw, W., Lorenz, D., Vlissides, J., Wegman, M.: Execution Patterns in Object-Oriented Visualization. In: Proceedings of the 69 4th USENIX Conference on Object-Oriented Technologies and Systems (COOTS), Santa Fe, NM, pp. 219–234 (1998)
7. Systä, T.: Understanding the Behaviour of Java Programs. In: Proceedings of the 7th Working Conference on Reverse Engineering, pp. 214–223 (2000)
8. Jerding, D., Rugaber, S.: Using Visualization for Architecture Localization and Extraction. In: Proc. of the 4th Working Conference on Reverse Engineering, pp. 219–234 (1997)
9. Wilde, N., Scully, M.C.: Software Reconnaissance: Mapping Program Features to Code. Software Maintenance: Research and Practice, 49–62 (1995)
10. Hamou-Lhadj, A., Lethbridge, T.: An Efficient Algorithm for Detecting Patterns in Traces of Procedure Calls. In: Proceedings of the 1st ICSE International Workshop on Dynamic Analysis (WODA), Portland, Oregon, USA (2003)
11. Weka 3: Data Mining Software in Java,
    http://www.cs.waikato.ac.nz/ml/weka/
12. Witten, I.H., Frank, E.: Data Mining: Practical Machine Learning Tools and Techniques with Java Implementations. Morgan Kaufmann (1999)
13. http://www.eclipse.org/tptp/

# Enhancement of the ROVER's Voting Scheme Using Pattern Matching

Milad AlemZadeh[1], Kacem Abida[1], Richard Khoury[2], and Fakhri Karray[1]

[1] Electrical and Computer Engineering Department, University of Waterloo,
Waterloo, Ontario, Canada
{malemzad,mkabida,karray}@uwaterloo.ca
http://www.uwaterloo.ca
[2] Software Engineering Department, Lakehead University,
Thunder Bay, Ontario, Canada
rkhoury@lakeheadu.ca
http://www.lakeheadu.ca

**Abstract.** Combining the output of several speech decoders is considered to be one of the most efficient approaches to reducing the Word Error Rate (WER) in automatic speech transcription. The Recognizer Output Voting Error Reduction (ROVER) is a well known procedure for systems' combination. However, this technique's performance has reached a plateau due to the limitation of the current voting schemes. The ROVER voting algorithms proposed originally rely on the frequency of occurrences and word level confidences, which leads to randomly broken ties and poor voting outcomes due to the unreliability of the decoder's confidence scores. This paper presents a pattern-matching-based voting scheme which has shown to reduce even further the WER.

**Keywords:** ROVER, combination, voting, pattern matching, WER.

## 1 Introduction

Speech decoders are becoming prominent in an increasing number of real-world applications, and are required to become more robust and more reliable. Nowadays, speech transcription is being used ubiquitously. The requirement for highly accurate decoders has increased substantially. Most researchers agree that one of the most promising approaches to the problem of reducing WER in speech transcription is to combine two or more speech decoders and compile a new composite more accurate output [1]. ROVER is the most widely used technique in this regard. However its performance has stagnated over the past few years mainly due to the limitations of the voting algorithms used. These shortcomings include the use of unreliable word level decoders' confidence scores, as well as randomly-broken ties when the frequency of occurrences is used during the voting. This paper proposes a novel voting scheme for the ROVER combination procedure. The new voting algorithm relies on a database of patterns to select the winner token from the different decoders' hypotheses. This paper is organized as follows. Section 2 details the ROVER procedure. Section 3 describes the

M. Kamel, F. Karray, and H. Hagras (Eds.): AIS 2012, LNCS 7326, pp. 167–174, 2012.

proposed approach and presents a working example. Experimental assessment then follows in section 4. The paper ends with a conclusion and future directions in section 5.

## 2    The ROVER Procedure

ROVER [2], is a system developed at NIST in 1997 to produce a composite of decoders' output when the outputs of multiple Automatic Speech Recognizers (ASR) are available. The goal is to produce a lower WER in the composite output. This is done through a voting mechanism to select the winner word from among the different decoders' outputs.ROVER is a two-step process. It starts by combining the multiple decoders' outputs into a single, minimal-cost word transition network (WTN) through dynamic programming. The resulting WTN is browsed by a voting process which selects the best output sequence. Three voting mechanisms have been presented in [2]. At each location in the composite WTN, a score is computed for each word using Equation 1, where $i$ is the current location in the WTN, $N_s$ is the total number of combined systems, $N(w, i)$ the frequency of word $w$ at the position $i$ and $C(w, i)$ is the confidence value for word $w$ at the position $i$.

$$S(w) = \alpha(\frac{N(w, i)}{N_s}) + (1 - \alpha)C(w, i) \tag{1}$$

The parameter $\alpha$ is set to be the trade-off between using word frequency and confidences. In the case there is an insertion or deletion, the NULL transition, noted as @, in the word transition network, will have the confidence $conf(@)$. A training stage is therefore needed to optimize both the $\alpha$ parameter and the NULL transition confidence value. This is commonly done through grid-based searching. The three voting schemes are frequency of occurrence, frequency of occurrence and average word confidence, and frequency of occurrence and maximum confidence. The first scheme only uses occurrences to select the winning word at each WTN slot. However, ties tend to occur frequently and they are randomly broken. The remaining two schemes rely on words confidences to overcome the problem of arbitrarily broken ties. The fact that ROVER's voting relies on confidences makes it a vulnerable technique. It is not safe to assume that the word confidences are reliable. Much research [3] is still underway into trying to come up with a robust and effective technique to provide a decent confidence measure for Large Vocabulary Continuous Speech Recognition (LVCSR). Even in [2], neither algorithm that relies on a confidence score achieved a significant error reduction compared to the frequency-based voting algorithm. Furthermore, ROVER is unable to outvote the erroneous ASR systems when only one single ASR is providing the correct output. The scoring schemes make it difficult to boost a single ASR output since both occurrence and confidence are used to score each word at a specific location.

# 3  Proposed Approach

In this section, the proposed new voting algorithm is presented, followed by a working example to illustrate the different steps of the proposed approach.

## 3.1  Novel Voting Algorithm

The proposed approach in this paper consists in relying on pattern matching in order to intelligently select the winner token at each location of the WTN. The assumption that underlies our work is that there exist multi-word statements that a speaker is likely to use, and that we can improve the voting stage within the ROVER procedure by recognizing these patterns. In this context, multi-word statements refer to a variety of entities, such as long nouns (e.g. the honorable member of parliament) or proper names (e.g. Harvard University).

Algorithm 1 presents the steps of our proposed approach. Basically, the algorithm browses the WTN in order to select the winner word at each slot. Once a winner is found, the token is tagged as *WINNER*.

---

**Algorithm 1** Novel Voting Algorithm

---
1: Build the WTN.
2: Tag all common tokens as *WINNER*.
3: Select all n-grams patterns, in which at least two consecutive words can be aligned with any consecutive slots of the WTN.
4: **while** ∃ slots without *WINNER* token **do**
5:    **while** ∃ $n$-grams patterns **do**
6:       Update $n$-grams list by keeping only patterns matching at least a WTN slot without the *WINNER* tag.
7:       Rank all $n$-grams patterns.
8:       Pick the highest scored pattern.
9:       Tag tokens matching the winner pattern as *WINNER*.
10:       Remove the highest scored pattern from the list.
11:    **end while**
12:    Use frequency of occurrences to select *WINNER* in the remaining slots, if any.
13: **end while**

---

**Step 1:** Align the recognition hypotheses to create the WTN.

**Step 2:** Browse the WTN to locate all slots in which the same token has been recognized by all speech decoders. Tag the token as *WINNER*. In other words, if all speech decoders agreed on the same token at a given slot, we suppose it has been correctly recognized and hence we tag it as *WINNER*.

**Step 3:** Given a corpus of patterns, this step consists of selecting all patterns that match any $n$-gram of consecutive tokens within the WTN. We set $n \geq 2$ given our requirement to match at least two consecutive tokens in the WTN. Moreover, at least one WTN slot aligned with the pattern should not already be marked as *WINNER*. These are the set of patterns from which new *WINNER* tokens can be discovered.

**Step 4:** Once the set of patterns relevant to the WTN has been collected, the WTN is browsed in a loop as long as there are still slots without a *WINNER*.

**Step 5:** As long as there are still relevant patterns in the set, new *WINNER* tags are attributed to slots in the WTN as explained in the next four steps.

**Step 6:** The list of patterns must be updated to keep patterns matching at least one WTN slot without a WINNER tag. These are the set of patterns from which new *WINNER* tokens can be discovered.

**Step 7:** All collected n-grams are ranked using equation 2,

$$W_p = (\sum_{w \in p} W_w)\frac{C_p - X_p}{N_p} \tag{2}$$

where $W_p$ is the weight of the pattern, $W_w$ is the occurrence of the word $w$ in the slot, $C_p$ is the number of consecutive words in the pattern $p$, $X_p$ is the number of false matches (i.e the number of words within the pattern which match tokens at slots where a *WINNER* tag is attributed to a different token), and $N_p$ is the total number of words in the pattern $p$. It is worth mentioning the impact of the NULL transition in equation 2. The NULL token has a weight, which is its occurrences at the slot, but it does not count as a word. It can thus contribute to the value of $W_w$, but not to $C_p$, or $N_p$.

**Steps 8, 9 and 10:** Once all the patterns' weights are computed, the highest-weighted pattern is selected to discover new *WINNER* tokens (i.e words in the pattern that also appear in at least one transition at the same slot).

**Step 12:** Once all patterns are exhausted, slots without a *WINNER* token may still exist. In order to select the winner token at each of these slots, the frequency of occurrence voting scheme is used. Basically, the token with the highest $W_w$ is the winner word. If ties occur, they are randomly broken.

One downside of our new algorithm compared to the classic ROVER voting algorithm is the additional processing it requires. For off-line applications, however, this should not be an issue. For real-time applications, the algorithm can be optimized in a number of ways, for instance by pruning the infrequent patterns from database, by caching frequently-used patterns, and by making use of the text indexing tools available in many commercial database systems.

## 3.2    Illustrative Example

In order to better illustrate the steps in Algorithm 1, we present a working example. Let us assume that a given speech waveform is transcribed by three different speech decoders. The resulting WTN is given in figure 1. Per step 2 of the algorithm, the token "The" at the first slot is a *WINNER*. Now let us assume for the sake of this example, that only four patterns are kept for this WTN. These patterns, along with their initial weights, are given in table 1. Notice that the first pattern matches three words, "united", "states", and "ink", but only the first two words are consecutive. Consequently, only "united states" is aligned with the WTN and "ink" is ignored. Notice also that the "boarding school" pattern gets

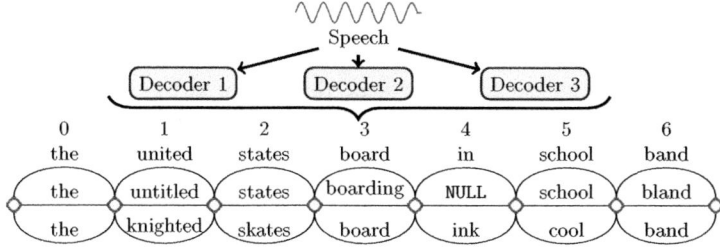

**Fig. 1.** Sample WTN

**Table 1.** Patterns List and Initial Weighting

| Patterns | Aligned Slots | Pattern Weight ($W_p$) |
|---|---|---|
| "united states of ink" | 1–2 | $(1+2)*(2-0)/4 = 1.5$ |
| "knighted skates inc." | 1–2 | $(1+1)*(2-0)/3 = 1.33$ |
| "states board of education" | 2–3 | $(2+2)*(2-0)/4 = 2$ |
| "boarding school" | 3–5 | $(1+1+2)*(2-0)/2 = 4$ |

an extra word weight because of the NULL transition between "boarding" and "school". Finally, note that none of the patterns cover the initial slot 0, the only position for which a *WINNER* token is identified, and consequently $X_p = 0$. At the first iteration, the highest-weighted pattern is "boarding school". Therefore the tokens "boarding", "NULL", and "school" at slots 3, 4, and 5 are tagged as *WINNER*. This pattern is removed from the list, before the second iteration begins.

The next iteration starts by updating the weights of the remaining three patterns. The updated list of patterns is shown in table 2. Only one pattern has

**Table 2.** Second Iteration: Patterns List and Weighting

| Patterns | Aligned Slots | Pattern Weight ($W_p$) |
|---|---|---|
| "united states of ink" | 1–2 | $(1+2)*(2-0)/4 = 1.5$ |
| "knighted skates inc." | 1–2 | $(1+1)*(2-0)/3 = 1.33$ |
| "states board of education" | 2–3 | $(2+2)*(2-1)/4 = 1$ |

its weight changed, i.e "states board of education", which mismatches a token at slot 3 (where "boarding" was tagged as *WINNER* during the previous iteration). Therefore the highest-weighted pattern is "united states of ink". The words "united" and "states" are thus tagged as *WINNER* at slot 1 and 2 respectively. The remaining patterns only match slots with winner tokens. Therefore, no more patterns are usable. The algorithm deals with slot 6 using frequency of occurrence, thus marking "band" as a *WINNER*. The composite final sentence is thus "the united states boarding school band".

# 4    Performance Evaluation

In this section, the experiments we carried out in order to assess the performance of the proposed voting scheme are presented. We compare WERs of ROVER with the original frequency of occurrence voting scheme, and ROVER with the novel voting algorithm. This section starts with the experimental framework, followed by the experimental results and interpretations.

## 4.1    Experimental Framework

The experiments were conducted using the transcription output of the Carnegie Mellon University Sphinx 4 decoder. In terms of data, we have considered the English Broadcast News Speech (HUB4) testing framework [5]. This corpus is composed of both speech data (LDC98S71) and transcripts (LDC98T28). It is a total of 97 hours of 16000 Hz recordings from radio and television news broadcasts. Transcriptions of this HUB4 corpus have been used to train the language model (named as LM-98T28 hereafter). Two other freely available language models were used: an open source model for broadcast news transcriptions from CMU [4], referred to as LM-BN99 hereafter, and another language model created from the English Gigaword corpus [6], referred to as LM-GIGA hereafter. In order to simulate speech decoders' outputs from several sites, we have loaded the Sphinx 4 decoder with different language models. A total of three configurations have been set up: Sphinx 4 loaded with LM-98T28 (s4-LM-98T28), Sphinx 4 loaded with LM-BN99 (s4-LM-BN99), and Sphinx 4 loaded with LM-GIGA (s4-LM-GIGA). All possible binary combinations were carried out. Table 3 reports the setup. Notice here that the order of combination matters. This is an inherent

**Table 3.** Decoders' Combinations IDs

| ID | Combination Configuration | ID | Combination Configuration |
|----|---------------------------|----|---------------------------|
| C1 | s4-LM-GIGA - s4-LM-98T28 | C4 | s4-LM-98T28 - s4-LM-BN99 |
| C2 | s4-LM-GIGA - s4-LM-BN99 | C5 | s4-LM-BN99 - s4-LM-98T2 |
| C3 | s4-LM-98T28 s4-LM-GIGA | C6 | s4-LM-BN99 - s4-LM-GIGA |

problem of ROVER's WTN building stage. Because of the use of different language models, a normalization step is required to standardize the output of the different speech decoders. In other words, the same word can be written in different ways and therefore all these variations have to be unified under a single form. Examples of these issues include acronyms such as CNN which can be written as c. n. n. (three distinct letters), c.n.n. (one single word), cnn (one single word), c n n (three letters), and capitalization such as Vote, voTe, vote, etc.

It is worth mentioning that the Sphinx 4 confidence scores were not reliable. Consequently, using these scores in the voting algorithm causes no change in the outcome. For this reason, we only used the frequency-based voting in the reminder our experiments.

In terms of patterns, we have used the list of all titles in Wikipedia that are two or more words in length. This created a set of 6, 479, 950 patterns ranging from bi-grams up to 45-grams, averaging 3-grams in length. A benefit of using Wikipedia is that these patterns cover most imaginable English topics [7–9].

### 4.2    Simulation Results and Analysis

The baseline performance is obtained through the use of the frequency of occurrence voting scheme. The ROVER baseline WER is given in table 4. The first

**Table 4.** ROVER Baseline WER

| Exp. | C1 | C2 | C3 | C4 | C5 | C6 |
|------|------|------|------|------|------|------|
| WER | 37.12 | 36.45 | 37.32 | 35.51 | 32.45 | 33.1 |

row in table 4 represents the different binary combinations previously defined in table 3. In order to illustrate the performance of the pattern matching-based voting scheme, the WER reduction is plotted. Figure 2 reports the WER reduction (absolute and relative) achieved with the novel voting algorithm. The WER

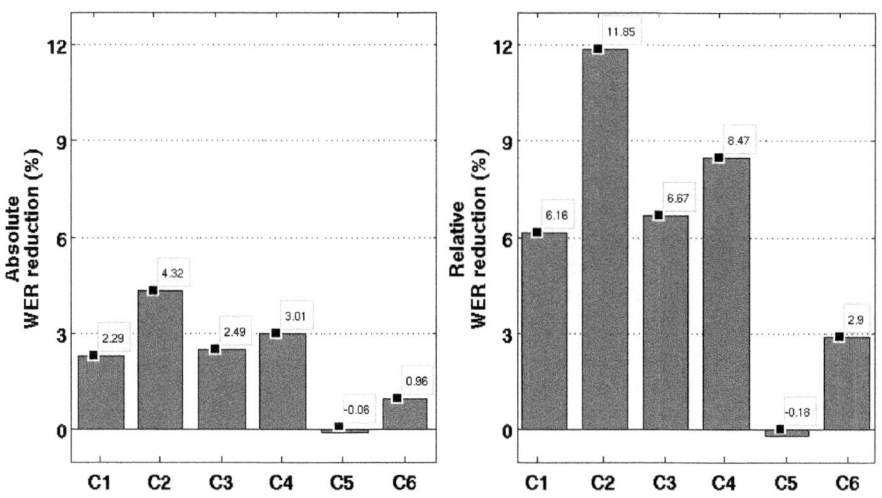

**Fig. 2.** Novel Voting Scheme WER Reduction

reduction reported in figure 2 shows that the novel voting scheme is outperforming the original frequency of occurrence scheme in almost all the experiments. The relative WER reduction ranged from 3% to up to 12%. This reduction is considered impressive since ROVER's performance has stagnated over the past few years, and it became more and more difficult to achieve even a small WER reduction. However, with the $C5$ binary combination, the pattern matching-based voting performs slightly worse than the original ROVER algorithm. This can be explained by the fact that the $C5$ baseline WER is the lowest (table 4). In other words, when the WER is low and decoders are highly optimized, the novel voting performs roughly the same as the original scheme.

## 5   Conclusion and Future Work

In this paper, a novel voting algorithm has been proposed to improve the performance of the ROVER procedure. The voting scheme relies on a database of patterns in order to select the winner token at each slot of the WTN. Experiments showed that this new approach outperforms the original voting scheme and achieve substantial WER reduction in most of the experiments. Future research can focus on enhancing our proposed algorithm in a number of ways. One possible direction is to expand it to handle more than pairs of decoders. Indeed, combining the output of three or more decoders in our system should improve the results compared to using only two decoders. We could also look at improving the technique used to break ties. The random decision our system makes at present is not optimal, and some kind of intelligent decision process should give better results.

## References

1. Abida, K., Karray, F.: Systems combination in large vocabulary continuous speech recognition. In: IEEE International Conference on Autonomous and Intelligent Systems, pp. 1–6 (2010)
2. Fiscus, J.: A post-processing system to yield reduced word error rates: Recognizer Output Voting Error Reduction (ROVER). In: IEEE Workshop on Automatic Speech Recognition and Understanding, pp. 347–352 (1997)
3. Jiang, H.: Confidence measures for speech recognition: a survey. J. Speech Communication 45(4), 455–470 (2005)
4. Huggins-Daines, D.: CMU Sphinx open source models (2008), http://www.speech.cs.cmu.edu/sphinx/models/
5. Fiscus, J., Garofolo, J., Przybocki, M., Fisher, W., Pallet, D.: 1997 English Broadcast News Speech, HUB4 (1998)
6. Graff, D., Cieri, C.: English Gigaword Corpus (2003)
7. Khoury, R.: The impact of Wikipedia on scientific research. In: 3rd International Conference on Internet Technologies and Applications, pp. 2–11 (2009)
8. Béchet, F., Charton, E.: Unsupervised knowledge acquisition for Extracting Named Entities from speech. In: ICASSP 2010, pp. 5338–5341 (2010)
9. Alemzadeh, M., Khoury, R., Karray, F.: Exploring Wikipedia's Category Graph for Query Classification. In: Kamel, M., Karray, F., Gueaieb, W., Khamis, A. (eds.) AIS 2011. LNCS, vol. 6752, pp. 222–230. Springer, Heidelberg (2011)

# A Novel Voting Scheme for ROVER Using Automatic Error Detection

Kacem Abida[1], Fakhri Karray[1], and Wafa Abida[2]

[1] Electrical and Computer Engineering Department, University of Waterloo,
Waterloo, Ontario, Canada
{mkabida,karray}@uwaterloo.ca
http://www.uwaterloo.ca
[2] Voice Enabling Systems Technology,
Waterloo, Ontario, Canada
wafa@vestec.com
http://www.vestec.com

**Abstract.** Recognizer Output Voting Error Reduction (ROVER), is a well-known procedure for decoders' output combination aiming at reducing the Word Error Rate (WER) in large vocabulary transcriptions. This paper presents a novel voting scheme to boost the current ROVER's performance. A contextual analysis is carried out to compute robust confidence scores, which are used during the voting stage. The confidence scores are collected from several automatic error detectors. Experiments have proven that it is possible to outperform the ROVER voting schemes.

**Keywords:** ROVER, Voting, Error Detection, WER, Combination.

## 1 Introduction

ROVER[1] is a two-step process, as shown in Figure 1. It starts by combining the multiple decoders' outputs into a single, minimal cost word transition network (WTN), through dynamic programming. The recognizers' confidence scores are between parenthesis in the WTN of figure 1. The resulting network is browsed by a voting process which selects the best output sequence. At each location in the composite WTN, a score is computed for each word using Equation 1, where $i$ is the current location in the WTN, $N_s$ is the total number of combined systems, $N(w,i)$ the frequency of word $w$ at the position $i$ and $C(w,i)$ is the confidence value for word $w$ at the position $i$. The parameter $\alpha$ is set to be the trade-off between using word frequency and confidences.

$$S(w) = \alpha(\frac{N(w,i)}{N_s}) + (1-\alpha)C(w,i) \tag{1}$$

The three voting schemes proposed in [1] are frequency of occurrence, frequency of occurrence and average word confidence, and frequency of occurrence and maximum confidence. The first scheme only uses occurrences to select the winning word at each WTN slot. However, ties tend to occur frequently and they

M. Kamel, F. Karray, and H. Hagras (Eds.): AIS 2012, LNCS 7326, pp. 175–183, 2012.
© Springer-Verlag Berlin Heidelberg 2012

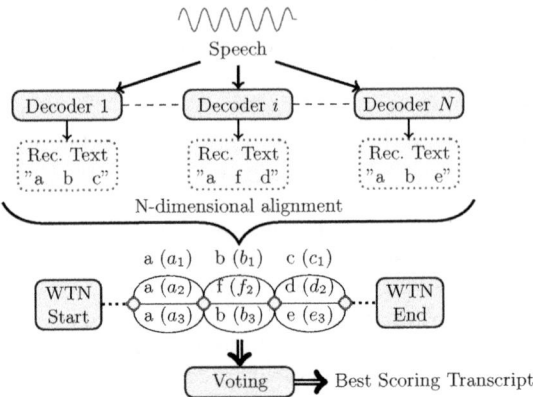

**Fig. 1.** ROVER Procedure

are randomly broken. The remaining two schemes rely on words confidences to overcome the problem of arbitrarily broken ties. The original ROVER voting mechanisms have several shortcomings. The fact that two of the ROVER schemes rely on the speech recognizers' confidences, make the whole procedure vulnerable. Furthermore, the frequency-based voting scheme proposed in [1] suffers from the random tiebreaker problem. In this research work, we propose a solution to overcome these issues. Our proposed voting scheme aims at collecting robust confidence scores through a contextual analysis, in order to resolve issues caused by the non reliable decoders' confidences. This contextual analysis is carried out through the use of automatic error detectors. These techniques aim at spotting contextual outliers within a transcription output. The probabilistic techniques are among the most promising approaches. These methods usually rely on thresholding a confidence value to decide whether a token is a recognition error. In this work, we propose to use these confidences in order to improve on the current ROVER voting schemes performance.

The remaining of this paper is organized as follows. In section 2, our approach to aggregate error filters is outlined, followed by the description of the novel voting scheme. A case study using two widely known error filters is presented. Section 3 reports the experimental results and analysis. The paper is concluded with a summary of findings.

## 2    Proposed Approach

The research work in this paper investigates a novel scoring mechanism for ROVER, in order to resolve some of the issues of the original ROVER voting schemes. The new voting scheme consists of using the confidence scores collected during the contextual analysis (error filtering analysis) to compute new word confidences at each slot of the WTN. The low recall and precision ratios of the current automatic error detectors in speech transcription led us to investigate ways to improve both of these ratios. The idea is to combine different error detection approaches in the hope that the new technique achieves higher recall, without degrading the precision.

## 2.1 Combination of Error Filters

Our goal is to improve both the precision and recall simultaneously upon the combination of the error detectors. The implicit assumption here is that the error detectors should have different performance in terms of these two ratios. That is when one approach achieves high recall and low precision, the other technique to be combined with it, needs to achieve high precision and low recall to ensure improvement in both ratios. The logic of our proposed approach is to preserve each technique's advantage and powerful characteristics during combination. Figure 2 describes the flow of the error detection combination approach [5]. The scale of each error detector's confidence score is different. Therefore a score

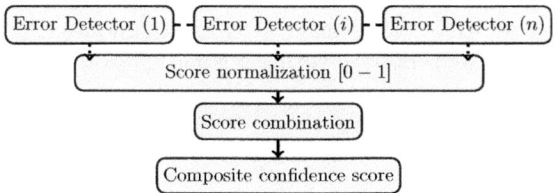

**Fig. 2.** Combination Procedure for $n$ Error Detectors

normalization stage is needed to standardize all confidence scores from various detection techniques to lie between zero and one. The scaling is done as follows: $X_{scaled} = \dfrac{X - min}{max - min}$, where $X$ is the score to be normalized, and $min$, respectively $max$, is the minimum, respectively maximum, value of the technique's confidence score. Once all the confidence scores have been normalized, a score combination formula is then applied to build a new score. Two score combination formulas were used, namely Weighted Average ($Score_{WA} = \sum\limits_{i=1}^{N} \alpha_i Score_i$) and Harmonic Mean ($Score_{HM} = N/\sum\limits_{i=1}^{N} \dfrac{1}{Score_i}$), where $Score_i$ refers to the confidence score of the $i$th error detection technique, $N$ is the total number of combined error detectors, and $\alpha_i$ are weighting scales to each technique in such a way $\sum\limits_{i=1}^{N} \alpha_i = 1$. The weighting factors play an important role in realizing a trade off between various detection techniques to optimize recall and precision ratios. These coefficients need to be optimized a priori during training.

## 2.2   New Voting Scheme

The new voting scheme consists of trading off between the original ROVER's scoring and the confidence scores obtained upon the combination of several error filters' scores. The original ROVER's scoring, given by Equation 1, is a weighted sum of both the frequency of occurrence and the speech decoder's confidence scores. We are proposing to alter this equation to take into account the confidence scores computed during the contextual analysis. Equation 2 is used to compute a new score for each word at each slot in the composite WTN.

$$S(w) = \beta \left[ \alpha(\frac{N(w,i)}{Ns}) + (1-\alpha)C(w,i) \right] + (1-\beta)Er(w,i) \qquad (2)$$

$Er(w,i)$ is the aggregated score from different error detectors assigned to the word $w$ at slot $i$, and $\beta$ is a parameter to balance between the original ROVER scores and the composite error filter scores. Similarly to the original $\alpha$ parameter, the $\beta$ factor needs to be optimized through training. The newly proposed voting algorithm is only applied on nodes with discrepancies. This is done to limit the incidence of false positives of the error filtering techniques. Figure 3 reports the overall procedure of the proposed voting scheme.

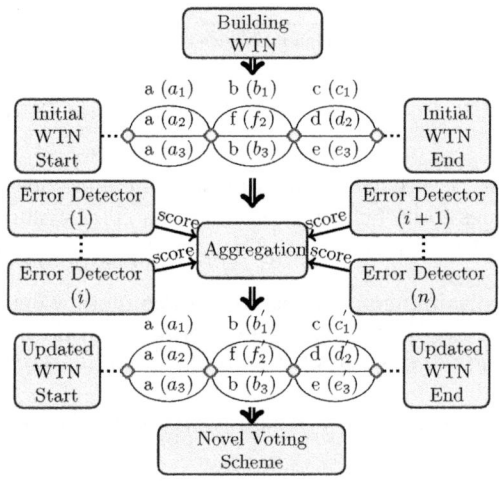

**Fig. 3.** Novel Voting Scheme Overview

Once the WTN is built, confidence scores from different error filters are aggregated together to create a composite confidence score for each token in the WTN. Then, the novel scoring formula is applied on the nodes with discrepancies. Notice in figure 3, how the confidences in the first slot have not been altered because there is no discrepancy. Tokens with the highest voting score are then selected for the final ROVER transcription. Our proposed voting scheme outperforms the original ROVER voting in two ways. First, if the speech decoders' confidence measures are unavailable, the scores from the contextual analysis (error classifiers) compensate for these missing confidences, and thereby considerably lower the risk of ties during voting. Second, even when decoders' confidences are available, our voting scheme lowers even further the risk of ties, as well as compensates the impact of using the unreliable decoders' confidences, because of the second set of confidences that are introduced from the contextual analysis. For these reasons, the new voting is more robust than the original ROVER voting schemes.

## 2.3 Case Study

The two selected filters used to assess the new voting scheme are Pointwise Mutual Information (PMI)[3] and Latent Semantic Indexing (LSI)[2] based error filters.

**PMI-Based Error Detection.** The PMI-based semantic similarity is a measure of how similar and how close in meaning $w_i$ and $w_j$ are. It is computed as follows: $PMI(w_i, w_j) = log\left(\dfrac{P(w_i, w_j)}{P(w_i).P(w_j)}\right)$, where $P(w_i) = \dfrac{c(w_i)}{N}$, $P(w_i, w_j) = \dfrac{c(w_i, w_j)}{N}$, and $c(w_i)$ and $c(w_i, w_j)$ are the frequency counts collected from a large textual corpus. The process of detecting an error using the PMI-based technique is described in algorithm 1.

---

**Algorithm 1.** PMI-based Error Detection

---

1: Identify the neighborhood $N(w)$.
2: Compute $PMI(w_i, w_j)$ $\forall(w_i, w_j)$ where $w_i \neq w_j$, including $w$. Scale up the PMIs so that $PMI(w_i, w_j) \geq 0$.
3: Compute Semantic Coherence $SC(w_i)$ $\forall w_i$ in $N(w)$, by aggregating the $PMI(w_i, w_j)$ of $w_i$ with all $w_j \neq w_i$.
4: Define $SC_{avg} = \dfrac{1}{|N(w)|} \displaystyle\sum_{i \in N(W)} SC(w_i)$.
5: Tag the word $w$ as an error if $SC(w) \leq K.SC_{avg}$.

---

In step 3 of the algorithm, the semantic coherence can be computed using different aggregation variants (averaging, summation, etc). The computed semantic coherence scores are used during our novel voting scheme. In the last step of the algorithm, the constant $K$ is a parameter to tune the error filtering.

**LSI-Based Error Detection.** The LSI procedure aims at extracting features that highlight the similarities between words. These features are obtained through the singular value decomposition of a large term-document matrix. The cosine similarity is then used to quantify similarities between terms. Two different aggregations have been used to compute the semantic similarity score of a given word in an utterance of length $M$: the mean semantic scoring (MSS),

$$MSS_i = \frac{1}{M} \sum_{j=1}^{M} cosine(w_i, w_j)$$ and the mean rank of the semantic scores (MR),

$$MR_i = \frac{1}{M} \sum_{j=1}^{M} RANK(cosine(w_i, w_j)).$$ The error detection applied on a word

$w_i$ in a given transcription output, is detailed in algorithm 2.

---

**Algorithm 2.** LSI-based Error Detection

---

1: Compute $\cos(w_i, w_j) \ \forall w_j$ in the transcription output.
2: Compute $MSS_i$ score or $MR_i$ score.
3: $w_i$ is an error if $MSS_i \leq K$ or $MR_i \leq K$.

---

The $MSS_i$ and $MR_i$ scores are collected and used by our proposed novel voting scheme. $K$ is a parameter used in order to adjust the error filtering.

## 3     Simulation Results and Analysis

Nuance v9.0 and Sphinx-4 decoders were used with the HUB4 standard testing framework[4]. Two language models have been used by Sphinx-4, which resulted in three different decoders' outputs. Experiments for all two-decoder (experiments C1 to C6) and three-decoder (experiments C7 to C12) combinations have been reported. The term-document matrix for the LSI-based detector, has been built using the latest Wikipedia XML dump. Google's one trillion-token corpus[4] was used to collect uni-gram and bi-gram frequencies required by the PMI-based error detector. The baseline WER performance is given by table 1.

Figure 4 reports the relative WER reduction, compared to ROVER's original voting, for all the different combinations of two-decoder outputs. In each experiment, relative WER reduction has been reported when confidence scores from individual error filters, as well as aggregated scores from two detectors, are used during the novel voting scheme. Figure 5 reports the relative WER reduction for all the different combinations of three decoders' outputs. The highest relative WER achieved for two decoders is 16.16%, specifically with experiment C6, when PMI and MR scores were combined using the weighted aggregation scheme (subplot (b)). However, in some cases, the new voting scheme increased the relative WER by almost 8%. When three decoders were combined together, the WER reduction didn't exceed 7%. In some experiments, there has been no change at all compared to the original ROVER scheme, and in some others, a slight increase in WER has been recorded. It is also worth mentioning that the combination of error detectors, through weighted and harmonic aggregations, has not achieved a statistically significant WER reduction compared to individual error filters. We also noticed that reducing the WER when three decoders are combined is more difficult. Our proposed voting scheme seemed vulnerable due to the use of unreliable confidence scores from error classifiers. These scores' fluctuations are certainly caused by the low recall and precision rates of the classifiers, where correct words have very low confidences, whereas erroneous tokens have high scores.

**Table 1.** ROVER's Baseline WER

| Exp. | C1 | C2 | C3 | C4 | C5 | C6 | C7 | C8 | C9 | C10 | C11 | C12 |
|------|------|------|------|------|------|------|------|------|------|------|------|------|
| WER | 18.94 | 19.55 | 19.56 | 20.36 | 24.67 | 24.45 | 16.79 | 16.92 | 18.21 | 18.78 | 19.02 | 18.56 |

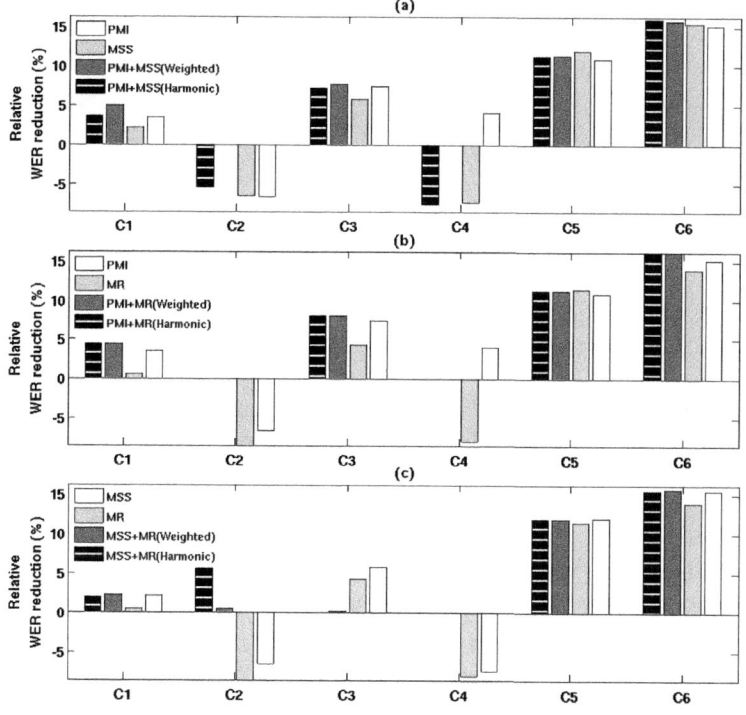

**Fig. 4.** Relative WER Reduction with 2-decoder Combination

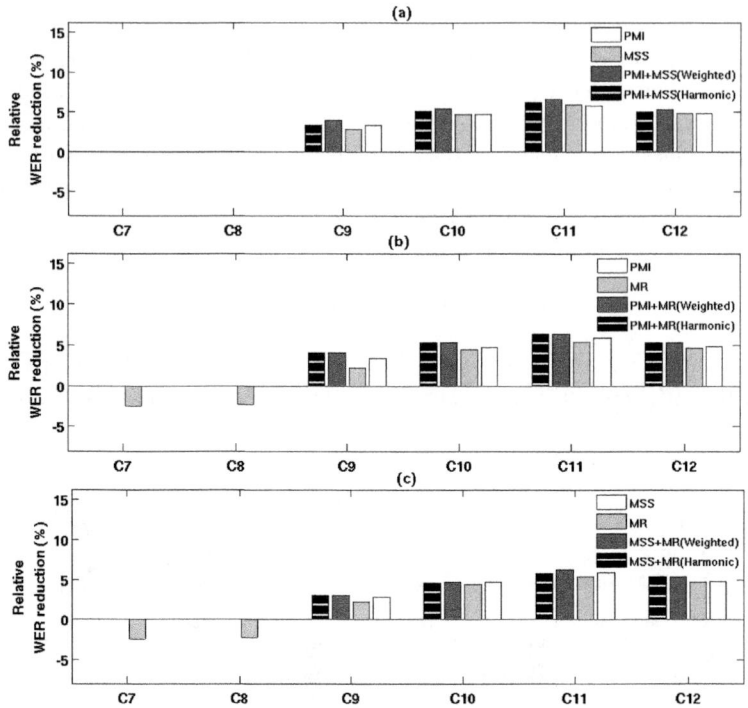

**Fig. 5.** Relative WER Reduction with 3-decoder Combination

## 4    Conclusion and Future Work

In this work, a novel ROVER's voting has been presented, which relies on confidences collected from error detectors. Aggregation of error filters' scores have been proposed. Experiments have shown it is possible to outperform the original voting, reaching in some cases up to 16% in relative WER reduction, but the proposed voting could not guarantee lower WER in all cases. Future work involves using machine learning to intelligently select the reliable error detectors (confidences) per token types (verbs, stopwords, etc).

## References

1. Fiscus, J.: A post-processing system to yield reduced word error rates: Recognizer Output Voting Error Reduction (ROVER). In: IEEE Workshop on Automatic Speech Recognition and Understanding, pp. 347–352 (1997)
2. Cox, S., Dasmahapatra, S.: High-level approaches to confidence estimation in speech recognition. IEEE Transactions on Speech and Audio Processing 10(7), 460–471 (2002)

3. Inkpen, D., Desilets, A.: Semantic Similarity for detecting recognition errors in automatic speech transcripts. In: HLT/EMNLP, pp. 49–56 (2005)
4. Abida, K., Karray, F.: ROVER enhancement with automatic error detection. In: 12th Annual Conference on the Int. Speech Communication Association (INTERSPEECH), pp. 2885–2888 (2011)
5. Abida, K., Abida, W., Karray, F.: Combination of Error Detection Techniques in Automatic Speech Transcription. In: Kamel, M., Karray, F., Gueaieb, W., Khamis, A. (eds.) AIS 2011. LNCS, vol. 6752, pp. 231–240. Springer, Heidelberg (2011)

# Features' Weight Learning towards Improved Query Classification

Arash Abghari, Kacem Abida, and Fakhri Karray

Electrical and Computer Engineering Department,
University of Waterloo,
Waterloo, Ontario,
Canada
{aabghari,mkabida,karray}@uwaterloo.ca
http://www.uwaterloo.ca

**Abstract.** This paper is an attempt to enhance query classification in call routing applications. We have introduced a new method to learn weights from training data by means of regression model. In this work, we have tested our method with *tf-idf* weighting scheme, but the approach can be applied to any weighting scheme. Empirical evaluations with several classifiers including Support Vector Machines (SVM), Maximum Entropy, Naive Bayes, and K-Nearest Neighbor (KNN) show substantial improvement in both macro and micro F1 measure.

**Keywords:** Query classification, weight learning, call routing, *tf-idf*.

## 1 Introduction

The purpose of this paper is to investigate a method to improve the accuracy of query classification for call routing applications. Call routing function is to route an incoming call to an appropriate destination. Human agents are doing this task very well, however, many enterprises try to reduce their reliance on human agents by utilizing self-service systems. To this end, touch tone systems have been used, but they are difficult to navigate and user must often go through several levels of menu. Natural language call routing is the alternative approach to overcome these limitations and problems. The prompt to the customers is general and open and they are expected to express their request freely. In this paper, we investigate a novel approach to enhance the performance of natural language based call routing applications by applying weights to the query terms. The weights which have been learned within training phase, represent the importance of each query term with respect to the destination.

This paper is organized as follows. In section 2, we review the previous work on weight learning approaches and call routing applications. Section 3 describes the proposed approach. Section 4 explains the specification of data sets and discusses experimental results. The paper ends with a conclusion and future directions in section 5.

M. Kamel, F. Karray, and H. Hagras (Eds.): AIS 2012, LNCS 7326, pp. 184–191, 2012.

## 2    Background Review

Term weighting has been well studied in the literature[1]. Term weighting has been applied to both Text Categorization (TC) [2,3] and Information Retrieval (IR) [4]. Gerard *et al.* [5] studied different methods of weighting suited for both document and query vectors. They investigated different forms of term frequency, collection frequency, and normalization components. Forman [3] replaced Inverse Document Frequency (IDF) with Bi-Normal Separation (BNS) in *tf-idf*. This weighting achieved the best result among other weighting methods including Information Gain (IG), Chi Squared, and Odds Ratio. Jin *et al.* [4] introduced a new scheme to automatically learn term weights based on the correlation between term frequency and document categories. This approach showed good performance in Information Retrieval task. Ullah *et al.* [6] investigated a soft computing approach to learn term weights. They grouped terms in the training queries into two groups with high and low weights, then applied genetic algorithm to find the weights.

Numerous work on call routing has been published [7,8,9,6]. Krasinski *et al.*[10] examined a method based on Hidden Markov Model (HMM), a generalized Probabilistic Descent (GPD) algorithm to deal "with non-keywords and word spotting". This approach showed good performance on small applications. Cox *et al.*[7] investigated different discriminative techniques. These techniques resulted in good performance while testing with training data but not with another set. Tyson *et al.*[8] introduced the Latent Semantic Index (LSI) enhanced with speech recognizer word confidence score. Sarikaya *et al.*[9] compared the performance of Deep Belief Network (DBN) against other classification techniques such as Maximum Entropy, Boosting, and Support Vector Machines (SVM). It showed that DBN can perform equally or better than other classifiers.

To the authors knowledge, none of those aforementioned weighting techniques has been applied to the realm of query classification and call routing application specifically. In this work, we are investigating an approach to incorporate weighting methods within query classification task.

## 3    Proposed Approach

The proposed approach in this paper consists of using features' weighting in order to improve the accuracy of the query classification problem. This technique has been successfully applied before in the context of document classification. However, porting the weighting concept to the query classification problem is not an easy task. This paper proposes a tailored algorithm to weight features in the context of query routing.

### 3.1    Features Weighting in a Query Collection

The term frequency–inverse document frequency, *tf-idf*, is the weighting scheme considered in this work. It is worth mentioning that the proposed algorithm is generic and independent of the weighting mechanism. The *tf-idf* is a commonly used numerical statistic in the field of information retrieval. In a nutshell, this metric reflects how important a word is to a document in a collection of documents. It is a product of the term frequency and the inverse document frequency. The term frequency is simply the number of occurrence of a term in a given document. The inverse document frequency of the term $t$ in the corpus $D$ is given by equation 1, where $|D|$ is the total number of documents in the collection and $|\{d \in D : t \in d\}|$ is the number of documents to which term $t$ belongs.

$$idf(t, D) = log\frac{|D|}{|\{d \in D : t \in d\}|} \qquad (1)$$

If we are given a set of documents and terms, it becomes straightforward to compute the *tf-idf* weights, as defined earlier. However, if instead we have only access to a set of labeled queries, the definition of the term frequency-inverse document frequency cannot be applied directly. In this regard, we propose reorganizing the collection of queries in order to compute the weights. The idea is to cluster all queries with the same label (i.e category) under a single set (i.e document). In other words, the number of documents actually becomes the number of categories. The previously defined *tf-idf* weighting scheme can now be applied on any labeled query collection, in which a document is defined as the set of queries labeled under the same category.

### 3.2   Query's Features Weighting

In section 3.1, we have detailed how to weight features or terms when a collection of labeled queries is available. However, in the case where we are only given one single query, it is impossible to compute the features' weights. In order to overcome this issue, we propose to learn features' weighting from a pre-weighted training query set. The idea is to learn a feature's weight based on the context from which the feature comes from. To this regard, we propose to use regression modeling in order to interpolate the computed weights during the training to the testing queries. Algorithm 1 explains the steps needed to prepare data for training the regression models. In summary, for each feature, there is a regression model estimating the weight with regards to the input query. In other word, we try to learn the relation between the weights in a query with their features. It is worth to mention that the regression model is our means to learn this relation and it can be substituted with any learning method. Algorithm 2 describes how to use the regression models to estimate the weights. A RBF Support Vector Regression(SVR) machine from LIBSVM [11] has been used for regression modeling.

### 3.3   Query Classification

After weighting the training query collection, and building the regression models (in order to weight single queries), the query classification procedure is carried out. The training procedure is detailed in figure 1. Given a collection of queries,

---

**Algorithm 1:** Data preparation and training regression models

---

**Input:**

$Q = \{q_1, \ldots, q_N\}$: Collection of $N$ training queries

$F = \{f_1, \ldots, f_K\}$: Feature space where each word in a training query can be a feature

$q = \{f_{1_q}, \ldots, f_{M_q}\}$: Each query $q$ consists of $M_q$ features where $M_q \leq K$

$w_q = \{\alpha_{1_q}, \ldots, \alpha_{M_q}\}$: Weights corresponded to the features of query $q$

$(x, y)$: A pair of training data where $x$ is the input and $y$ is the output

**Output:**

$R = \{r_1, \ldots, r_K\}$: Set of regression models. Each regression model $r_j$ is corresponded to the feature $f_j$ where $j = 1, \ldots, K$.

**begin**

    `// preparing training data for regression models`

    **foreach** $q \in Q$ **do**

        **foreach** $f_{j_q} \in q$ **do**

          | add $(q, \alpha_{j_q})$ to the training set of $r_{j_q}$

        **end**

    **end**

    `// training regression models`

    **foreach** $r_j \in R$ **do**

        | train $r_j$ using a regression algorithm such as SVR

    **end**

**end**

---

**Algorithm 2:** Estimating weights using regression models

---

**Input:**

$F = \{f_1, \ldots, f_K\}$: Feature space

$q = \{f_{1_q}, \ldots, f_{M_q}\}$: Input query $q$ consists of $M_q$ features where $M_q \leq K$

$R = \{r_1, \ldots, r_K\}$: Set of trained regression models. Each regression model $r_j$ is corresponded to the feature $f_j$ where $j = 1, \ldots, K$.

**Output:**

$w_q = \{\alpha_{1_q}, \ldots, \alpha_{M_q}\}$: Weights corresponded to the features of query $q$

**begin**

    **foreach** $f_{j_q} \in q$ **do**

        | $\alpha_{j_q} = r_{j_q}(q)$

    **end**

**end**

---

the first step is to extract the most important words in the collection (features extraction). This step includes for instance the elimination of all the stop words. Once the set of features has been identified, the features' weighting procedure is carried out, which results in weighted set of feature vectors, along with the regression models. The final training step consists of feeding these updated feature vector to a classification technique for training. Once a classifier is trained using

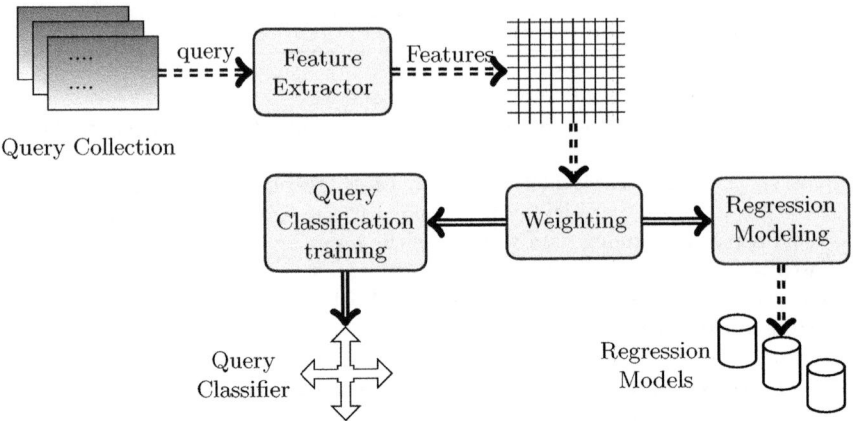

**Fig. 1.** Query Classification Training Procedure

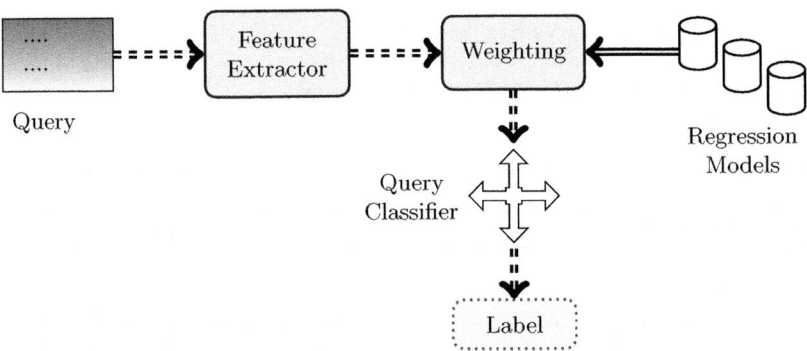

**Fig. 2.** Query Classification Testing Procedure

the weighted query collection set, the testing is carried out as shown in figure 2. Note that weightings are now estimated by the regression models. Given a query, the first step consists of selecting the features and weighting them using the regression models trained earlier. The weighted feature vector is then fed to the trained classification model for routing.

## 4   Experiments and Results

### 4.1   Experimental Setup

The proposed approach has been assessed on queries collected from four major telecommunication companies' call center. Every query is tagged by a human expert to one single category. Table 1 reports details about the four data sets. Four different classifiers including Linear Support Vector Machine (SVM) from

**Table 1.** Data sets Specification

| Data Set | Total Queries | Number of Categories | Number of Words |
|----------|---------------|----------------------|-----------------|
| Telecom 1 | 2511 | 18 | 1043 |
| Telecom 2 | 4200 | 26 | 1550 |
| Telecom 3 | 13364 | 28 | 3357 |
| Telecom 4 | 116275 | 100 | 3208 |

LIBSVM [11], Maximum Entropy (MaxEnt) and Naive Bayes (NB) probabilistic from MALLET [12], and N-Nearest Neighbor (KNN) classifier have been used to assess the impact of features' weighting on query classification.

Two metric measures were used to quantify the classification performance, namely micro F1-measure and macro F1-measure. They provide different view of the classifier performance and may give different results. The micro F1-measure put more weight on more frequent categories while macro F1-measure highlights also the performance of a classifier on categories with the small number of positive training data. To find out if the improvement or regression provided by the proposed method is statistically significance, we performed student t-test on the results.

## 4.2 Results and Interpretations

All the experiments have been carried out using four-fold cross validation. Each fold has been repeated ten times and the average has been taken as a final result. For KNN classifier the numbers are the average taken for different number of neighborhoods in the range of 1 to 10. A pre-processing task including stemming and stop-words removal has been applied for both training and testing data. We have also chosen only top 15% most important features ranked by Information Gain to train the classifier and regression models. To compare the performance of our approach, all classifiers have been also trained and tested with unweighted data which means each feature is assigned a binary weight. Figure 3 depicts the macro F1-measure, micro F1-measure, and the significance test results. The proposed features' weighting approach has improved both macro and micro F1-measures in case of Naive Bayes and KNN classifiers for all data sets. One can observe high variance with KNN compared to other classifiers. The reason is that these numbers are the average of macro and micro F1-measures over different neighborhoods (from 1 to 10). Nevertheless, the weighting approach is able to reduce this variance substantially in most of the cases. The macro and micro F1-measures have been increased by applying our weighting method to Maximum Entropy classifier except for data set 4 and micro F1-measure of data set 1 where some performance degradation is observed. It is interesting to see that there is no improvement in case of SVM classifier except for the second testing set, where some improvement has been achieved on macro F1-measure. It seems that SVM classifier is weighting-resistant somehow. In other words, SVM

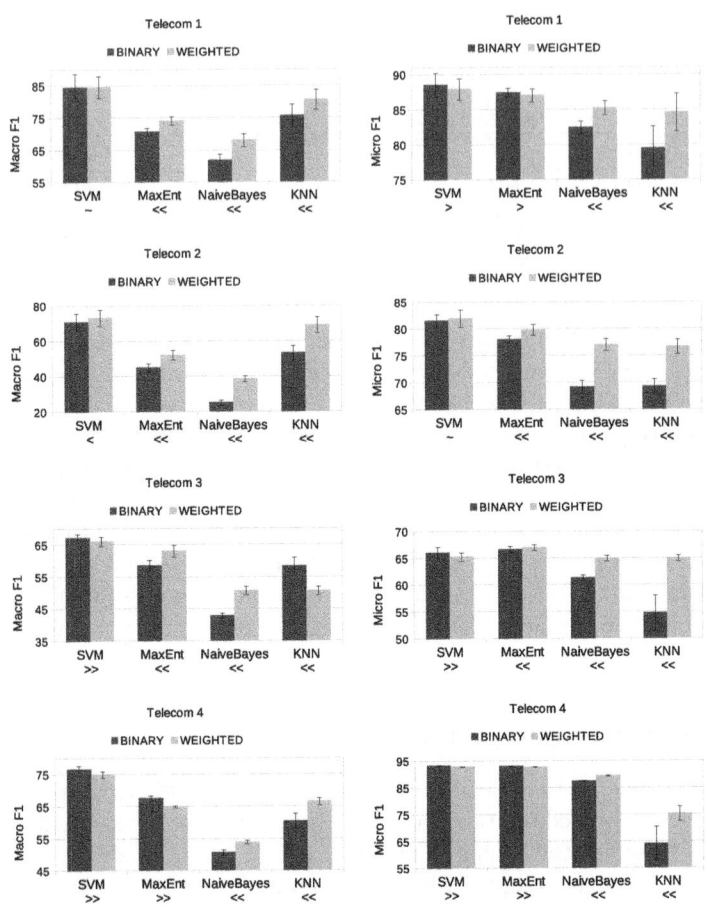

**Fig. 3.** Macro F1-measure, micro F1-measure and significance test results
"≪" or "≫" means P-value ≤ 0.01
"<" or ">" means 0.01 < P-value ≤ 0.05
"∼" means P-value > 0.05

classification performance is not affected much by applying weighting schemes. This finding is in concurrence with other works in the literature[2,3]. Having said this, Forman [3] has shown that Bi-Normal Separation weighting can improve the performance of SVM classifier. The significance test results shows that the improvements obtained by the proposed method is reliable and statistically significant.

## 5   Conclusion

In this paper, we have introduced a novel approach to assign weights to the features in a query classification task. By learning weights from training data using regression modeling, we have shown that both micro and macro F1 metrics can be substantially improved. Experiments have been carried out using the *tf-idf* weighting technique. However, our proposed approach is weighting scheme agnostic, and therefore can be generalized to any technique. As a future work, we are going to investigate other weighting schemes like Bi-Normal Separation since it has been reported to work well with SVM classifiers [3]. Furthermore, we are intending to implement alternatives modeling procedures, such as Stacked Denoising Autoencoders (SDA) [13] in order to learn the weights from the query collection.

# References

1. Forman, G., Guyon, I., Elisseeff, A.: An extensive empirical study of feature section metrics for text classification. J. Machine Learning Research 3, 1289–1305 (2003)
2. Debole, F., Sebastiani, F.: Supervised term weighting for automated text categorization. In: 18th ACM Symposium on Applied Computing, pp. 784–788. ACM Press (2003)
3. Forman, G.: BNS feature scaling: an improved representation over tf-idf for SVM text classification. In: 17th ACM Conference on Information and Knowledge Management, pp. 263–270 (2007)
4. Jin, R., Chai, J.Y., Si, L.: Learn to weight terms in information retrieval using category information. In: 22nd International Conference on Machine Learning, pp. 350–360. ACM Press (2005)
5. Salton, G., Buckley, C.: Term-weighting approaches in automatic text retrieval. In: Information Processing and Management, pp. 513–523 (1988)
6. Ullah, S., Karray, F., Abghari, A., Podder, S.: Soft computing-based approach for natural language call routing systems. In: 9th International Symposium Signal Processing and Its Applications (ISSPA), pp. 1–4 (2007)
7. Cox, S.: Discriminative techniques in call routing. In: IEEE Conference on Acoustics, Speech, and Signal Processing, ICASSP (2003)
8. Tyson, N., Matula, V.C.: Improved LSI-based natural language call routing using speech recognition confidence scores. In: IEEE conference on Computational Cybernetics (ICCC), pp. 409–413 (2004)
9. Sarikaya, R., Hinton, E., Ramabhadran, B.: Deep belief nets for natural language call-routing. In: IEEE Conference on Acoustics, Speech and Signal Processing (ICASSP), pp. 5680–5683 (2011)
10. Krasinski, D.J., Sukkar, R.A.: Automatic speech recognition for network call routing. In: 2nd IEEE Workshop Interactive Voice Technology for Telecommunications Applications, pp. 157–160 (1994)
11. Chang, C., Lin, C.: LIBSVM: a library for support vector machines. ACM Transaction on Intelligent Systems and Technology, 1–27 (2011), Software, http://www.csie.ntu.edu.tw/~cjlin/libsvm
12. McCallum, A.: A machine learning for language toolkit (2002), http://mallet.cs.umass.edu
13. Vincent, P., Larochelle, H., Bengio, Y., Manzagol, P.: Extracting and composing robust features with denoising autoencoders. In: 25th International Conference on Machine Learning (ICML), pp. 1096–1103 (2008)

# A Novel Template Matching Approach to Speaker-Independent Arabic Spoken Digit Recognition

Jiping Sun[1], Jeremy Sun[1], Kacem Abida[2], and Fakhri Karray[2]

[1] Voice Enabling Systems Technology,
Waterloo, Ontario, Canada
{jiping,jeremy}@vestec.com
http://www.vestec.com

[2] Electrical and Computer Engineering Department, University of Waterloo,
Waterloo, Ontario, Canada
{mkabida,karray}@uwaterloo.ca
http://www.uwaterloo.ca

**Abstract.** In this paper we propose a quantized time series algorithm for spoken word recognition. In particular, we apply the algorithm to the task of spoken Arabic digit recognition. The quantized time series algorithm falls into the category of template matching approach, but with two important extensions. The first is that instead of selecting some typical templates from a set of training data, all the data is processed through vector quantization. The second extension consists of a built-in temporal structure within the quantized time series to facilitate the direct matching, instead of relying on time warping techniques. Experimental results have shown that the proposed approach outperforms the time warping pattern matching schemes in terms of accuracy and processing time.

**Keywords:** spoken digits recognition, template matching, vector quantization, time series, Arabic digits.

## 1 Introduction

Voice recognition may be categorized into two major approaches. The first approach applies multiple layers of processing. Typically in this approach the input audio is first mapped into sub-word units such as phonemes by hidden Markov modeling. Then phonemes are connected into words using word models. This approach usually involves sophisticated implementations of large software and requires lots of data. These systems aim at solving large vocabulary, speaker independent continuous speech recognition tasks. Further details can be found in [1]. The second approach aims at simpler tasks. A typical example is digit recognition for voice dialing, voice print verification, etc. Most typical approach is to model the user voice by a set of templates and recognize inputs by template matching. The work in [2] and [9] describes in more details such an approach

M. Kamel, F. Karray, and H. Hagras (Eds.): AIS 2012, LNCS 7326, pp. 192–199, 2012.

to voice recognition. We need to mention at this point that common to both approaches there is a signal processing front-end stage, which converts audio data to feature vectors, most common of which are the Mel Frequency Cepstral Coefficients (MFCC) features [2]. Each feature vector in the sequence represents a short-term frame of time duration about 10 milliseconds.

There are two common issues in the template matching approach for voice recognition. The first is the selection of good templates from a set of training data. As human speech varies from time to time and from situation to situation, each instance of a word can be slightly different from other instances. It is therefore often difficult to decide which one of these instances should be used as the model. The second issue is the accuracy. This is because the few selected templates usually have missing information to match various testing utterances. As a result, template matching-based systems are often used in speaker-dependent applications to avoid inter-speaker variations. Authors in [3] described an approach by which all training data available is used in the creation of "templates" to increase the information content of the models. Therefore their templates are extended by data quantization so that the units in the templates are no longer feature vectors from individual audio inputs but instead are quantization vectors. A quantization vector combines the information of many data vectors through the process of averaging. Similar ideas of applying Dynamic Time Warping (DTW) on quantized time series can be found in [8].

In this paper we propose a different quantization-based template approach which extends to speaker-independent tasks. This model has an increased complexity in the template structures in order to retain more information from spoken words variations. The remainder of this paper is organized as follows. Section 2 details the proposed model structures and algorithms. Section 3 presents the experimental setup and the results. The paper is concluded with discussion and future directions in section 4.

## 2   The Algorithm of Quantized Time Series

Given a training set of labeled vectors $\{x_i^c\}, x_i \in \Re^d, c = 1, ..., k$ where each vector $x_i$ is a member in $d$-dimensional real valued Euclidean space, each class $c$ has $n_c$ instance vectors, the following sub-algorithm steps, illustrated by figure 1, are executed in sequence to obtain a set of models, $Q^c, c = 1, ..., k$, in the form of quantized time series. For both training and testing, the data is normalized per each dimension. Second, a statistical procedure is executed to find the distribution of adjacent vector distances, that is, the distance between neighbor vectors in each time series. Based on this distribution a merge threshold is defined and used to merge neighbor vectors. This step shrinks the lengths of vector time series to reduce the processing time of ensuing steps as well as to reflect the fact that adjacent similar frames in speech, can be considered as staying in the same states. Following this step is the main learning step of vector quantization, which applies an extension of the time series vectors by adding an additional dimension of time. This step results in a set of quantized time series models,

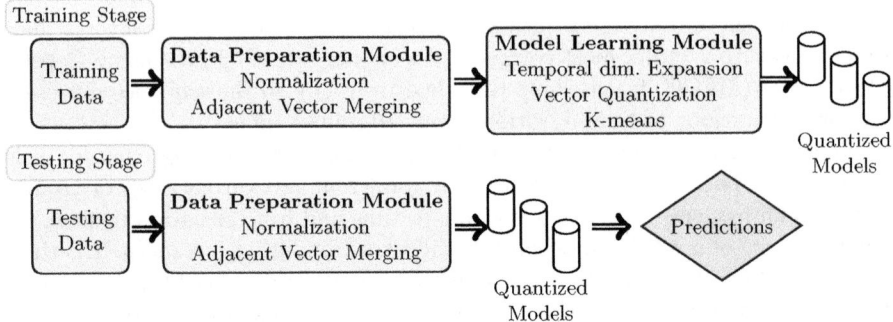

**Fig. 1.** Quantized Time series Training and Testing Procedures

each for one class. Finally a rather simple template matching algorithm is used for predicting the class of an input testing sequence. The next sections provide details of each step of the illustrated procedure.

## 2.1   Data Normalization

The values in each dimension of the time series may have different scales and with large differences in range. This will affect the performance of distance-based classifiers if direct distance calculation such as Euclidean distance is used. In order to effectively measure the similarities between vectors, all the dimensions are normalized to lie within the $[-1, +1]$ range. Algorithm 1 describes the normalization procedure. Once normalized, similarities between vectors is performed through Euclidean distance directly.

---

**Algorithm 1.** Data Normalization

---
1: Set upper-bound to $ub = +1$, and lower-bound to $lb = -1$.
2: **for** each dimension $m = 1, ..., d$ **do**
3:     Find the maximum and minimum of all values $x_i^c[m]$, $\max_m$ and $\min_m$
4:     Convert all $x_i^c[m]$ to $lb + (ub - lb)\dfrac{x_i^c[m] - \min_m}{\max_m - \min_m}$
5: **end for**

---

## 2.2   Merging Adjacent Feature Vectors

All vectors of each class $c$ appear in certain variable-length sequences (or time series) corresponding to utterances. Let $X_j^c = \{x_1^c...x_n^c\}$ be such a sequence, where $j$ ranges from one to $n_c$. This instance is usually formed once the audio is processed by the front-end signal processor in order to obtain feature vectors. In this paper, the $X_j^c$ is a spoken instance $j$ of a digit $c$. Some of these adjacent vectors $x_i^c$ can be similar especially when they belong to the same phoneme.

---

**Algorithm 2.** Adjacent Vector Merging Algorithm

---

1: **for all** $X_j^c$, $j \in [1, n_c]$, $c \in [1, k]$ **do**
2:      Compute Euclidean distances, $dist_{i,i+1}(x_{ji}^c, x_{ji+1}^c)$, of all adjacent vector pairs.
3:      Organize $dist_{i,i+1}$ in 100 histogram bins within $[0, (\max_{dist_{i,i+1}} - \min_{dist_{i,i+1}})]$
4:      Set $GoldenRatio = 1.61$
5:      Set the merging threshold $dis_{merge}$ to the $t^{th}$ bin's right boundary where
$$\frac{\sum_{a=1}^{100} count(his[a])}{\sum_{a=1}^{t} count(his[a])} = GoldenRatio$$
6: **end for**
7: **for all** $X_j^c$, $j \in [1, n_c]$, $c \in [1, k]$ **do**
8:      **if** $dist_{i,i+1}(x_{ji}^c, x_{ji+1}^c) \le dis_{merge}$ **then**
9:          Merge $x_{ji}^c$ and $x_{ji+1}^c$ into one vector, by averaging values at each dimension.
10:      **end if**
11: **end for**

---

Therefore, we can reduce the vector sequence $X_j^c$ length by merging these similar feature vectors. The adjacent vector merging procedure is detailed in algorithm 2. The algorithm first aims at finding the merging distance, below which adjacent feature vectors must be combined. In step 2, all distances between adjacent feature vectors $x_{ji}$ are computed. The $j$ index here refers to the instance $X_j^c$. The maximum and minimum of these distances are then determined. The range $[0, (\max_{dist_{i,i+1}} - \min_{dist_{i,i+1}})]$ is then divided into 100 histogram bins in step 3. Each distance $dist_{i,i+1}$ is placed in the appropriate bin, in such a way it is less or equal than the bin's right boundary. Once step 3 is achieved, the $dist_{i,i+1}$ become scattered across all bins with different distribution. Step 5 determines the merging distance. The idea is to locate the bin in which the ratio of the cumulative distances by the total number of distances is equal to the empirical "Golden Ratio". The merging distance, $dis_{merge}$, is then selected as the right boundary of this specific bin. The remaining steps consists of merging any adjacent feature vectors if their distances is less than $dis_{merge}$.

## 2.3  Vector Quantization with Extended Temporal Dimension

Once the data is normalized and adjacent vectors in every sequence are merged, the vectors in the training data go through a vector quantization process to reduce the data into a smaller number of centroids. Before quantization, each feature vector of each sequence is extended by one extra dimension to reflect its relative position in the sequence. Since the range of every dimension is normalized between $-1$ and $+1$, this extended dimension will take the value between 0 and 2. The extra feature is called temporal dimension. The vector quantization and model sequence learning technique is detailed in algorithm 3. The algorithm is carried out separately for each class, therefore resulting in a set of centroids for

---

**Algorithm 3.** Vector Quantization Algorithm

---

1: Set centroid radius $r$, merging distance $d_m$, radius delta $r_\Delta$, put the first vector in
    the data as a member of centroids $Ctnds$.
2: **for** each training vector $x_i^c$ in a class **do**
3:    Find the nearest centroid $ctn$ in $Ctnds$.
4:    **if** $dist(x_i^c, ctn) < r$ **then**
5:        merge $x_i^c$ with $ctn$ by averaging at each dimension.
6:    **else**
7:        Set $x_i^c$ as a new member of $Ctnds$.
8:    **end if**
9:    **for** Every pair of centroids $ctn_i$ and $ctn_j$ in $Ctnds$ **do**
10:      **if** $dist(ctn_i, ctn_j) < d_m$ **then**
11:         merge $ctn_i$ and $ctn_j$ by averaging at each dimension;
12:      **end if**
13:    **end for**
14:    **for** each centroid $ctn$ in $Ctnds$ **do**
15:      **if** $data\_count(ctn) \geq mean(data\_count(Ctnds))$ **then**
16:        $r(c) += r_\Delta$
17:      **else**
18:        $r(c) -= r_\Delta$
19:      **end if**
20:    **end for**
21: **end for**

---

each class. In the case of Arabic digit recognition, this resulted in ten separate sets of learned centroids, each for one digit (zero to nine). This online vector quantization algorithm was adapted from the Self Learning Vector Quantization algorithm in [4]. After vector quantization, the centroids are further refined by the general $k$-means algorithm in order to adjust the values of each centroid so that the centroids better model the statistical properties of the training data. Finally the centroids are sorted by the extended temporal dimension into sequences. These sequences are the quantized time series models, one for each class (i.e. digit). This is represented as $Q^c, c = 1, ..., k$, for $k$ classes.

## 2.4   Matching Test Sequence to Quantized Time Series Models

The time series models are sequences of learned centroids sorted by the temporal dimension. To match a test vector sequence to a time series model, the temporal dimension is used to project a vector in the test sequence to a (smaller) set of vectors in the model having similar temporal locations. This is illustrated in Figure 2. The matching algorithm is detailed in algorithm 4. In step 6 of the algorithm, the $\alpha$ parameter is a user defined threshold for projecting an instance test vector into a subset of model's vectors. In our experiments of Arabic spoken digit recognition, the $\alpha$ has been set to 0.4. This template-matching algorithm has a fundamental difference from the DTW-based matching algorithm or similar algorithms. In the later, there is always a time-warping factor involved,

Quantized time series model

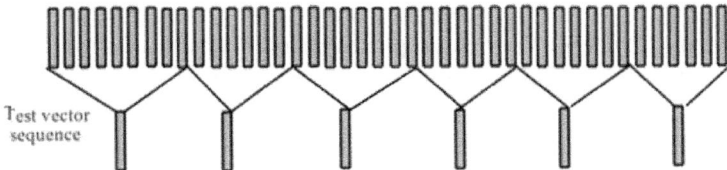

Test vector
sequence

**Fig. 2.** Matching a test vector sequence to a quantized time series model

---

**Algorithm 4.** Algorithm of vector sequence matching

---
1: Given a set of quantized time series models $Q^c, c = 1, ..., k$ and a test vector sequence $X_t$,
2: **for all** $Q^c, c = 1, ..., k$ **do**
3:    Set sum $S_c$ to 0.0
4:    **for all** $x_i$ in $X_t$ **do**
5:        Carry out Normalization, and Adjacent Vector Merging.
6:        Add the temporal dimension.
7:        Project $x_i$ to a subset $O$ of $Q^c$ in such a way that $\forall j, abs(o_j[0] - x_i[0]) \leq \alpha$
8:        Find the nearest centroid $o$ to $x_i$ in $O$;
9:        Calculate similarity between $x_i$ and $o$ using $sim_{x_{j,o}} = e^{-x_{j,o}}$
10:        $S_c = S_c + sim_{x_{j,o}}$
11:    **end for**
12: **end for**
13: Prediction: class of $X_t = c, if S_c > S_i, \forall i, i \neq c$

---

which requires the computation of much more values and the execution of much complex steps, while in our algorithm the matching is rather direct as both the model and the input vectors are linked by a timing condition. Therefore the computational complexity of our algorithm is lower than standard DTW-based algorithms.

## 3   Experiments and Results

In this section we highlight the experiments in which we have applied the quantized time series algorithm to the task of spoken Arabic digit recognition. Procedures of data preparation, model learning and testing novel sequences from the test set are described. Following that results obtained from the experiments are reported .

### 3.1   Data Preparation and Model Learning

The Spoken Arabic Digit data set[5] consists of 66 training speakers (33 male speakers and 33 female speakers) and 22 testing speakers (11 male speakers and

11 female speakers). Each speaker spoke each digit from 0 to 9 ten times. Therefore there are 6600 training utterances which after signal processing front-end became 6600 MFCC vector sequences (time series). Similarly 2200 testing utterances became 2200 MFCC vector sequences (time series). In the training set, the original 6600 vector sequences contained 263256 vectors. After data normalization (algorithm 1) and adjacent vector merging (algorithm 2, $dis_{merge} = 0.43$), 106792 vectors remained in the training set. Vector quantization and $k$-means operations (algorithm 3, $r = mpd/2.2$, $d_m = r/24.0$, $r_\Delta = r/14.0$, where $mpd$ is the mean pairwise distance between all vectors in the data) generated 10 quantized time series models containing 2475 vectors in total. Each vector is a 13-dimension MFCC feature vector for both the testing data and the training data as provided by the original UCI data set. Thus the model achieved a 1:106.36 model to data compression ratio.

## 3.2   Experiment and Results

Both DTW and the proposed methods are applied on the 22 speakers in the test set. As the speakers in the test data set are different from those in the train set, the task is speaker-independent voice recognition. To compare the performance of the proposed algorithm, two DTW model sets are selected. In the first, one sequence is selected for each digit from the data of the first speaker in the training set. This formed a model of 10 templates. To increase the performance of the DTW method, another set of templates were selected: For every sixth speaker in the training set one template for each digit is selected. This included 11 speakers of the training set, forming a model of 110 templates in total, 11 templates for each digit. This model set was used on the test speakers by finding the closest match as the winner. As a result, the accuracy has improved but since the matching task was much heavier, the processing time was much longer. The results of these tests are given in Table 1. As the templates used in the second

**Table 1.** Testing Set: Arabic Spoken Digits Recognition

| No.time series | Algorithm | Model size | Accuracy | CPU Time |
|---|---|---|---|---|
| 2200 | DTW | 10 templates | 65.23% | 1075 seconds |
| 2200 | DTW | 110 templates | 85.31% | 6480 seconds |
| 2200 | Quantized time series | 10 templates | 93.32% | 903 seconds |

task are from the first speaker of the training set, a speaker-dependent task was performed (third task) to verify that the templates are valid. The results show that the proposed algorithm achieved a much higher accuracy compared to standard DTW-based template matching algorithm in the task of speaker-independent voice recognition, in particular, for recognizing spoken Arabic digits. Results also showed that the quantized time series technique is much faster than both DTW-based pattern matching procedures.

## 4    Conclusion and Future Work

In this paper we have proposed a novel algorithm for template-matching based voice recognition. Through a set of simple processing steps, the training data set is transformed into a set of quantized time series models. Experiments recognizing spoken Arabic digits in a speaker-independent setting have demonstrated the effectiveness of the proposed algorithm. Since the computation and data structure used in this algorithm are both simple and have very small memory requirements, it falls well within the template matching category of voice recognition and can be readily implemented in embedded or mobile platforms.

To further improve the performance of the proposed algorithm, the models centroids can be further refined through discriminative learning, that is by examining error patterns and adjusting the values of each mean vector. Furthermore, soft subspace weights can be assigned to the dimensions of centroids[6,7]. We want to investigate whether adjusting such weights by examining error patterns can lead to higher accuracies.

## References

1. Young, S.: Statistical modeling in Continuous Speech Recognition (CSR). In: International Conference on Uncertainty in Artificial Intelligence, pp. 562–571. Morgan Kaufmann, Seattle (2001)
2. Muda, L., Begam, M., Elamvazuthi, I.: Voice recognition algorithms using Mel Frequency Cepstral Coefficient (MFCC) and Dynamic Time Warping (DTW) Techniques. J. of Computing 2(3), 138–143 (2010)
3. Zaharia, T., Segarceanu, S., Cotescu, M., Spataru, A.: Quantized Dynamic Time Warping (DTW) algorithm. In: 8th International Conference on Communications, Bucharest, pp. 91–94 (2010)
4. Rasanen, O.J., Laine, U.K., Altosaar, T.: Self-learning vector quantization for pattern discovery from speech. In: International Speech Communication Association, pp. 852–855. ISCA, Brighton (2009)
5. Frank, A., Asuncion, A.: UCI Machine Learning Repository. University of California, School of Information and Computer Science, Irvine, CA (2010), http://archive.ics.uci.edu/ml
6. Domeniconi, C., Gunopulos, D., Ma, S., Yan, B., Al-Razgan, M., Papadopoulos, D.: Locally adaptive metrics for clustering high dimensional data. J. Data Min. Knowl. Discov. 14(1), 63–97 (2007)
7. Karray, F., De Silva, C.: Soft computing and Intelligent systems design. Pearson Education Limited, Canada (2004)
8. Martin, M., Maycock, J., Schmidt, F.P., Kramer, O.: Recognition of Manual Actions Using Vector Quantization and Dynamic Time Warping. In: Graña Romay, M., Corchado, E., Garcia Sebastian, M.T. (eds.) HAIS 2010, Part I. LNCS, vol. 6076, pp. 221–228. Springer, Heidelberg (2010)
9. Chanwoo, K., Seo, K.: Robust DTW-based recognition algorithm for hand-held consumer devices. J. of IEEE Transactions on Consumer Electronics 51(2), 699–709 (2005)

# An Heterogeneous Particle Swarm Optimizer with Predator and Scout Particles

Arlindo Silva[1], Ana Neves[1], and Teresa Gonçalves[2]

[1] Escola Superior de Tecnologia do Instituto Politécnico de Castelo Branco
{arlindo,dorian}@ipcb.pt
[2] Universidade de Évora
tcg@uevora.pt

**Abstract.** We present a new heterogeneous particle swarm optimization algorithm, called scouting predator-prey optimizer. This algorithm uses the swarm's interactions with a predator particle to control the balance between exploration and exploitation. Scout particles are proposed as a straightforward way of introducing new exploratory behaviors into the swarm. These can range from new heuristics that globally improve the algorithm to modifications based on problem specific knowledge. The scouting predator-prey optimizer is compared with several variations of both particle swarm and differential evolution algorithms on a large set of benchmark functions, selected to present the algorithms with different difficulties. The experimental results suggest the new optimizer can outperform the other approaches over most of the benchmark problems.

**Keywords:** swarm intelligence, particle swarm optimization, heterogeneous particle swarms.

## 1 Introduction

The particle swarm optimization algorithm (PSO) was initially based on a metaphor of social interaction between individuals in flocks of birds [6]. The result was a population based stochastic optimization algorithm, where individuals, called particles, are represented as vectors of real numbers in a multidimensional space. A second vector represents each particle's velocity. A particle is attracted both by the best position it has previously found in the search space and by the best position found by its neighbors in the swarm. The intensity of these attractions is used to compute the particles' velocities. The algorithm searches for optima by iteratively changing the velocity of each particle and, as a result, its position, thus defining a trajectory through the search space.

Since its introduction, the PSO algorithm has been successfully applied to problems in different areas, including antenna design, computer graphics visualization, biomedical applications, design of electrical networks and many others [10]. Amongst the qualities that led to this popularity are its conceptual simplicity, ease of implementation and low computational costs. The algorithm works well with small swarms and tends to converge fast to a solution. In addition, particle swarm optimization has proven itself to be easily adaptable to new domains

M. Kamel, F. Karray, and H. Hagras (Eds.): AIS 2012, LNCS 7326, pp. 200–208, 2012.
© Springer-Verlag Berlin Heidelberg 2012

of application and to work well when hybridized with other approaches. While there might be other algorithms in the evolutionary computation field that can claim better optimization results, the combination of simplicity, flexibility and performance of the PSO algorithm remains very appealing to practitioners.

Notwithstanding its success and popularity, the basic PSO has some drawbacks that have been identified almost since its introduction [1,13]. Controlling the balance between exploration and exploitation is one of the main difficulties. Maintaining some level of diversity in the swarm after it has converged is also a problem, since the basic PSO does not have a mechanism equivalent to mutation in evolutionary algorithms, capable of introducing diversity even after the algorithm has stagnated. Other issues include a performance degradation when optimizing non-separable functions and difficulties in fine-tuning a solution, after a reasonably good position is found and the swarm has converged to it.

Many changes and additions to the original PSO have been proposed since its introduction, to try to eliminate its perceived weaknesses [11]. Some of the more promising variations in use today are hybrid approaches, which use one or more mutation operators to overcome some of those weaknesses [5]. Memetic variants have also been proposed, allying the advantages of local search algorithms with the global exploratory capabilities of the PSO [9].

Recently, it has increased the interest in heterogeneous particle swarm optimizers, where different particles within the same swarm can have different behaviors and/or properties [3,7]. These approaches include the use of different neighborhood topologies or update equations for different particles. Heterogeneity allows for different parts of the swarm to be tuned for different aspects of the problem being optimized or for different phases of the exploration process.

In this article we present a new heterogeneous particle swarm algorithm, called scouting predator-prey optimizer (SPPO). Besides the normal particles in the swarm, the SPPO uses an extra particle, called a predator, and a subset of the swarm is updated with alternative update rules. The particles in this subset are called scout particles. The heterogeneous particles improve different aspects of the standard PSO, which leads to improved performance when compared to state of the art PSO and differential evolution (DE) based algorithms [15]. The empirical comparison is made over 16 carefully chosen optimization benchmark functions, which present the algorithms with a diverse set of challenges and difficulties, in an effort to better highlight their strengths and weaknesses.

## 2   The Algorithms

In particle swarm optimization each swarm member is represented by three $m$-size vectors, assuming an optimization problem $f(\mathbf{x})$ in $\mathbb{R}^m$. For each particle $i$ we have a $\mathbf{x_i}$ vector that represents the current position in the search space, a $\mathbf{p_i}$ vector storing the best position found so far and a third vector $\mathbf{v_i}$ corresponding to the particle's velocity. For each iteration $t$ of the algorithm, the current position $\mathbf{x_i}$ of every particle $i$ is evaluated by computing $f(\mathbf{x_i})$. Assuming a minimization problem, $\mathbf{x_i}$ is saved in $\mathbf{p_i}$ if $f(\mathbf{x_i}) < f(\mathbf{p_i})$, i.e. if $\mathbf{x_i}$ is the best solution found by the particle so far. The velocity vector $\mathbf{v_i}$ is then computed with equation 1 and used to update the particle's position (equation 2).

$$\mathbf{v_i^{t+1}} = w\mathbf{v_i^t} + \mathbf{u}(0, \phi_1) \otimes (\mathbf{p_i} - \mathbf{x_i^t}) + \mathbf{u}(0, \phi_2) \otimes (\mathbf{p_g} - \mathbf{x_i^t}) \qquad (1)$$

$$\mathbf{x_i^{t+1}} = \mathbf{x_i^t} + \mathbf{v_i^t} \qquad (2)$$

In equation 1 $(\mathbf{p_i} - \mathbf{x_i^t})$ represents the distance between a particle and its best position in previous iterations and $(\mathbf{p_g} - \mathbf{x_i^t})$ represents the distance between a particle and the best position found by the particles in its neighborhood, stored in $\mathbf{p_g}$. $\mathbf{u}(0, \phi_1)$ e $\mathbf{u}(0, \phi_2)$ are random number vectors with uniform distributions between 0 e $\phi_1$ and 0 e $\phi_2$, respectively. $w$ is a weigh that decreases linearly with $t$ and $\otimes$ is a vector component-wise multiplication.

## 2.1 The Predator Effect

One of the limitations of the standard particle swarm algorithm is its inability to introduce diversity in the swarm after it has converged to a local optimum. Since there is no mechanism similar to a mutation operator, and changes in $\mathbf{x_i}$ are dependent on differences between the particles' positions, as the swarm clusters around a promising area in the search space, so does velocity decreases and particles converge. This is the desirable behavior if the optimum is global, but, if it is local, there is no way to increase velocities again, allowing the swarm to escape to a new optimum, i.e. there is no way to go back to a global exploration phase of search after the algorithm entered a local search (exploitation) phase.

We introduced the predator-prey idea [14] to alleviate this problem. Here we present a simplified and updated version of the predator-prey optimizer. The predator particle's velocity $\mathbf{v_p}$ is updated using equation 3, oscillating between the best particle's best and current position. This update rule makes the predator effectively chase the best particle in the search space.

$$\mathbf{v_p^{t+1}} = w\mathbf{v_p^t} + \mathbf{u}(0, \phi_1) \otimes (\mathbf{x_g^t} - \mathbf{x_p^t}) + \mathbf{u}(0, \phi_2) \otimes (\mathbf{p_g} - \mathbf{x_p^t}) \qquad (3)$$

The role of the predator particle in the SPPO algorithm is to introduce a perturbation factor in the swarm and to guarantee that this disturbance increases as the swarm converges to a single point. To achieve this, we add a perturbation to a particles's velocity in dimension $j$, as defined by equation 4, where $u(-1, 1)$ and $u(0, 1)$ are random numbers uniformly distributed between the arguments, $x_{max}$ and $x_{min}$ are, respectively the upper and lower limit to the search space and $r$ is the user defined perturbation probability.

$$v_{ij}^t = v_{ij}^t + u(-1, 1)|x_{max} - x_{min}|, \text{if } u(0, 1) < r \exp^{-|x_{ij} - x_{pj}|} \qquad (4)$$

From equation 4 follows that a random perturbation is added to the velocity value in dimension $j$ with a probability that depends on the particles's distance to the predator in that dimension. This probability is maximum $(r)$ when that distance is 0 but rapidly decreases if the particle escapes the predator. Since the predator chases the best particle, the perturbation in the swarm is more likely when all the particles are very near, i.e. during the exploitation phase, and becomes almost inexistent when the particles are far apart. This mechanism allows for particles to escape and find new optima far from the current attractor even in the last phases of exploitation.

## 2.2   Scout Particles

The predator-prey optimizer is a an heterogeneous particle swarm algorithm, since the predator particle is updated using a different equation. We propose a new level of heterogeneity by including scout particles into the swarm. Scout particles, or scouts, are a subset of the swarm that implements different exploration strategies. They can be used to introduce improvements to the global algorithm, e.g. a local search sub-algorithm, or to implement problem dependent mechanisms to better adapt the algorithm to a specific problem. Since interaction with the main swarm is basically done by sharing the best value found by the particles, the introduction of scouts doesn't disrupt the global swarm behavior. In this sense, it is a very flexible way to introduce new capabilities to the optimization algorithm, or even to hybridize it with a different optimizer altogether.

To illustrate this idea, we use two scout particles to improve the performance of the predator-prey algorithm in continuous optimization problems. For the first scout we choose the best particle in the swarm at a given iteration and perform a random mutation on one of its dimensions $j$ using equation 5, where $n(0, \sigma^2)$ is a a random number drawn from a normal distribution with average 0 and standard deviation $\sigma$. $\mathbf{p_g}$ is updated to the new $\mathbf{p'_g}$ only if $f(\mathbf{p'_g}) < f(\mathbf{p_g})$. $\sigma$ starts at $x_{max}/10$ and is updated during the run using the $1/5$ rule borrowed from evolution strategies [2]. After every 100 generations $\sigma$ is doubled if the mutation success rate is over $1/5$ and is halved otherwise. This mutation mechanism allows for a local search to be made around $\mathbf{p_g}$ over time.

$$p'_{gj} = p_{gj} + n(0, \sigma) \tag{5}$$

The second scout particle uses an update rule inspired by opposition based learning (OBL) [16]. The basic idea behind OBL is that sometimes it might be useful to search in the opposite direction of the current position. Opposition based extensions of particle swarm [17] and differential evolution [12] have improved on the results of the corresponding base algorithms. For this second scout, we use the particle with worst evaluation, $\mathbf{p_w}$, based on the heuristic that in the opposite of the worst particle might be a more promising search region. We compute the opposite particle position $\mathbf{p'_w}$ using equation 6 for each dimension $j$, with $a^t_j$ and $b^t_j$ being, respectively, the maximum and minimum value for $p_{ij}$ at generation $t$. Again, $\mathbf{p_w}$ is only updated to the new $\mathbf{p'_w}$ if $f(\mathbf{p'_w}) < f(\mathbf{p_w})$.

$$p'_{wj} = a^t_j + b^t_j - p_{wj} \tag{6}$$

Scout particles are updated prior to the main update cycle of the swarm, where they can cumulatively be updated using equations 1 and 4. To save objective function evaluations, we update the best particle but not the worst one.

## 3   Experimental Results

In our experimental setup we used 16 benchmark functions carefully chosen from the optimization bibliography to pose different difficulties to the algorithms. $f_1$-$f_3$ are all unimodal functions, but $f_2$ is non-separable and in $f_3$ the optimum is in a very flat zone with little information to guide the algorithms. $f_4$-$f_{10}$ are multimodal functions with many local optima, but all have strong symmetries around the global optimum, located at $\mathbf{x_i^*} = 0$. $f_6$,$f_7$,$f_9$ and $f_{10}$ are non-separable.

**Table 1.** Benchmark function parameters

| Function | Name | Range | $f(\mathbf{x}^*)$ | $x_i^*$ | Displacement |
|---|---|---|---|---|---|
| $f_1$ | Sphere | $[-100 : 100]$ | 0 | 0 | 25 |
| $f_2$ | Rotated Ellipsoid | $[-100 : 100]$ | 0 | 0 | 25 |
| $f_3$ | Zhakarov's | $[-5 : 10]$ | 0 | 0 | 1.25 |
| $f_4$ | Rastrigin's | $[-5.12 : 5.12]$ | 0 | 0 | 1.28 |
| $f_5$ | Ackley's | $[-32 : 32]$ | 0 | 0 | 8 |
| $f_6$ | Griewangk's | $[-600 : 600]$ | 0 | 0 | 150 |
| $f_7$ | Salomon's | $[-100 : 100]$ | 0 | 0 | 25 |
| $f_8$ | Kursawe's | $[-1000 : 1000]$ | 0 | 0 | 250 |
| $f_9$ | Shaffer's | $[-100 : 100]$ | 0 | 0 | 25 |
| $f_{10}$ | Levy-Montalvo's | $[-500 : 500]$ | 0 | 1 | 125 |
| $f_{11}$ | Rosenbrock's | $[-5 : 10]$ | 0 | 1 | 0 |
| $f_{12}$ | Michalewicz's | $[0 : \pi]$ | na | $\pi$ | 0 |
| $f_{13}$ | Shubert's | $[-10 : 10]$ | na | na | 0 |
| $f_{14}$ | Schwefel's | $[-500 : 500]$ | 0 | 420.97 | 0 |
| $f_{15}$ | Rana's | $[-520 : 520]$ | -512.75 | 514.04 | 0 |
| $f_{16}$ | Whitley's | $[-10 : 10]$ | 0 | 1 | 0 |

To avoid a possible bias of the algorithms towards the origin of the search space, the global optimum of the functions $f_1$-$f_{10}$ was displaced by the amount shown on table 1. Functions $f_{11}$-$f_{16}$ are also multimodal, but don't have the symmetries of the previous group. They present different difficulties, including local optima very far apart, global optima near the borders of the search space and/or surrounded by very noisy regions. Additionally, only $f_{12}$ and $f_{14}$ are separable. Functions were optimized in 40 dimensions. Table 1 lists names and parameters for the benchmark functions, including the optimum value $f(\mathbf{x}^*)$ and its position $x_i^*$. Table 2 presents the actual functions.

We tested the proposed predator-prey algorithm against PSO and differential evolution (DE) based approaches. We included DE algorithms since differential evolution shares many of the PSO advantages, namely in terms of simplicity,

**Table 2.** Benchmark functions

$$f_1(\mathbf{x}) = \sum_{i=1}^{n} x_i^2$$

$$f_2(\mathbf{x}) = \sum_{i=1}^{n} \left( \sum_{j=1}^{i} x_j \right)^2$$

$$f_3(\mathbf{x}) = \sum_{i=1}^{n} x_i^2 + \left( \sum_{i=1}^{n} 0.5 i x_i \right)^2 + \left( \sum_{i=1}^{n} 0.5 i x_i \right)^4$$

$$f_4(\mathbf{x}) = 10n + \sum_{i=1}^{n} \left( x_i^2 - 10\cos(2\pi x_i) \right)$$

$$f_5(\mathbf{x}) = -20 \exp \left( -0.2 \sqrt{\frac{1}{n} \sum_{i=1}^{n} x_i^2} \right) - \exp \left( \frac{1}{n} \sum_{i=1}^{n} \cos(2\pi x_i) \right) + 20 + e$$

$$f_6(\mathbf{x}) = \frac{1}{4000} \sum_{i=1}^{n} x_i^2 - \prod_{i=1}^{n} \frac{\cos(x_i)}{\sqrt{i}} + 1$$

$$f_7(\mathbf{x}) = 1 - \cos(2\pi \|\mathbf{x}\|) + 0.1\|\mathbf{x}\|$$

$$f_8(\mathbf{x}) = \sum_{i=1}^{n} \left( |x_i|^{0.8} + 5\sin(x_i^3) \right)$$

$$f_9(\mathbf{x}) = \sum_{i=1}^{n-1} s(x_i, x_{i+1}) + s(x_n, x_1), \quad s(x, y) = 5 + \frac{\sin^2\left(\sqrt{x^2+y^2}\right) - 0.5}{(1 + 0.001(x^2+y^2))^2}$$

$$f_{10}(\mathbf{x}) = 0.1 \left( \sin^2(3\pi x_1) + \sum_{i=1}^{n-1} (x_i - 1)^2 \left( 1 + \sin^2(3\pi x_{i+1}) \right) + (x_n - 1)^2 \left( 1 + \sin^2(2\pi x_n) \right) \right)$$

$$f_{11}(\mathbf{x}) = \sum_{i=1}^{n-1} \left( 100 \left( x_i^2 - x_{i+1} \right)^2 + (1 - x_i)^2 \right)$$

$$f_{12}(\mathbf{x}) = -\sum_{i=1}^{n} \sin(x_i) \left( \sin\left( \frac{i x_i^2}{\pi} \right) \right)^{20}$$

$$f_{13}(\mathbf{x}) = \prod_{i=1}^{n} \left( \sum_{j=1}^{5} j \cos\left((j+1)x_i + j\right) \right)$$

$$f_{14}(\mathbf{x}) = 418.98291n + \sum_{i=1}^{n} \left( -x_i^2 \sin(\sqrt{|x_i|}) \right)$$

$$f_{15}(\mathbf{x}) = \frac{\sum_{i=1}^{n-1} r(x_i, x_{i+1}) + r(x_n, x_1)}{n},$$

$$r(x, y) = x \sin\left( \sqrt{|y + 1 - x|} \right) \cos\left( \sqrt{|y + 1 + x|} \right)$$

$$+ (y + 1) \cos\left( \sqrt{|y + 1 - x|} \right) \sin\left( \sqrt{|y + 1 + x|} \right)$$

$$f_{16}(\mathbf{x}) = \sum_{i=1}^{n} \sum_{j=1}^{n} \left( \frac{(100(x_i^2 - x_j)^2 + (1 - x_j)^2)^2}{4000} - \cos\left( 100(x_i^2 - x_j)^2 + (1 - x_j)^2 \right) + 1 \right)$$

flexibility and performance. We used a recent state-of-the art hybrid PSO (HM-RPSO) variant [4], which reported very good results in the optimization of a large set of benchmark functions when compared with several other PSO variants and evolutionary algorithms. The DE approach used for comparison, which also showed promising experimental results, is called free search differential evolution (FSDE) [8]. Our implementations were first tested on the benchmarks reported in the original papers to minimize implementation discrepancies. Standard versions of the PSO and DE algorithms were also included in our experiments.

**Table 3.** Experimental results: average and standard deviations of the best values found over 50 runs of an algorithm for each function

| | PSO | DE | HMRPSO | FSDE | SPPO |
|---|---|---|---|---|---|
| $f_1$ | 1.65169e-26 | 6.25047e-23 | 3.38079e-23 | 0.242096 | **7.57306e-31** |
| | (5.54796e-26) | (4.4143e-22) | (1.4217e-23) | (0.174053) | **(3.02794e-30)** |
| $f_2$ | 316.126 | 56.2572 | 0.00226426 | 61.5534 | **2.52261e-09** |
| | (254.974) | (397.747) | (0.00212582) | (33.1706) | **(2.40497e-09)** |
| $f_3$ | 15.8846 | 0.000692142 | 5.19511e-09 | 1.13648 | **6.72887e-10** |
| | (9.53778) | (0.0016271) | (4.86953e-09) | (0.807662) | **(5.23577e-10)** |
| $f_4$ | 55.8172 | 148.852 | 17.8104 | 7.72839 | **0.0198992** |
| | (17.0791) | (84.0428) | (5.6274) | (11.3974) | **(0.140708)** |
| $f_5$ | 0.476326 | 5.47737 | 1.27917e-12 | 0.150284 | **3.96305e-14** |
| | (2.91565) | (6.9252) | (2.26036e-13) | (0.0844456) | **(8.27882e-15)** |
| $f_6$ | 0.0134364 | 0.0861482 | **0.00777161** | 0.290238 | 0.0524945 |
| | (0.0169437) | (0.282405) | **(0.0112705)** | (0.191083) | (0.0498488) |
| $f_7$ | 0.639873 | 0.389874 | 0.553874 | **0.375873** | 1.19387 |
| | (0.137024) | (0.194044) | (0.0973316) | **(0.169706)** | (0.219842) |
| $f_8$ | -41.2553 | -76.2049 | -24.9186 | 171.819 | **-150.326** |
| | (50.7622) | (67.5841) | (25.9715) | (215.74) | **(16.6553)** |
| $f_9$ | 5.15659 | 15.976 | 6.77497 | 12.0213 | **0.860774** |
| | (2.34525) | (0.391461) | (1.32032) | (1.81459) | **(0.424196)** |
| $f_{10}$ | 0.0181309 | 0.0621908 | 5.56247e-22 | 55.1298 | **5.86288e-28** |
| | (0.126894) | (0.216245) | (3.83363e-22) | (78.7154) | **(3.38157e-28)** |
| $f_{11}$ | 56.518 | 41.2635 | 37.1949 | **0.0592396** | 4.31659 |
| | (37.0099) | (42.3479) | (28.3189) | **(0.0552312)** | (13.8899) |
| $f_{12}$ | -33.667 | -16.4953 | -36.3654 | -23.9191 | **36.9508** |
| | (1.29863) | (0.953367) | (1.52178) | (2.08068) | **(1.04873)** |
| $f_{13}$ | -6.50704e+44 | -3.24599e+31 | -5.44541e+45 | -3.48103e+45 | **-7.44425e+45** |
| | (8.53809e+44) | (1.58113e+32) | (3.81935e+45) | (7.0238e+45) | **(2.21794e+45)** |
| $f_{14}$ | 3206.31 | 4523.58 | 579.052 | 2317.55 | **71.0635** |
| | (571.308) | (1102.55) | (581.897) | (1343.43) | **(92.6734)** |
| $f_{15}$ | -325.619 | -466.38 | -401.814 | -417.29 | **-475.215** |
| | (15.4024) | (19.8534) | (27.1564) | (51.9677) | **(24.1802)** |
| $f_{16}$ | 357.564 | 404.922 | 181.022 | **3.3508** | 44.7104 |
| | (248.078) | (264.003) | (196.041) | **(18.8189)** | (68.9411) |

The parameters for the algorithms are the ones proposed by the respective authors. For the scouting predator-prey optimizer we used $\phi_1 = \phi_2 = 1.6$, $w$ was decreased from 0.4 to 0.2 and $r = 0.0008$. Population size was set to 20 and the algorithms were run for 1e4 iterations or until 2e5 objective function evaluations were performed. In table 3 we present averages and standard deviations for the best values found for each test function, taken over 50 runs of each pair algorithm/function. The random number generator was initiated to ensure that all algorithms started with the same population for corresponding runs.

From the results in table 3 we can conclude that the SPPO algorithm obtained the best average results in 12 of the 16 benchmark functions. The FSDE algorithm obtained best results for 3 of the benchmark function and the HM-RPSO algorithm obtained the best average result for only 1 function. It should be noted that the hybrid algorithms all performed better than the corresponding base algorithms (PSO and DE) in almost all of the benchmark functions. In the group of unimodal functions, the SPPO found better average results for all of the 3 functions, indicating the faster convergence rate of all the tested algorithms. For the group of multi-modal functions with significant symmetries ($f_4$-$f_{10}$), the SPPO algorithm performed better for 5 of the 7 functions, being outperformed by both FSDE and HMRPSO in one of the remaining functions. In the last group of multimodal functions, again the SPPO algorithm achieved better results in the majority of the functions (4 out of 6), with the FSDE algorithm obtaining best average results for the remaining 2. If we take into account only the 9 non-separable functions, the SPPO loses some of its supremacy, but it still obtains better results for 5 benchmark functions, against FSDE with 3 functions and HMRSO, which achieves the best average value for just one function.

## 4   Conclusions

In this paper we described a new heterogeneous particle swarm optimizer algorithm and compared it experimentally with state of the art variations of both particle swarm and differential evolution optimizers. The new heterogeneous PSO uses a two-fold approach to improve the performance of the standard particle swarm optimizer. A predator particle, and its interaction with the swarm, is used to oppose premature convergence and control the balance between exploration and exploitation. Scout particles are used to add new exploratory behaviors to the algorithm, without perturbing the general dynamic of the swarm.

The algorithms were tested using a set of benchmark functions that included unimodal, multimodal with strong symmetries, multimodal with different difficulties, separable and non-separable functions. The experimental results, where the SPPO found better average results for 12 out 16 benchmark functions, suggest that the new algorithm can be competitive over a large set of very different optimization problems. Both the ease with which new behaviors were added to the swarm, and the experimental results obtained, illustrate the power of heterogeneous particle swarm algorithms in general and of the use of scout particles in particular.

# References

1. Angeline, P.J.: Evolutionary Optimization Versus Particle Swarm Optimization: Philosophy and Performance Differences. In: Porto, V.W., Waagen, D. (eds.) EP 1998. LNCS, vol. 1447, pp. 601–610. Springer, Heidelberg (1998)
2. Beyer, H.-G., Schwefel, H.-P.: Evolution strategies - a comprehensive introduction. Natural Computing 1, 3–52 (2002)
3. Engelbrecht, A.: Heterogeneous Particle Swarm Optimization. In: Dorigo, M., Birattari, M., Di Caro, G.A., Doursat, R., Engelbrecht, A.P., Floreano, D., Gambardella, L.M., Groß, R., Şahin, E., Sayama, H., Stützle, T. (eds.) ANTS 2010. LNCS, vol. 6234, pp. 191–202. Springer, Heidelberg (2010)
4. Gao, H., Xu, W.: A new particle swarm algorithm and its globally convergent modifications. IEEE Transactions on Systems, Man, and Cybernetics, Part B: Cybernetics 41(5), 1334–1351 (2011)
5. Gao, H., Xu, W.: Particle swarm algorithm with hybrid mutation strategy. Applied Soft Computing 11(8), 5129–5142 (2011)
6. Kennedy, J., Eberhart, R.: Particle swarm optimization. In: Proceedings. IEEE International Conference on Neural Networks, vol. 4, pp. 1942–1948 (1995)
7. Montes de Oca, M., Pena, J., Stutzle, T., Pinciroli, C., Dorigo, M.: Heterogeneous particle swarm optimizers. In: IEEE Congress on Evolutionary Computation, CEC 2009, pp. 698–705 (May 2009)
8. Omran, M.G.H., Engelbrecht, A.P.: Free search differential evolution. In: Proceedings of the Eleventh Conference on Congress on Evolutionary Computation, CEC 2009, pp. 110–117. IEEE Press, Piscataway (2009)
9. Petalas, Y., Parsopoulos, K., Vrahatis, M.: Memetic particle swarm optimization. Annals of Operations Research 156, 99–127 (2007)
10. Poli, R.: Analysis of the publications on the applications of particle swarm optimisation. J. Artif. Evol. App., 4:1–4:10 (January 2008)
11. Poli, R., Kennedy, J., Blackwell, T.: Particle swarm optimization. Swarm Intelligence 1, 33–57 (2007)
12. Rahnamayan, S., Tizhoosh, H., Salama, M.: Opposition-based differential evolution. IEEE Transactions on Evolutionary Computation 12(1), 64–79 (2008)
13. Shi, Y., Eberhart, R.: Empirical study of particle swarm optimization. In: Proceedings of the 1999 Congress on Evolutionary Computation, CEC 1999, vol. 3, p. 3 vol. (xxxvii+2348) (1999)
14. Silva, A., Neves, A., Costa, E.: An Empirical Comparison of Particle Swarm and Predator Prey Optimisation. In: O'Neill, M., Sutcliffe, R.F.E., Ryan, C., Eaton, M., Griffith, N.J.L. (eds.) AICS 2002. LNCS (LNAI), vol. 2464, pp. 103–110. Springer, Heidelberg (2002)
15. Storn, R., Price, K.: Differential evolution – a simple and efficient heuristic for global optimization over continuous spaces. Journal of Global Optimization 11, 341–359 (1997)
16. Tizhoosh, H.R.: Opposition-based learning: A new scheme for machine intelligence. In: Proceedings of the International Conference on Computational Intelligence for Modelling, Control and Automation, CIMCA 2005, vol. 01, pp. 695–701. IEEE Computer Society Press, Los Alamitos (2005)
17. Wang, H., Li, H., Liu, Y., Li, C., Zeng, S.: Opposition-based particle swarm algorithm with cauchy mutation. In: IEEE Congress on Evolutionary Computation, CEC 2007, pp. 4750–4756 (September 2007)

# Job Shop Scheduling with Transportation Delays and Layout Planning in Manufacturing Systems: A Multi-objective Evolutionary Approach

Kazi Shah Nawaz Ripon, Kyrre Glette, Mats Hovin, and Jim Torresen

Department of Informatics, University of Oslo, Norway
{ksripon,kyrrehg,matsh,jimtoer}@ifi.uio.no

**Abstract.** The job shop scheduling problem (JSSP) and the facility layout planning (FLP) are two important factors influencing productivity and cost-controlling activities in any manufacturing system. In the past, a number of attempts have been made to solve these stubborn problems. Although, these two problems are strongly interconnected and solution of one significantly impacts the performance of other, so far, these problems are solved independently. Also, the majority of studies on JSSPs assume that the transportation delays among machines are negligible. In this paper, we introduce a general method using multi-objective genetic algorithm for solving the integrated problems of the FLP and the JSSP considering transportation delay having three objectives to optimize: makespan, total material handling costs, and closeness rating score. The proposed method makes use of Pareto dominance relationship to optimize multiple objectives simultaneously and a set of non-dominated solutions are obtained providing additional degrees of freedom for the production manager.

## 1 Introduction

Scheduling exists almost everywhere in real-world manufacturing situations. It can be defined as the assigning of shared resources (machines, people, etc) efficiently over time to competing activities (operations) in such a manner that some goals can be achieved economically and a set of constraints can be satisfied. The job shop scheduling problem (JSSP), a popular model in scheduling theory, is widely acknowledged as one of the most difficult NP-hard problems [1]. The JSSP can be characterized as one in which a finite set $\{J_j\}_{1 \leq j \leq n}$ of $n$ independent jobs, each comprising a specified sequence of operations to be performed on a finite set $\{M_k\}_{1 \leq k \leq m}$ of $m$ machines and requiring certain amounts of time, are to be processed. The target is to find the optimum processing sequence of jobs so that one or more performance measure(s) can be optimized. The JSSP has been widely studied over the last four decades, and has great importance to manufacturing industries with an objective to minimize the production cost.

In a schedule for any particular job, a time delay between two successive operations is necessary for transporting materials from one machine to another, especially if the machines are not adjacent. This fact stems mainly from the

M. Kamel, F. Karray, and H. Hagras (Eds.): AIS 2012, LNCS 7326, pp. 209–219, 2012.

observation that in real-world manufacturing industries, particularly producing several products simultaneously, time delays between consecutive operations on different machines are very usual. However, in most JSSPs, transportation delay is neglected while generating the final schedule. In this work, we deal with a variant of the JSSP considering transporting delay.

A facility is an entity in a manufacturing system such as a machine tool, a department, a workstation, a manufacturing cell, or a warehouse, which assists in a dedicated task and makes it possible to produce goods easily. The facility layout planning (FLP) is concerned with the most efficient physical arrangement of a number of interacting facilities on the factory floor of a manufacturing system in order to meet one or more objectives. Traditionally, the FLP has been presented as a quadratic assignment problem (QAP), which assigns $p$ facilities to $p$ equal area locations with the constraint that each facility is restricted to one location and one facility should choose only one location. Layout planning in a manufacturing company is an important economical consideration, and it is NP-hard as well [2]. A proper layout will help any company improve its business performance and can reduce the total operating expenses by 50% [3].

Both JSSPs and FLPs are inevitable in industrial plants. They appear frequently in manufacturing industries, as well as engineering design contexts, including automobile, chemical, computer, semiconductor, printing, pharmaceutical and construction industries. Recognizing their importance, a diverse spectrum of researchers ranging from management scientists to production workers has developed a wide variety of approaches in their search for fast and optimal solutions to these problems. A comprehensive survey of the current research work on JSSPs and FLPs can be found in [4],[5]. Traditionally, JSSPs and FLPs are performed independently, where JSSPs are executed separately after layouts have been generated. Therefore, it is possible that layouts so generated may not be optimal from the scheduling point of view, or vice versa. However, considering the fact that these two problems are strongly dependent on each other, it is necessary to integrate them more tightly so that the global optimization for the entire manufacturing system can be achieved.

Targeting the autonomy and global optimization of manufacturing environment, the adaptation of the new production process has brought an urgent need to study and develop integrated FLP and JSSP approaches so that the placement of facilities (where the machines will be placed) and the job shop assembly operations assigned to these facilities can be co-optimized as much as possible. Since the requests of job orders deeply affect the transportation sequence of materials among the facilities, finding an optimum layout for shared resources is very important. In practice, the choice of layout for facilities and the scheduling of jobs significantly impact the performance of the other [6]. The importance of an integrated JSSP and FLP involves reducing costs and responding quickly to customer demands to stay alive in the intense competition of global market. There are several other advantages of this integration, for example maximizing the profitability, flexibility, and productivity of manufacturing plant, as well as the performance of a production system.

FLPs and JSSPs are computationally difficult tasks. In a $p$–facility layout problem, we would have to evaluate $p!$ different layouts. For the standard JSSP of $n$–jobs and $m$–machines, the search space will be $(n!)^m$. Within the context of an integrated JSSP and FLP, the size of the search space is $(n!)^m p!$. Accordingly, it is computationally infeasible to try every possible solution. For solving problems of such combinatorial nature, optimal algorithms require a large computational effort and extensive memory requirements, even when the problem size is small. For this reason, there has been much research on heuristic and meta-heuristic approaches to find near-optimal solutions. Comprehensive surveys on these approaches are available in [4],[7]. Among these approaches, the genetic algorithm (GA) [8] has found wide application in research intended to solve the combinatorial optimization problems. Generally speaking, GAs outperform other heuristic and meta-heuristic methods due to their ability to generate feasible solutions in a minimum amount of time [5],[7].

Real-world manufacturing systems are multi-objective by nature and they require the decision maker to consider a number of criteria before arriving at any conclusion. A solution that is optimal with respect to a given criterion might be a poor candidate for where another is paramount. Hence, the trade-offs involved in considering several criteria provide useful insights for the decision maker. It is only recently that some attempts have been made for the multi-objective industrial problems [2]. However many approaches usually optimize a single objective and treat other objectives as constraints [9]. Alternatively, they combine multiple objectives into a single objective function using a weighted linear combination of all objectives and then a single-objective optimization algorithm is used to find a single solution at a time [2]. However, relative preferences require prior domain knowledge and the solution quality is sensitive to the relative preferences used. To cope with this, Pareto-optimality has become an alternative. In the presence of multiple objectives, there is no single "optimum" solution and the resulting optimization problem gives rise to a set of approximately efficient solutions — Pareto-optimal solutions. A solution is called Pareto-optimal (non-dominated) if it is not possible to improve the value of one objective without worsening the value of the other. From a system designer's point of view, it is very desirable to obtain a set of non-dominated solutions providing the flexibility of considering all the objectives simultaneously.

All these motivate us to propose a method for solving the integrated problems of JSSP and FLP using multi-objective GA that presents the final solutions as a set of Pareto-optimal solutions optimizing multiple objectives simultaneously. Additionally, we consider transportation delay for the JSSP to reflect the real-world manufacturing scenario. The solution would answer two questions simultaneously: the schedule of each job and the layout for facilities (machines) considering the scheduling of jobs among the facilities; with the goals of minimizing the makespan (maximum completion time of all jobs) for the JSSP, as well as minimizing the total material handling (MH) cost (sum of the distances between all facilities multiplied by the corresponding flows) and maximizing the closeness rating (CR) score (the desired relative "closeness" requirement for two facilities to be next to each other) for the FLP.

## 2    The Model

There are basically two combinatorial optimization problems that we need to consider for the proposed method:

1. Layout for the facilities (where the machines will be placed) which is dependent on the schedule.
2. Schedules for the jobs which are dependent on the corresponding layout.

Throughout this paper, we use facility and machine interchangeably. For obvious reason, the number of machines will be equal to the number of facilities. The main assumptions for the JSSP are described as follows:

- The processing of job $J_j$ on machine $M_k$ is called the operation $O_{jk}$.
- Every job has a unique sequence of operations on machines — it's technological sequence.
- All jobs are available for processing at time zero.
- Operation $O_{jk}$ can be processed by only one machine $M_k$ at a time.
- Each machine performs operations one after another.
- Every operation $O_{jk}$ which has begun on a machine $M_k$, must be uninterrupted for the duration $t_{jk}$ — it's processing time.
- The machines are not identical and perform different operations.

**Table 1.** Problem instance for a $6 \times 6$ JSSP with (machine, processing time) pair

| Job-n | (k,t) | (k,t) | (k,t) | (k,t) | (k,t) | (k,t) |
|-------|-------|-------|-------|-------|-------|-------|
| Job-1 | 3,1 | 1,3 | 2,6 | 4,7 | 6,3 | 5,6 |
| Job-2 | 2,8 | 3,5 | 5,1 | 6,1 | 1,1 | 4,4 |
| Job-3 | 3,5 | 4,4 | 6,8 | 1,9 | 2,1 | 5,7 |
| Job-4 | 2,5 | 1,5 | 3,5 | 4,3 | 5,8 | 6,9 |
| Job-5 | 3,9 | 2,3 | 5,5 | 6,4 | 1,3 | 4,1 |
| Job-6 | 2,3 | 4,3 | 6,9 | 1,1 | 5,4 | 3,1 |

As described earlier, almost all existing JSSP approaches assume that the transportation delay among consecutive machines is negligible. However, this assumption is impractical for real-world manufacturing systems where JSSPs and FLPs are closely related. We can describe this with an arbitrary integrated example of a $6 \times 6$ JSSP (Table 1) and the corresponding example of a $3 \times 2$ FLP (Fig. 1). In this JSSP, the first two consecutive operations of job–1 require the transportation of materials from machine–3 to machine–1. However, the layout presented in Fig. 1 shows that the facilities, where these machines (3 & 1, within circle) are located, are far apart. Therefore, it is natural that there must be some transportation delays between two consecutive operations, at least if they are processed by two machines which are not adjacent. Furthermore, for layouts

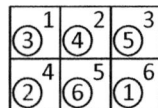

**Fig. 1.** An example for $3 \times 2$ FLP

**Fig. 2.** An example for $3 \times 5$ FLP

with more facilities like the $3 \times 5$ one presented in Fig. 2, the transportation delay is a key factor affecting the performance of the JSSP a lot. Consequently, it affects the performance of the manufacturing systems.

Simultaneously, solutions of JSSPs also influence the final layouts. We can describe this using the solution of above JSSP (Table 1) presented in Fig. 3. It is a traditional schedule without considering transportation delays between distant facilities. In practice, it is beneficial to place the machines adjacently on which the consecutive operations of the same job are scheduled immediately without any time delay. From Fig. 3, we can find that the first two operations for job–1 are scheduled instantly from machine–3 to machine–1. If these two machines are placed together, the MH cost will be reduced significantly. This is because the distance between these two facilities will be reduced. For every job, we can find similar pairs of machines. Accordingly, we can find a matrix which will give the priority for placing every pair of facilities. Fig. 4 presents the priority matrix (PM) for the JSSP solution presented in Fig. 3. A cell without any value in the matrix indicates that there is not any such priority. There will be different PMs corresponding to different solutions of the same JSSP. That is why, the design of the layout is also dependent on the JSSP solution.

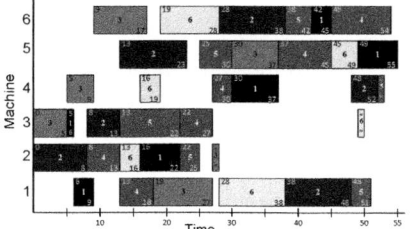

**Fig. 3.** Solution for the JSSP shown in Table 1

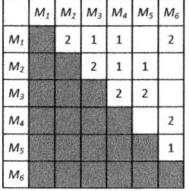

|       | $M_1$ | $M_2$ | $M_3$ | $M_4$ | $M_5$ | $M_6$ |
|-------|-------|-------|-------|-------|-------|-------|
| $M_1$ |       | 2     | 1     | 1     |       | 2     |
| $M_2$ |       |       | 2     | 1     | 1     |       |
| $M_3$ |       |       |       | 2     | 2     |       |
| $M_4$ |       |       |       |       |       | 2     |
| $M_5$ |       |       |       |       |       | 1     |
| $M_6$ |       |       |       |       |       |       |

**Fig. 4.** Corresponding PM for the solution presented in Fig. 3

## 3   Solution Methodology

In this method, we use the non-dominated sorting genetic algorithm 2 (NSGA-2) [10] as the multi-objective GA. The crowding-distance assignment and the non-dominated sorting strategy to achieve the elitisms for the next generation are the same as in NSGA-2.

## 3.1   Chromosome Representation

In this work, chromosomes are composed of two parts. In the first part (for the JSSP), an indirect representation interpreted by the schedule builder is applied. The genes for this portion are represented as a string of integers of length $m \times n$, where each job integer ($n$ job) is repeated $m$ times ($m$ machines). This representation is known as "permutation with repetition" [11]. By scanning the permutation from left to right, the $k$–th occurrence of a job number refers to the $k$–th operation in the technological sequence of this job as depicted in Fig. 5. The advantage of this representation is that it requires a very simple schedule builder because all generated schedules are legal.

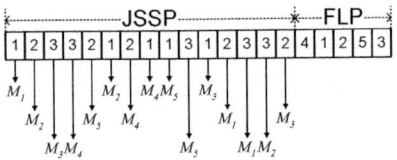

**Fig. 5.** Chromosome for a 3–job, 5–machine, and 5-facility problem

For the FLP part, a direct representation is used. The solution is represented as a string of integers of length $p$ (no. of facilities). The integers denote the facilities and their positions in the string denote the positions of the facilities in the layout. As mentioned earlier, the number of facilities will be equal to the number of machines. Fig. 5 presents a chromosome for a 3–job, 5–machine, and 5–facility problem (the technological sequence of each job is shown arbitrarily).

## 3.2   Schedule Builder

In indirect representation for the JSSP, the chromosome encodes a sequence of preferences (simple ordering of jobs in a machine). Hence, a schedule builder is required to decode the chromosome into a schedule. The schedule builder also performs the evaluation procedure and should be chosen with respect to the performance measure of optimization. In this method, we employed a modified Giffler and Thompson (G&T) algorithm [12] to build the schedule. The schedules built by this version of algorithm are active schedules. A schedule is active if no *permissible left shift* can be applied to that schedule. A permissible left shift is the reassigning of an operation to the left without delaying other jobs to reduce the makespan of any schedule. Active schedules are good in average and at the same time an optimal schedule is always active so the search space can be safely limited to the set of all active schedules. Here, we modified the G&T algorithm so it takes into account the transportation delays considering the positions of facilities in the layout constructed from the same chromosome. If the consecutive operations are to be processed by the facilities which are not adjacent, transportation delays are added before starting the next operation.

## 3.3   Formulation of the Objective Functions

We use three objectives simultaneously: one for the JSSP and two for the FLP. Each objective is calculated considering only the genes of the respective portion of the chromosome (as shown in Fig. 5). In the proposed method, however, each objective is subject to the formulation of its counterpart.

For JSSPs, minimization of makespan is considered as the main optimization criterion [1]. It can be expressed by the following mathematical model where $C_i$ is the completion time of job–$i$ incorporating the transportation delay between two distant facilities considering the complete solution for the integrated problem.

$$Makespan = max[C_i] \tag{1}$$

Minimization of the total MH cost is the most common objective in FLPs. However, the CR score which aims at placing facilities that utilize common materials, personnel, or utilities adjacent to one another, while separating facilities for the reasons of safety, noise, or cleanliness is also important in FLPs [13]. We modified the mathematical modeling of total MH costs considering the adjacency of the facilities on which consecutive operations of the same job are scheduled immediately. It can be found by the corresponding JSSP portion of the same solution.

$$Total\ MH\ costs = \sum_{i=1}^{p}\sum_{j=1}^{p}\sum_{k=1}^{p}\sum_{l=1}^{p} f_{ik} W_{ijkl} X_{ij} X_{kl} \tag{2}$$

$$CR\ scores = \sum_{i=1}^{p}\sum_{j=1}^{p}\sum_{k=1}^{p}\sum_{l=1}^{p} C_{ijkl} X_{ij} X_{kl} \tag{3}$$

subject to

$$\sum_{i=1}^{p} X_{ij} = 1, \quad j = 1, 2, \ldots, p \tag{4}$$

$$\sum_{j=1}^{p} X_{ij} = 1, \quad i = 1, 2, \ldots, p \tag{5}$$

$$W_{ijkl} = \begin{cases} R_{ik}\ ; & \text{if } PM \text{ has an entry for facilities } i \text{ and } k \\ d_{jl}\ ; & \text{otherwise} \end{cases} \tag{6}$$

$$R_{ik} = \begin{cases} d_{jl}\ ; & \text{if facilities } i \text{ and } k \text{ are adjacent} \\ PM_{ik} d_{jl}\ ; & \text{otherwise} \end{cases} \tag{7}$$

$$C_{ijkl} = \begin{cases} cl_{ik}\ ; & \text{if locations } j \text{ and } l \text{ are neighbors} \\ 0\ ; & \text{otherwise} \end{cases} \tag{8}$$

Where, $i$, $k$ are facilities; $j$, $l$ are locations in the layout; $f_{ik}$ is the flow from facility $i$ to $k$; $d_{jl}$ is the distance from location $j$ to $l$; $X_{ij} = 1$ or $0$ for locating $i$ at $j$; $PM_{ik}$ is the value in PM, $cl_{ik}$ is the CR value when facilities $i$ and $k$ are neighbors with common boundary, and $p$ is the number of facilities in the layout. The numerical values used for the CR values are: absolutely necessary=6, essentially important=5, important=4, ordinary=3, un-important=2, and undesirable=1.

### 3.4   Genetic Operators

The chromosome representation in this approach is different from that of the conventional GA. As a result, the direct application of traditional genetic operators may create an illegal solution. Hence, some problem-specific genetic operators are required. To apply crossover, first we randomly choose the crossover points for the two parent chromosomes with the restriction that both points should be on the same portion. Based on the crossover points, we apply two types of crossover operation: (i) the improved precedence preservation crossover (IPPX) [1] for the JSSP portion, and (ii) the crossover operation proposed by Suresh et al. [14] for the FLP portion. In the traditional permutation based crossover techniques for JSSPs, there exist redundancies at the tail of a chromosome. A schedule, decoded from a chromosome sequence of redundant genes, has no effect on the final solution. It also imposes additional time complexity. The IPPX resolves this issue and also reduces the execution time. The details of IPPX can be found in [1]. For the FLP portion, we follow the crossover technique described in [14]. This crossover technique maintains the partial structure of parents to a larger extent than by the existing crossover operations such as PMX, OX and CX [13].

For mutation, we use swap mutation with the same restriction of selecting two genes randomly from the same portion of a chromosome. If the genes are from the FLP part, we use traditional swap mutation; while in the case of the JSSP portion, we use swap mutation with additional restriction of choosing two non-identical genes.

## 4   Performance Evaluation

### 4.1   Description of Test Data

There are no benchmark datasets published for the combined areas of FLP and JSSP considering the transportation delay, even without transportation delay, as of now. In order to test our proposed method, we have created test problems by combining the data sets available in the literature for JSSPs and FLPs. The problem data for the JSSP are *mt06*, *abz7*, *tai02*, *tai10*, *la38*, *la39*, and *la40*. The description of these data is available in [15]. We choose test problems for the FLP (*nug6*, *nug15*) from published literature [13] to match the number of machines with the number of facilities. Due to the lack of practical data, we set the transportation delay to 1 unit whenever necessary.

### 4.2   Test Setup

The experiments are conducted using population of 200 chromosomes and 400 generations. The probabilities of crossover and mutation are [0.7, 0.8, 0.9] and [0.1, 0.2, 0.3], respectively. We use traditional tournament selection with tournament size of 2. Each problem is tested for 30 times with different seeds and different (random) combinations of crossover and mutation rates. Then each of the final generations is combined and a non-dominated sorting is performed to produce the final non-dominated solutions.

**Table 2.** Results for multiple objectives ($J$ = jobs, $M$ = machines, $F$ = facilities)

| JSSP | J | M | FLP | F | Makespan, LB (w/o tr. time) | MH Cost, best known | Makespan | | MH Cost | | CR Score | |
|------|---|---|-----|---|------|------|------|------|------|------|------|------|
| | | | | | | | Best | Avg | Best | Avg | Best | Avg |
| $mt6$ | 6 | 6 | $nug6$ | 6 | 55 | 43 | 58 | 62.6 | 46 | 49.2 | 44 | 40.60 |
| $tai02$ | 15 | 15 | $nug15$ | 15 | 1244 | 575 | 1263 | 1327.6 | 583 | 702.05 | 168 | 156.25 |
| $tai10$ | 15 | 15 | $nug15$ | 15 | 1241 | 575 | 1260 | 1318.2 | 580 | 691.5 | 166 | 156.40 |
| $abz7$ | 20 | 15 | $nug15$ | 15 | 654 | 575 | 681 | 716.2 | 580 | 690.25 | 184 | 174.65 |
| $la38$ | 15 | 15 | $nug15$ | 15 | 1184 | 575 | 1209 | 1285.1 | 586 | 704.3 | 166 | 154.85 |
| $la39$ | 15 | 15 | $nug15$ | 15 | 1233 | 575 | 1250 | 1326.8 | 583 | 707.5 | 170 | 156.75 |
| $la40$ | 15 | 15 | $nug15$ | 15 | 1222 | 575 | 1238 | 1381.7 | 584 | 706.2 | 168 | 156.20 |

## 4.3  Experimental Analysis

The values provided in Table 2 show the best and average values obtained by the proposed method in the context of makespan, total MH costs, and CR score. This table also shows the lower bounds for the JSSP benchmark problems. Since, all these benchmark problems assume that there is no transportation delay, it is not reasonable to compare with the lower bounds. However, we mention the lower bound as a reference. As mentioned earlier, we solve the JSSP and the FLP as an integrated manufacturing design problem. Thus, while computing the makespan, we take into account the transportation delay between two distant facilities considering the corresponding layout obtained by the same chromosome. Therefore, the makespan may be larger that the usual. Even so, the makespan obtained by our proposed approach is satisfactory enough.

Similarly, the total MH cost is also dependent on the corresponding schedule obtained by the same chromosome. Accordingly, the total MH cost may be a little higher than the best known values achieved by some existing FLP approaches [13]. Despite that it can be seen in Table 2 that the total MH costs obtained by the proposed method are also satisfactory enough. This table also shows the performance statistics of the proposed method in the context of CR score. However, for the lack of existing results, we are unable to present any comparison for CR score. It is worthwhile to mention that, all the existing approaches are designed for either JSSPs or FLPs, and the main goal of our proposed method is to find trade-off solutions for the integrated problems of JSSPs and FLPs optimizing multiple objectives. Also, according to the Pareto-optimal theory, the final and average values may be influenced by the presence of other objectives. Considering all, the overall performance of the proposed method is very promising.

Fig. 6 demonstrates the convergence behavior of the proposed method over generations for $la38$. These figures also justify that this method clearly optimizes all the objectives with generations. From the figures, it can be found that from first generations to last generations, the proposed method is able to minimize makespan and total MH cost, as well as to maximize CR score successfully.

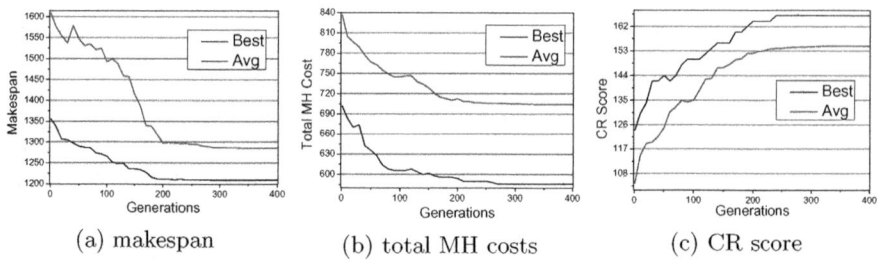

(a) makespan          (b) total MH costs          (c) CR score

**Fig. 6.** Optimization of objectives over generations for *la*38

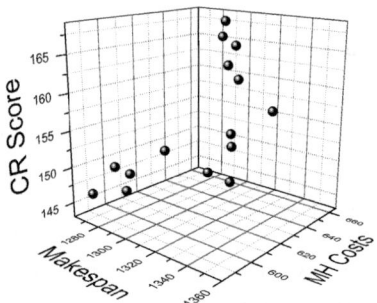

**Fig. 7.** Final solutions for *tai*02

To further demonstrate the convergence and diversity of the final solutions, non-dominated solutions of the final generation produced by the proposed method for *tai*02 is presented in Fig. 7. It should be mentioned that in all cases, many of the final solutions are Pareto-optimal. In the figure, the occurrences of the same non-dominated solutions are plotted only once. From the figure, it can be observed that the final solutions are well spread and convergent. Also, the proposed method is capable of finding extreme solutions. Thus, it can provide a wide range of alternative choices for the designers. Therefore, the designers have the flexibility of choosing among the trade-off solutions via simultaneously considering all the objectives and taking decision based on the current requirements.

## 5   Conclusion

Traditionally, layout planning and scheduling are performed sequentially in manufacturing system, where scheduling is executed after layout for facilities are designed. In practice, the choice of layout for facilities and the scheduling of jobs significantly impact the performance of each other, and their coordination is essential to achieve global optimization for the entire production process. Still these two combinatorial optimization problems are handled independently. Then again, transportation delay between two consecutive operations is neglected while solving JSSPs. This paper presents a method using multi-objective genetic algorithm for solving the FLP and the JSSP considering transportation delay as an integrated problem and presents the final solutions as a set of Pareto-optimal solutions, which better represent the real-world manufacturing scenario. The experimental results justify that this method is capable of finding a set of non-dominated solutions optimizing multiple objectives simultaneously. Thus, it provides a wide range of alternative choices, allowing decision makers to be more flexible and to make better decisions based on market circumstances. Future research may be on the performance of this method to solve real-world case studies, as well as the use of different objective functions to investigate its performances in different manufacturing situations.

# References

1. Ripon, K.S.N., Siddique, N., Torresen, J.: Improved Precedence Preservation Crossover for Multi-Objective Job Shop Scheduling Problem. Evol. Syst. 2(2), 119–129 (2011)
2. Sarker, R., Ray, T., Fonseca, J.: An Evolutionary Algorithm for Machine Layout and Job Assignment Problems. In: 2007 IEEE Congress on Evolutionary Computation (CEC 2007), pp. 3991–3997 (2007)
3. Tompkins, A.: Facilities Planning, 2nd edn. John Wiley & Sons, New York (2003)
4. Lawrynowic, A.: A Survey of Evolutionary Algorithms for Production and Logistics Optimization. Res. Logist. 1(2), 57–91 (2011)
5. Drira, A., Pierreval, H., Hajri-Gabouj, S.: Facility Layout Problems: A Survey. Annu. Rev. Control. 31(2), 255–267 (2007)
6. Wang, L., Keshavarzmanesh, S., Feng, H.-Y.: A Hybrid Approach for Dynamic Assembly Shop Floor Layout. In: 2010 IEEE Conference on Automation Science and Engineering (CASE), pp. 604–609 (2010)
7. Aytug, H., Khouja, M., Vergara, F.E.: Use of Genetic Algorithms to Solve Production and Operations Management Problems: A Review. Int. J. Prod. Res. 41(17), 3955–4009 (2003)
8. Holland, J.: Adaptation in Natural and Artificial Systems. University of Michigan Press, Ann Arbor (1975)
9. Grieco, A., Semeraro, Q., Tolio, T.: A Review of Different Approaches to the FMS Loading Problem. Int. J. Flex. Manuf. Sys. 13(4), 361–384 (2001)
10. Deb, K., Pratap, A., Agarwal, S., Meyarivan, T.: A Fast and Elitist Multiobjective Genetic Algorithm: NSGA-II. IEEE T. Evolut. Comput. 6(2), 182–197 (2002)
11. Bierwirth, C.: A Generalized Permutation Approach to Job Shop Scheduling with Genetic Algorithms. OR Spectrum 17, 87–92 (1995)
12. Varela, R., Serrano, D., Sierra, M.: New Codification Schemas for Scheduling with Genetic Algorithms. In: Mira, J., Álvarez, J.R. (eds.) IWINAC 2005, Part II. LNCS, vol. 3562, pp. 11–20. Springer, Heidelberg (2005)
13. Ripon, K.S.N., Glette, K., Mirmotahari, O., Høvin, M., Tørresen, J.: Pareto Optimal Based Evolutionary Approach for Solving Multi-Objective Facility Layout Problem. In: Leung, C.S., Lee, M., Chan, J.H. (eds.) ICONIP 2009, Part II. LNCS, vol. 5864, pp. 159–168. Springer, Heidelberg (2009)
14. Suresh, G., Vinod, V.V., Sahu, S.: A Genetic Algorithm for Facility Layout. Int. J. Prod. Res. 33(12), 3411–3423 (1995)
15. OR-Library, http://people.brunel.ac.uk/~mastjjb/jeb/info.html

# Adapting Strategies to Opponent Models in Incomplete Information Games: A Reinforcement Learning Approach for Poker

Luís Filipe Teófilo[1,2], Nuno Passos[2],
Luís Paulo Reis[1,3], and Henrique Lopes Cardoso[1,2]

[1] LIACC – Artificial Intelligence and Computer Science Lab., University of Porto, Portugal
[2] FEUP – Faculty of Engineering, University of Porto, DEI, Portugal
[3] EEUM – School of Engineering, University of Minho, DSI, Portugal
{luis.teofilo,ei08029,hlc}@fe.up.pt,
lpreis@dsi.uminho.pt

**Abstract.** Researching into the incomplete information games (IIG) field requires the development of strategies which focus on optimizing the decision making process, as there is no unequivocal best choice for a particular play. As such, this paper describes the development process and testing of an agent able to compete against human players on Poker – one of the most popular IIG. The used methodology combines pre-defined opponent models with a reinforcement learning approach. The decision-making algorithm creates a different strategy against each type of opponent by identifying the opponent's type and adjusting the rewards of the actions of the corresponding strategy. The opponent models are simple classifications used by Poker experts. Thus, each strategy is constantly adapted throughout the games, continuously improving the agent's performance. In light of this, two agents with the same structure but different rewarding conditions were developed and tested against other agents and each other. The test results indicated that after a training phase the developed strategy is capable of outperforming basic/intermediate playing strategies thus validating this approach.

**Keywords:** Incomplete Information Games, Opponent Modeling, Reinforcement Learning, Poker.

## 1 Introduction

One of the fields with large focus on AI research is games. Because games have a limited set of well-defined rules, studying them allows for easy testing of new approaches making it possible to accurately measure their degree of success. This is done by comparing results of many games played against programs based on other approaches or against human players, meaning that games have a well-defined metric for measuring the development progress [1]. It is then possible to determine with more accuracy whether if the solution is optimal to solve a given problem. Also, the fact that games have a recreational dimension and present an increasing importance for the entertainment industry today motivates further research on this domain.

M. Kamel, F. Karray, and H. Hagras (Eds.): AIS 2012, LNCS 7326, pp. 220–227, 2012.
© Springer-Verlag Berlin Heidelberg 2012

Remarkable results were achieved in games research, such as the well-known Deep Blue Computer, which was the first computer to ever defeat a human chess champion [2]. However such success has not yet been achieved for incomplete information games. This is because that their game state is not fully visible which means that there are hidden variables/features. Therefore decision making in these games is more difficult, because predictions about the missing data must be made. This makes it almost impossible to obtain an optimal solution. Poker is a very popular game that presents these characteristics because players do not know the cards of their opponents.

The research on Computer Poker has been active in the past years. Several Poker playing agents were developed but none of them has reached a level similar to a expert human player. In order to overcome the limitations found in previously developed agents, a new agent has been developed. This approach tries to mimic human players by combining opponent models used by expert players and a reinforcement learning method. The usage of reinforcement learning in the conception of the agent's strategy allowed for good adaption of the agent to several pre-defined opponent types. Two different versions of this agent were implemented and they differ (only) in what regards the reward calculation.

The rest of the paper is organized as follows. Section 2 briefly shows this paper's background by presenting the game of Poker and basic opponent modeling. Section 3 describes related work, with particular emphasis on Computer Poker. Section 4 presents this paper's approach to create a Poker agent. Section 5 describes the validation process of this approach by indicating the experimental procedure and results. Finally, some conclusions are drawn and future research recommendations are as well suggested in section 6.

# 2    Background

Poker is a generic name for hundreds of games with similar rules [3], which are called variants. This work is focused on a simplified version of the Texas Hold'em variant, which is probably the most popular nowadays. Hold'em rules also have specific characteristics that allow for new developed methodologies to be adapted to other variants with reduced effort [4].

The game is based upon the concept of players betting that their current hand is stronger than the hands of their opponents. All bets throughout the game are placed in the pot and, at the end of the game, the player with the highest ranked hand wins. Alternatively, it is also possible to win the game by forcing the opponents to fold their hands by making bets that they are not willing to match. Thus, since the opponents' cards are hidden it is possible to win the game with a hand with lower score This is done by bluffing - convincing the opponents that one's hand is the highest ranked one.

## 2.1    Hand Ranking

A Poker hand is a set of five cards that defines the player's score. The hand rank at any stage of the game is the score given by the 5 card combination composed by player cards and community cards that has the best possible score. The possible hand ranks are (from stronger to weaker): Straight Flush (sequence of same suit), Four of a

Kind (4 cards with same rank), Full House (Three of a Kind + Pair), Straight (sequence), Three of a Kind (3 cards with same rank), Two Pair, One Pair (2 cards with same rank) and Highest Card (when not qualified to other ranks).

## 2.2    Rules of Simplified Texas Hold'em

The rules of the Poker variant used in this study represent a subset of the rules of Texas Hold'em, but only the initial round of the game is considered and it only allows for two players. In each game one of the players posts a mandatory minimum bet and the other one must bet half of that value. These players are called respectively big-blind and small-blind. After that, each player receives two cards – pocket cards – that are only seen by the player. The first player to act is the small-blind player and then the players play in turns. In each turn they can either match the highest bet (Call), increase that bet (Raise) or forfeit the game and lose the pot (Fold). When one of the players calls (except on the first big-blind player turn) five community cards are drawn from the deck and both players show their cards. The winning player is the one with the highest hand rank. If one of the players folds the other one wins the pot.

## 2.3    Opponent Modeling

One way of classifying opponents in this Poker variant is through VPIP and the aggression factor (AF) of the player [3]. The VPIP is the percentage of games in which the players raises at least one time. The aggression factor is the ration between the number of 'aggressive' actions and the number of 'passive' actions (equation 1).

$$AF = \frac{NumRaises}{NumCalls} \tag{1}$$

Poker experts classify opponents according to table 1, using the aforementioned indicators.

**Table 1.** Classifying Poker players using Agression Factor and VPIP

|            | $VPIP < 0,28$   | $VPIP \geq 0,28$ |
|------------|-----------------|------------------|
| $AF \geq 1$ | TightAgressive  | LooseAgressive   |
| $AF < 1$   | TightPassive    | LoosePassive     |

# 3    Related Work

The first approach to build Poker agents was a rule-based approach which involves specifying the action that should be taken for a given game state [1]. The following approaches were based on simulation techniques [1, 5-7], i.e. generating random instances in order to obtain a statistical average and decide the action. These approaches led to the creation of agents that were able to defeat weak human opponents.

The great breakthrough in Poker research began with the use of Nash's equilibrium theory [8, 9]. Since then, several approaches based on Nash Equilibrium emerged: Best Response [10], Restricted Nash Response [1, 11] and data-biased response [12]. Currently, the best Poker agent Polaris [12] uses a mixture of these approaches.

Other recent methodologies were based on pattern matching [13, 14] and on the Monte Carlo Search Tree algorithm [14, 15].

A successful work closely related to this approach is [16]. It presents a reinforcement learning methodology to another simplified version of Poker – 1 card Poker. This approach uses Q-Learning to learn how to play against several opponent types.

Despite all the breakthroughs achieved, there is no known approach in which the agent has reached a level similar to a competent human player.

# 4    Proposed Approach

This section describes the structure of the two agents. These agents are similar in every aspect except reward conditions.

## 4.1    Common Structure

The agents were developed with a Q-Table containing the state-action pairs. The state ($\sigma$) is defined as:

— **G**: A value representing a pair of cards that make the player's hand. This is useful since many hands have the same relative value (e.g. $\{2\clubsuit, 4\heartsuit\}$ and $\{2\diamondsuit, 4\clubsuit\}$).
— **P**: The player's seat on the table (big-blind or small-blind).
— **T**: A value representing the opponent type (Tight Aggressive, Tight Passive, Loose Aggressive and Loose Passive).
— **A**: A value representing the last action before the agent's turn (Call, Raise).

Each state has a direct correspondence to tuple (**C** – call weight, **R** – raise weight) as described in equation 2.

$$\sigma(G, P, T, A) \to (C, R) : C + R \leq 1;$$
$$G \in \{0 \dots 127\}; P \in \{'Big,' Small'\}; T \in \{'TA',' TP',' LA',' LP'\}; \quad (2)$$
$$A \in \{'Call',' Raise'\}; C, R \in [0,1]$$

The Q-Table is initially empty and the weights are filled up with random numbers as there is need for them. The value of the weights stabilizes as the games proceed, so as to choose the option which maximizes profit. However convergence to stable weight values is not guaranteed because the game state to action mapping may not be sufficient to fully describe the defined opponent types.

When the agent plays, it searches the Q-Table to obtain the values of C and R so as to decide on the action to take. After retrieving these values, a random number ($N \in [0,1]$) is generated. The probability of choosing an action is as on equation 3:

$$Action = \begin{cases} Call, N \in [0, C] \\ Raise, N \in ]C, C + R] \\ Fold \ otherwise \end{cases} \quad (3)$$

The flowchart present on figure 1 describes the complete process of update and usage of the Q-Table.

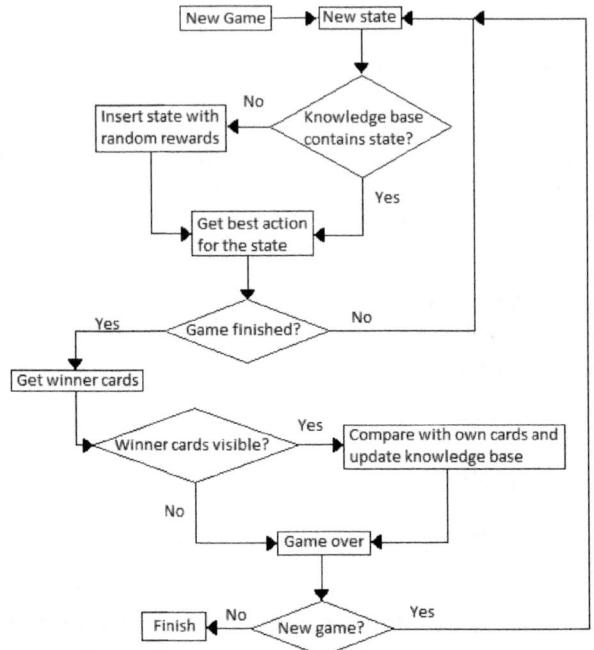

**Fig. 1.** Structure of the agent's behavior

## 4.2    Differences between Agents

Two agents with this structure were implemented: WHSLearner and WHLearner. The only difference between them resides on the reward calculation. Whilst WHSLearner updates the rewards based on the evaluation of the adequacy of the decision, WHLearner considers the actual outcome of the game. Table 2 shows how C and R variables are updated.

**Table 2.** Decision matrix for WHSLearner agent

| Agent | | Agent Action | | |
|---|---|---|---|---|
| **WHSLearner** | **WHLearner** | **Fold** | **Call** | **Raise** |
| Good Choice | Game Won | $C\downarrow, R\downarrow$ | $C\uparrow, R\uparrow\uparrow$ | $C\downarrow, R\uparrow\uparrow$ |
| Bad Choice | Game Lost | $C\uparrow, R\uparrow\uparrow$ | $C\downarrow, R\downarrow\downarrow$ | $C\uparrow, R\downarrow\downarrow\downarrow$ |

## 5    Tests and Results

To validate this paper's approach several tests were done. In the following subsections the experimental procedure and the obtained results are presented.

## 5.1    Procedure

All tests were performed in a simulated environment. The simulator used was Open Meerkat [17], which is an open source simulator that provides an API that facilitates the implementation and test of Poker agents. Open Meerkat was specially modified to implement the rules described on section 2. The agent was tested against four agents:

— AlwaysCallAgent – an agent that matches every bet;
— AlwaysAllInAgent – an agent that always bets its full bankroll;
— MegaBot [13] – an agent that combines several tactics with a simple heuristic to choose the most appropriate tactic against each kind of opponent;
— SimpleBot [1] – this agent uses opponent modeling and calculates the expected utility of each action based on a Bayesian analysis from simulations.

There were two stages to the agent testing phase. The first stage allowed the agent to build the Q-Table before actually testing its capabilities. Several simulations were run until the weights of actions for each state stabilized. The second stage was meant to test the agent against other agents as described above. There were 100.000 game simulations for each opponent, with seat permutation, to reduce the variance of results, following [1]. The results are displayed in profit evolution charts (figure 2-6).

## 5.2    Experiments

In the first experiments the opponents were AlwaysCallAgent and AlwaysAllInAgent.

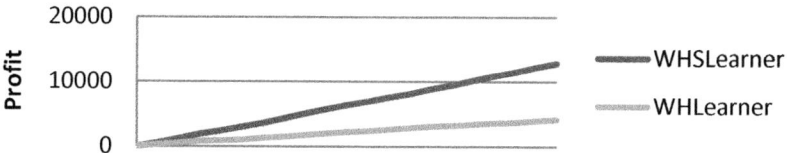

**Fig. 2.** Profit of both agents when facing the AlwaysCallAgent

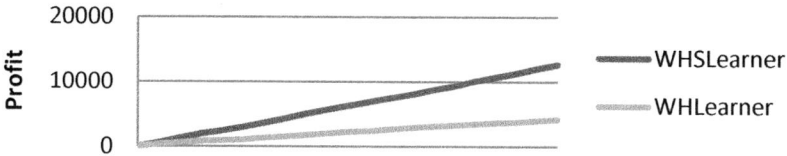

**Fig. 3.** Profit of both agents when facing the AlwaysAllInAgent

The results presented on Figure 2 and 3 show that this agent's approach is capable of easily beating very basic agents. In both experiments, the WHSLearner performed better. Figures 4 and 5 present the results of experiments where the agent played against more competitive opponents – MegaBot and SimpleBot. As can be seen, against more competitive opponents, the agents still got a positive profit. An interesting fact is that WHLearner performed better than WHSLearner against SimpleBot. This could be due to WHLearner having taken more advantage of the training stage than WHSLearner because of the number of wins of the first was higher than the number of good decisions of the former.

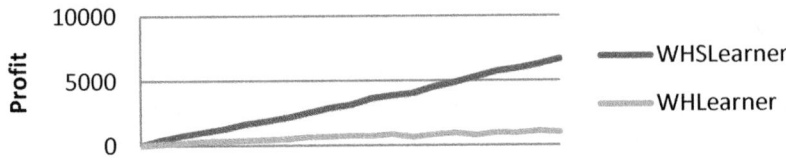

**Fig. 4.** Profit of both agents when facing the MegaBot

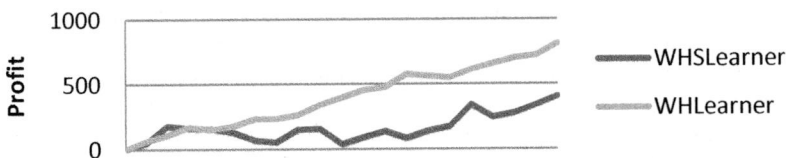

**Fig. 5.** Profit of both agents when facing the SimpleBot

Finally, the last experiment opposed both agents. Results are shown on figure 6.

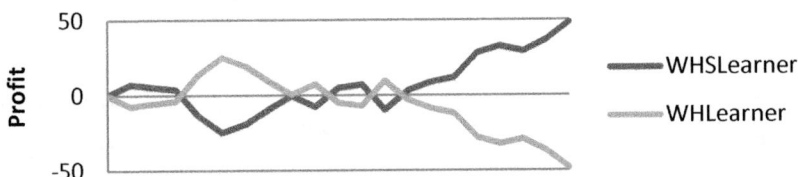

**Fig. 6.** Profit of both agents when facing each other

Like in most past tests, WHSLearner performed better. This means that adapting rewards using decision assessment is probably better than using the matches' outcome.

## 6     Conclusions and Future Work

Results showed that this approach is a valid starting point to create a complete Texas Hold'em agent, since the agent outperformed every opponent in all experiments. Another important conclusion can be extracted for the differences between the performance of WHSLearner and WHLearner. In most experiences, WHSLearner performed better, which means that rewarding good decisions maybe better than rewarding good outcomes in reinforcement learning algorithms.

Future work in this project should focus on developing an agent that considers the whole set of rules of Texas Hold'em. Moreover, this approach should be tested with human players for a more proper assessment. Finally more variables can be introduced to better represent the game state.

**Acknowledgments.** I would like to thank Fundação para a Ciência e a Tecnologia for supporting this work by providing my Ph.D. Scholarship SFRH/BD/71598/2010.

# References

1. Billings, D.: Algorithms and Assessment in Computer Poker. Ph.D. University of Alberta, Edmonton, Alberta, Canada (2006)
2. Newborn, M.: Kasparov versus Deep Blue: Computer Chess Comes of Age, 1st edn. Springer (1996)
3. Sklansky, D.: The Theory of Poker: A Professional Poker Player Teaches You How to Think Like One, 4th edn. Two Plus Two (2007)
4. Billings, D.: Computer Poker. M.Sc. University of Alberta, Canada (1995)
5. Davidson, A.: Opponent Modeling in Poker: Learning and Acting in a Hostile and Uncertain Environment. M.Sc. University Alberta, Edmonton, Alberta, Canada (2002)
6. Schauenberg, T.: Opponent Modeling and Search in Poker. M.Sc. University Alberta, Edmonton, Alberta, Canada (2006)
7. Frank, I., Basin, D., Matsubara, H.: Finding optimal strategies for imperfect information games. In: Proceedings 15th National/10th Conference on Artificial Intelligence/Innovative Applications of Artificial Intelligence, pp. 500–507. American Association for Artificial Intelligence, Menlo Park (1998)
8. Johanson, M.: Robust Strategies and Counter-Strategies: Building a Champion Level Computer Poker Player. M.Sc. University Alberta, Edmonton, Alberta, Canada (2007)
9. Gilpin, A., Sandholm, T.: A competitive Texas Hold'em poker player via automated abstraction and real-time equilibrium computation. In: Proceedings 5th International Joint Conference on Autonomous Agents and Multiagent Systems, Hakodate, Japan, pp. 1453–1454 (2006)
10. Gilpin, A., Sandholm, T.: Better automated abstraction techniques for im-perfect information games, with application to Texas Hold'em poker. In: Proceedings 6th International Joint Conference on Autonomous agents and Multiagent Systems. Article 192, Honolulu, Hawaii, United States, 8 pages (2007)
11. Billings, D., Burch, N., Davidson, A., Holte, R.C., Schaeffer, J., Schauenberg, T., Szafron, D.: Approximating game-theoretic optimal strategies for full-scale poker. In: Proceedings 18th International Joint Conference on Artificial Intelligence, Acapulco, Mexico, pp. 661–668 (2003)
12. Johanson, M., Bowling, M.: Data Biased Robust Counter Strategies. Journal of Machine Learning Research 5, 264–271 (2009)
13. Teófilo, L.F., Reis, L.P.: Building a No Limit Texas Hold'em Poker Agent Based on Game Logs Using Supervised Learning. In: Kamel, M., Karray, F., Gueaieb, W., Khamis, A. (eds.) AIS 2011. LNCS, vol. 6752, pp. 73–82. Springer, Heidelberg (2011)
14. Kleij, A.A.J.: Monte Carlo Tree Search and Opponent Modeling through Player Clustering in no-limit Texas Hold'em Poker. M.Sc. University of Groningen, Netherlands (2010)
15. Van den Broeck, G., Driessens, K., Ramon, J.: Monte-Carlo Tree Search in Poker Using Expected Reward Distributions. In: Zhou, Z.-H., Washio, T. (eds.) ACML 2009. LNCS, vol. 5828, pp. 367–381. Springer, Heidelberg (2009)
16. Dahl, F.A.: A Reinforcement Learning Algorithm Applied to Simplified Two-Player Texas Hold'em Poker. In: Flach, P.A., De Raedt, L. (eds.) ECML 2001. LNCS (LNAI), vol. 2167, pp. 85–96. Springer, Heidelberg (2001)
17. Open Meerkat Poker Testbed (2012), http://code.google.com/p/opentestbed/

# Face Recognition with Weightless Neural Networks Using the MIT Database

K. Khaki and T.J. Stonham

Brunel University, Engineering Department
Uxbridge, Middlesex, UK

**Abstract.** In this paper we propose a new face recognition method based on the weightless neural network system [1]. The algorithm uses 5-pixel n-tuples to map images, which passes through a ranking transform to obtain a binary n-tuple state. A digital neural network correlates the recurring states obtained from the current input pattern to those extracted from the test set. The data used in this paper is from the MIT-CBCL facial database [2], and the training data and testing data set each consist of 10 individual persons, with 100 examples of each subject. An error rate of 0.1% FAR and 0.1% FRR was achieved on data which was totally independent of the training set.

## 1 Introduction

In the recent years, many algorithms have been proposed to solve the problem of face recognition and detection [3][4][5]. However, these solutions have unacceptable prerequisites; for example, the images may require computationally intensive pre-processing, leading to excessive computational overheads for real-time operation. Also, error free results are rarely encountered in face recognition, and most researchers in the area underestimate the potential diversity of humanly perceived classes of images. The solution presented in this paper offers real-time recognition processing, and a single pass training facility, and has produced results which are significantly better than other methods applied to this data, with error rates two orders of magnitude less than hitherto reported.

The methodology has also been tested on other public domain databases, with error free results. However, the MIT database has been chosen for this paper because it offers more challenging recognition problems. Each image contains a face, but it is unconstrained in terms of position, background, lighting and clothing. The database was first used by MIT, where, they converted each image into a 3D computer representation, and achieved an overall recognition percentage of 88% [2]. This paper reports on an algorithm that results in 99.9% accuracy.

## 2 Algorithm

Conventional methods rely on feature extraction from images, and correlating them with unseen inputs. However, feature extraction is a complex problem that is not entirely error-free, and thus compromises the subsequent recognition performance.

M. Kamel, F. Karray, and H. Hagras (Eds.): AIS 2012, LNCS 7326, pp. 228–233, 2012.

This then prevents the successful application of face-based biometrics for high security applications such as airports. Modern technology allows a direct interface between a camera and a digital neural network, with information being captured at the pixel level, and this allows the network to evolve its functionality. Hence, using this methodology offers the potential to remove the problematic feature extraction stage, resulting in enhanced performance.

## 2.1 Initial Stages

The 'test' folder of the MIT-CBCL database consists of 10 image sets; each contains 200 images of an individual. The images vary in size, ranging from 100x100 to 115x115 pixels. Resizing is done initially to bring all images to a constant 100x100 pixels. This is done using subsample normalization, which does not change the values of any of the pixels in the image; rather it deletes rows and columns of pixels to reduce the overall size of the image. The database was then divided into a set of 100 images of each of the 10 subjects which were used solely for training the neural network. The remaining set of similar size was used solely for testing.

## 2.2 Methodology

The algorithm developed and used in this work is a random n-tuple mapping on grayscale images, where each image is divided into 200 sets of 5 pixels groups. An n-tuple state is then obtained by using a ranking transform which eliminates the requirement for pixel thresholding operation [1]. The digital neuron then correlates the recurring states obtained from the training phase with the current input from the test pattern. The algorithm is stated below:

Binary System

Images $\xrightarrow{\text{Threshold}}$ Input Vector: $I_{i}.. I_{i*j}$     $\text{Function} = \sum(n_1, n_2, n_3......n_t)$

where : $I_i \varepsilon (0,1)$     where: $n_i$ - $i^{th}$ state of sample at

position $I_j .. I_n$

n sample $\longrightarrow$ subset of $I$ - $I_j, I_k... I_n$     No. of Functions $= \dfrac{N}{n}$     where: $N = i*j$

where: $j, k..$ - random numbers

$n = 5$

To avoid the threshold, for each n-tuple $I_j$ to $I_n$ is ranked.

Ranks: $I_j \geq I_k.... \geq I_n$ is state 0

through to

$I_j \leq I_k.... \leq I_n$ is state $(n! -1)$

A visual representation is below in Fig 1.

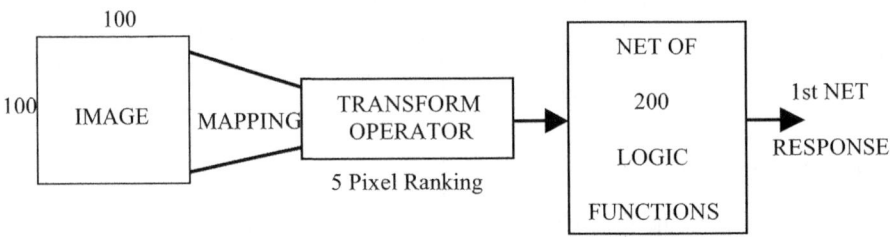

**Fig. 1.** Overall Training Methodology

### 2.3     Neural Network

The algorithm makes use of a single layer weightless neural network, with no hidden layers. The training requires a single pass through the training set; therefore no error back propagation is required. This results in very fast training, of the order of 10 frames a second in simulation. The number of neurons in the network depends on the size of the image, and is calculated from the number of pixels divided by the n-tuple size.

## 3     MIT-CBCL Data Set

Sample images of the 10 classes used in training and testing are displayed in Fig 2.

**Fig. 2.** Two samples of the 200 images available for each person. These represent the diversity of the pattern classes.

## 4     Results

The training set and testing set consisted of 10 individuals, each with 100 unique images. The algorithm had a training and testing rate of 10 images per second in a computer simulation, and produced a recognition accuracy of 99.9% overall, with a 0.1% False Rejection Rate (FRR) and 0.1% False Acceptance Rate (FAR). This compares with the 88% accuracy published by MIT [2] and it also shows significant savings in processing time. Additionally it should be noted that the memory requirements for a single neuron which is in effect implementing a 6 variable logic function, is 128 bits, and as 2000 neurons are used per net, the system has an efficient hardware implementation capability with a total storage requirement of 250kB; this is independent of the size of the data set being processed. Fig 3 shows the overall recognition results of the method presented in this paper.

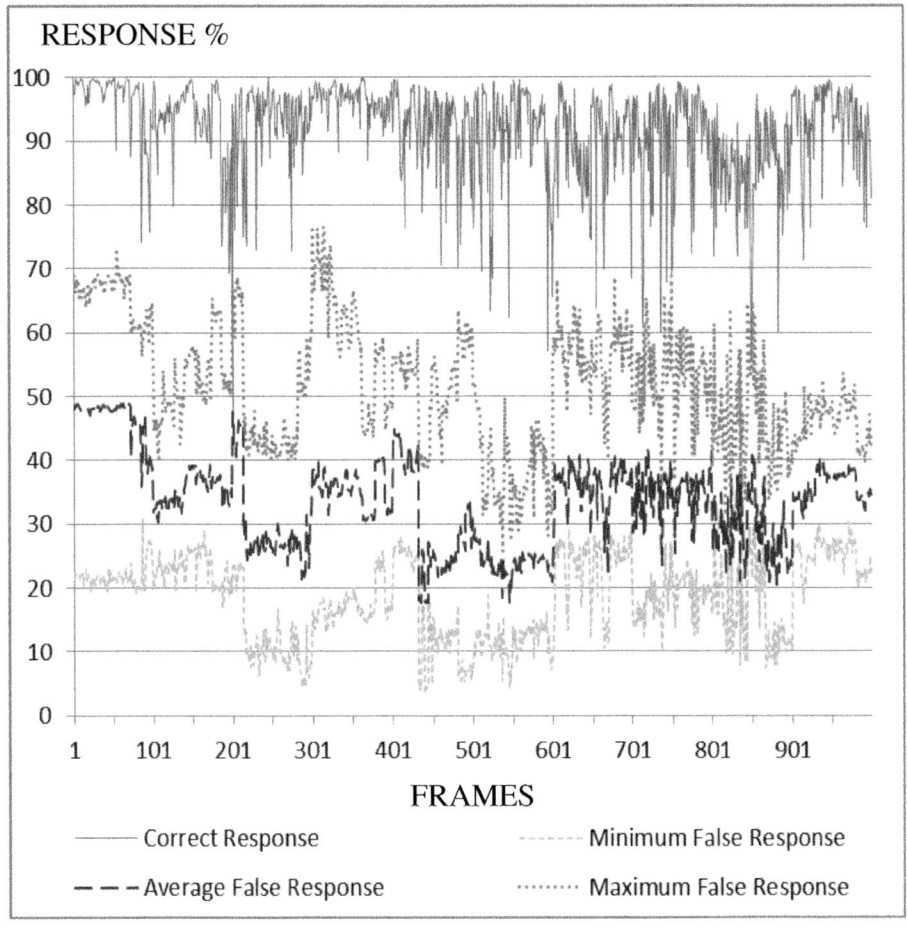

**Fig. 3.** MIT-CBCL Test Result – See Examples in Fig. 2

The y-axis of the graph represents the percentage response of the net to the frame, and the x-axis represents the frame being tested on. This graph is a culmination of the results of the ten trained nets on the 1000 untrained unique images. The graph key on the bottom of the graph represents the four different curves in it – namely:

1. The correct response – relates to the net that the frame is actually supposed to belong to, and it's response to that frame. This means that for the first 100 images, the correct response relates to the response of net 1, and for the next 100 images, it is the response of net 2, etc.
2. The minimum false response – relates to the minimum response out of all the other 9 nets that the frame does not belong to.
3. The average false response – refers to the average of the responses of the 9 nets that the frame being tested on does not belong.
4. The maximum false response – refers to the maximum response out of all the other 9 nets that the frame does not belong to.

Therefore, by subtracting the maximum false response from the correct response for each frame, the confidence of the system can be seen in terms of the separation of the accepted and rejected images. This is displayed in Fig 4:

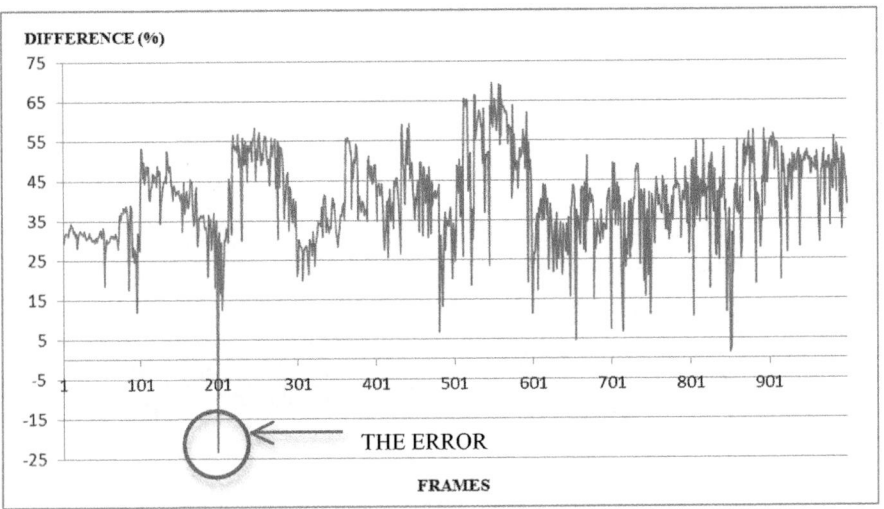

**Fig. 4.** Confidence graph between correct response and maximum false response

By observing the graph, the accuracy of the results can be seen. The difference in Fig.4 must be positive for correct recognition and the higher the difference, the greater the confidence in the recognition system. The 1 error in 1000 tests is in frame 201. The average difference in the responses between accepted and rejected is 40.3%. This shows the method can establish a significant differential between images which have subtle differences, where the commonality of all the images in the database is that they are human faces, and consequently there is a high perceptual similarity between any two human faces.

## 5    Conclusion

The 99.9% accuracy with the MIT database has resulted in one error in 1000 images. It should be noted that these results have been obtained from an un-optimised system.

The training and feature sampling strategies have been randomly specified. It is therefore expected that robust error free performance can be achieved in an optimised system. It may be noted that error free classification has been achieved with other public domain databases, namely AT&T and Yale.

**Acknowledgment.** Credit is hereby given to the Massachusetts Institute of Technology and to the Center for Biological and Computational Learning for providing the database of facial images.

# References

[1] Lauria, S., Mitchell, R.J.: Weightless Neural Nets for Face Recognition: a Comparison. In: IEEE Signal Processing Society Workshop (1998)
[2] Weyrauch, B., Huang, J., Heisle, B., Blanz, V.: Component-based Face Recognition with 3D Morphable Models (2004)
[3] Samer Charifa, M., Suliman, A., Bikdash, M.: Face Recognition Using a Hybrid General Backpropagation Neural Network. In: 2007 IEEE International Conference on Granular Computing (2007)
[4] Nazeer, S.A., Omar, N., Khalid, M.: Face Recognition using Artificial Neural Networks Approach. In: ICSCN 2007, pp. 420–425. MIT Campus, Anna University (2007)
[5] Bojkovic, Z., Samcovic, A.: Face Detection Approach In Neural Network Based Method For Video Surveillance. In: NEUREL 2006. IEEE (2006)

# P300 Speller Efficiency with Common Average Reference

Mohammed J. Alhaddad, Mahmoud Kamel, Hussein Malibary,
Khalid Thabit, Foud Dahlwi, and Anas Hadi

King AbdulAziz University, Jeddah, Saudi Arabia
{Malhaddad,miali,hmalibary,drthabit,
fdehlawi,ahadi008}@kau.edu.sa

**Abstract.** P300 detection is known to be challenging task, as P300 potentials
are buried in a large amount of noise. In standard recording of P300 signals, activity at the reference site affects measurements at all the active electrode sites.
Analyses of P300 data would be improved if reference site activity could be separated out. This step is an important one before the extraction of P300 features.
The essential goal is to improve the signal to noise ratio (SNR) significantly,
i.e. to separate the task-related signal from the noise content, and therefore is
likely to support the most accurate and rapid P300 Speller. Different techniques
have been proposed to remove common sources of artifacts in raw EEG signals.
In this research, twelve different techniques have been investigated along with
their application for P300 speller in three different Datasets. The results as a
whole demonstrate that common average reference CAR technique proved best
able to distinguish between targets and non-targets. It was significantly superior
to the other techniques.

**Keywords:** EEG, P300, Common Average Reference, Classification, Montaging, Re-reference.

## 1   Introduction

Ideally, the voltage readings should represent a pure measure of activity at the recording site. The difficulty is that voltage is a relative measure that necessarily compares
the recording site with another – reference- site. If there is any activity at the reference
site, this will contribute equally to the resulting voltage recording [3]. According to
this, one of the major problems in EEG recording is to find a region in the human body
whose bio-potential activity can be considered as neutral as possible. Nowadays, it is
well known that it is impossible to find a "zero-potential" site on the human body [8].
The most common way of performing EEG recordings is by using as a common reference (CR) an electrode placed somewhere on the head. Starting from this Common
Reference, several other re-reference technique or montages can be constructed for
interpretation or processing purposes [5]. There are several different recording
reference electrode placements such as vertex (Cz), linked-ears, linked-mastoids,
ipsilateral-ear, contralateral-ear, C7 reference, bipolar references, and tip of the nose.
Reference-free techniques are represented by common average reference, weighted
average reference, and source derivation. Each technique has its own set of advantages

M. Kamel, F. Karray, and H. Hagras (Eds.): AIS 2012, LNCS 7326, pp. 234–241, 2012.

and disadvantages. The choice of reference may produce topographic distortion if relatively electrically neutral area is not employed. Linking reference electrodes from two earlobes or mastoids reduces the likelihood of artificially inflating activity in one hemisphere. Nevertheless, the use of this method may drift away "effective" reference from the midline plane if the electrical resistance at each electrode differs. Cz reference is advantageous when it is located in the middle among active electrodes, however for close points it makes poor resolution. Reference-free method do not suffer from problems associated with an actual physical reference [4].

CAR has been proposed as a method for providing an inactive reference. The underlying principle is that electrical events produce both positive and negative poles. The integral of these potential fields in a conducting sphere sums to exactly zero [3].

In previous studies, McFarland et al. [6] discussed the selection of spatial filter for EEG-based communication. The Comparison were among standard ear-reference, (CAR), small Laplacian and a large Laplacian . CAR and large Laplacian methods proved best able to distinguish between top and bottom targets [6]. Another study with other results was performed by Ng and Raveendran [2]. According to that study CAR produced poor results, since only the channels over the motor cortex region were considered [2]. Both studies were on motor imagery paradigm.

In P300 paradigm, channel referencing were discussed by Krusienski et al [7], they conclude that there is no significant difference between CAR and ear-reference, and this is true only in the case of selecting best channels subset [7].

A solution to the reference problem proposed by Hu et al. consists in identifying the reference signal by constraining the Blind Source Separation (BSS) model to particular mixing system which implies that the non-zero reference signal is independent from all other measures. This approach is based on the hypothesis that the reference electrode placed on the scalp is not influenced by the intracranial measures. If for intra-cranial measures this hypothesis (although not proven) can be employed, in a cephalic referenced scalp EEG context it cannot hold, as the reference electrode itself records a noisy mixture of cerebral and extra-cerebral sources [5]. Improvements of this work with more accurate and generalized model were done by R. Salido-Ruiz et al [8] but still have the same problem.

Still that reference technique has engendered ongoing debate [3]. And they need more and more practical experiments. Improving the reference technique will positively affect all of the BCI research and P300 is not an exception.

## 2    Re-reference Methods

Montaging or re-reference methods are used to enhance the signal to noise ratio SNR. Twelve different montage methods have been tested, most of them were used by [2] in motor imagery paradigm.

1. Common Reference: No re-montaging is done
2. Common average reference: The mean of all the electrodes is removed for all the electrodes.
   *E.g.Cz CAR=Cz-(Fp1+AF3+F7+......+Fz+MA1+MA2)/34*

3. Surface Laplacian (4 adjacent): The weighted mean (depends on the distance) of the 4 adjacent electrodes is removed from the central electrode.
   *E.g. CzSL4=Cz-0.7(FC1+ FC2+ CP1+CP2)/4*

4. Surface Laplacian (8 adjacent): The weighted mean (depends on the distance) of the 8 surrounding electrodes is removed from the central electrode.
   *E.g. CzSL8=Cz-[0.7(FC1+FC2+CP1+CP2)+0.5(Fz+Pz+ C3+C4)]/4*

5. Bipolar (front to back): The difference of an electrode with the one behind it .
   *E.g. CzBFB= Cz-Pz*

6. Bipolar (front to back skip 1): The difference of 2 electrodes that lies in front and also behind that electrode.
   *E.g. CzBS1= Cz-Oz*

7. Bipolar (Symmetrical): The difference of 2 electrodes that is symmetrical to one another.
   *E.g. C3SYM= C3-C4*

8. Bipolar (left to right): The difference of an electrode with the one right to it.
   *E.g. C3LR= C3-Cz*

9. Bipolar (right to left): The difference of an electrode with the one left to it.
   *E.g. CzRL= Cz-C3*

10. Using T7, T8 channels : The mean of  T7,T8 channels  is removed for all the electrodes .
    *E.g.CzT7T8=Cz-(T7+T8)/2*

11. Common average reference without mastoid channels: The mean of all the electrodes without mastoid channels is removed for all the electrodes .
    *E.g.Cz CAR=Cz-(Fp1+AF3+F7+......+Fp2+Fz+AF4)/32*

12. Reference estimation: Here we apply the work of *R. Ranta et al.* [8] by the estimation of the reference *^r*
    *E.g.Cz est = Cz +^r*

## 3     Application of Re-reference Techniques to the P300 Speller

The above 12 techniques were applied on 3 different datasets for different subjects:

- **Dataset 1:** P300 dataset provided by Hoffmann et al. [1].
- **Dataset 2:** P300 dataset from the BCI competition 2003 [10].
- **Dataset 3:** P300 dataset obtained from our BCI Lab at KAU hospital. The data were recorded using gUSBamp amplifier and digitized at 256 Hz with 8 electrodes.

**Table 1.** Comparison between these three datasets

|  | Dataset 1 | Dataset 2 | Dataset 3 |
|---|---|---|---|
| **Sampling rate** | 2048Hz | 240Hz | 256Hz |
| **Subjects Number** | 8 | 2 | 1 |
| **Filtering** | 1.0Hz-12Hz | 0.1Hz-60Hz | 0.1-60Hz |
| **Electrodes no.** | 32 | 64 | 8 |
| **Electrodes Locations** | Fp1, AF3, F7, F3, FC1, FC5, T7, C3, CP1, CP5, P7, P3, Pz, PO3, O1, Oz, O2, PO4, P4, P8, CP6, CP2, C4, T8, FC6, FC2, F4, F8, AF4, Fp2, Fz, and Cz | FC5, FC3, FC1, FCz, FC2, FC4, FC6, C5, C3, C1, Cz, C2, C4, C6, Cp5, Cp3, Cp1, Cpz, Cp2, Cp4, Cp6, Fp1, Fpz, Fp2, AF7, AF3, AFz, AF4, AF8, F7, F5, F3, F1, Fz, F2, F4, F6, F8, Ft7, Ft8, T7, T8, T9, T10, Tp7, Tp8, P7, P5, P3, P1, Pz, P2, P4, P6, P8, PO7, PO3, POz, PO4, PO8, O1, Oz, O2, and Iz | F3, F4, T7, C3, Cz, C4, T8, and Pz |

## 3.1    Dataset 1

The techniques were first applied to Dataset 1, and we found that CAR gives best classification accuracy. Comparison based on classification accuracy between CAR and T7,T8 methods is shown in Fig.1. This accuracy is obtained with Bayesian Linear Discriminant Analysis (BLDA), averaged over all subjects and sessions, plotted against time, for all electrode configurations. According to the accuracy achieved in Fig.1 its clear that CAR outperforms T7, T8 method.

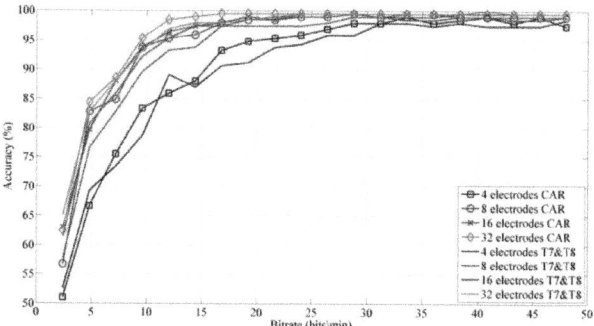

**Fig. 1.** T7,T8 & CAR comparison based on classification accuracy, for electrode configurations(4,8,16 and 32)

Comparison based on classification accuracy of other methods - mentioned in section 2- can be found in Fig.2.

Fig.2 shown the comparison between CAR and the different re-reference methods based on classification accuracy. In Fig.2.a, the CAR illustrates slightly better than Common Reference, whereas in Fig.2.b clear deference between CAR and Surface Laplacian with 4 & 8 adjacent electrodes and this is due the small number of

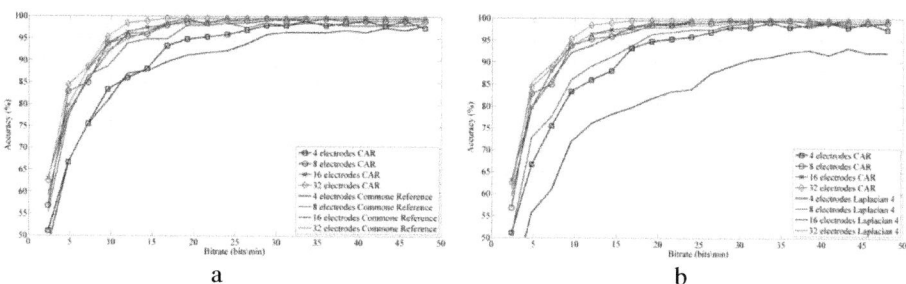

**Fig. 2.** Comparison between the different re-reference methods based on classification accuracy, for electrode configurations (4, 8, 16 and 32), obtained with BLDA, averaged over all subjects and sessions,  plotted against time, (A) Common Reference, (B) Laplacian 4, (C) Laplacian 8, (D) Bipolar Front to back, (E) Bipolar Front to back skip 1, (F) Bipolar Symmetrical, (G) Bipolar Left to Right, (H) Bipolar Right to Left, (I) CAR without Mastoid, (J) Reference estimation

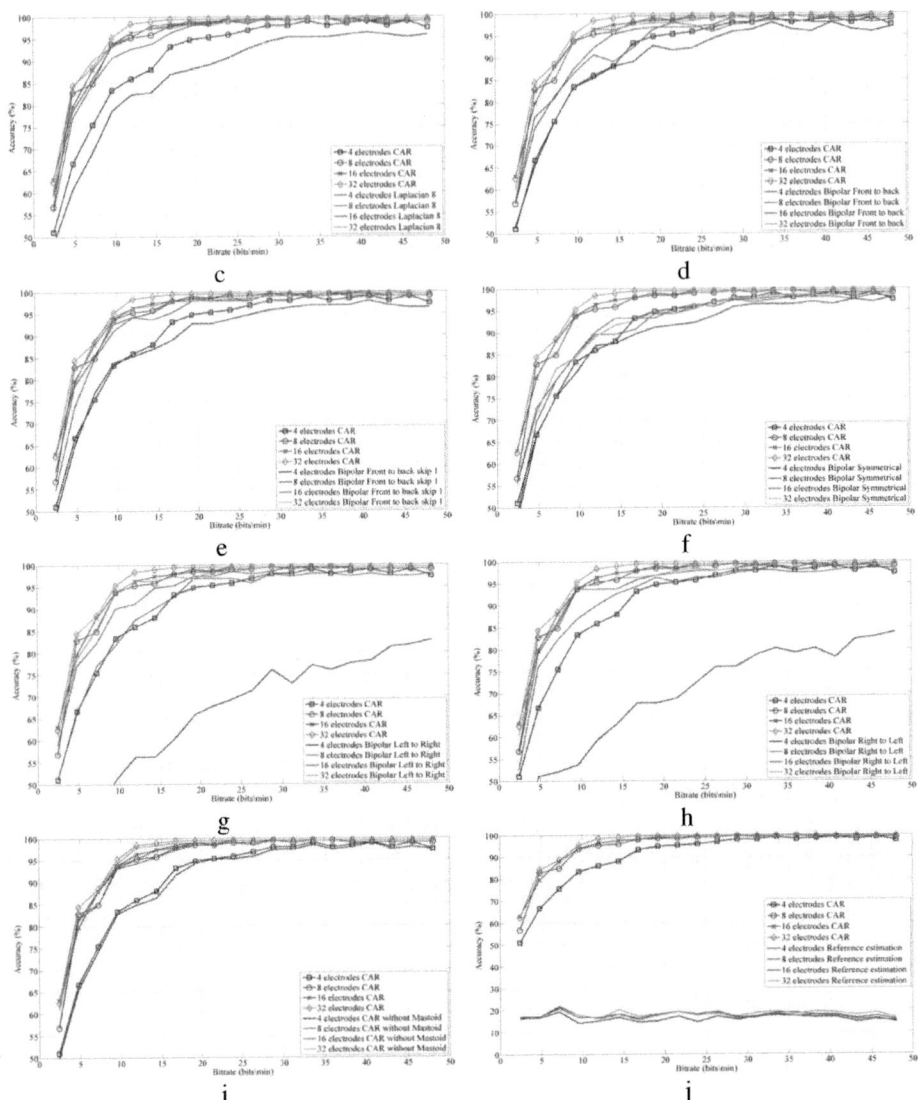

**Fig. 2.** (*Continued*)

electrodes has been used, nevertheless, Fig.2.c demonstrates Laplacian method but not as good as CAR. However, In Bipolar techniques we have five methods which are front to back as shown in Fig.2.d, front to back skip 1 in Fig.2.e, Symmetrical in Fig.2.f, left to right in Fig.2.g and, right to left in Fig.2.h, it obvious that left to right and right to left are the worst among the five methods, in the other hand we can find that front to back skip 1 gives the best; but again not as good as CAR.

Fig.2.i illustrates CAR method but without mastoid electrodes, and we can find the effect of removing such electrodes on the accuracy. Finally, reference estimation method Fig2.j provides poor results and this support the hypothesis that the reference electrode is a noisy mixture of cerebral and extra-cerebral sources, so it is dependent on the other electrodes, which conflict with the BSS model assumption.

## 3.2    Dataset 2

After that we again applied the techniques to Dataset 2, and the result where compared to each other using Correlation Coefficients method. Where the ensample average is taken over number of trials and correlation coefficients are calculated regarding the whole p300 ensample average. And as shown in Fig.3 that CAR was the pest method.

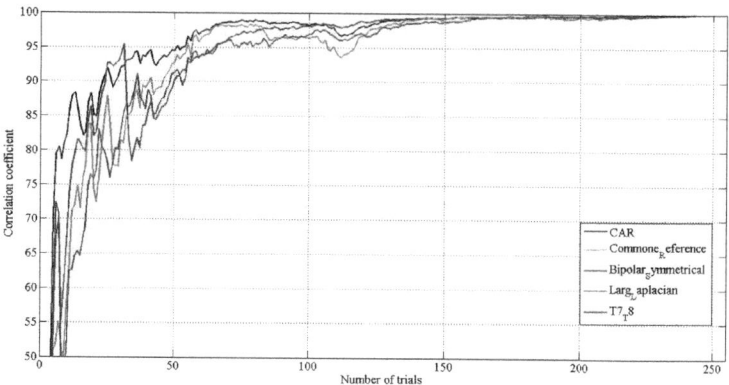

**Fig. 3.** Correlation coefficients with whole p300 ensample average VS Number of trials

## 3.3    Dataset 3

Finally we applied CAR in Dataset 3, and we found that the accuracy is improved when we use CAR as a rereference technique. In Fig.4 we can see a comparison based on classification accuracy between CAR and CR (right ear lobe) methods.

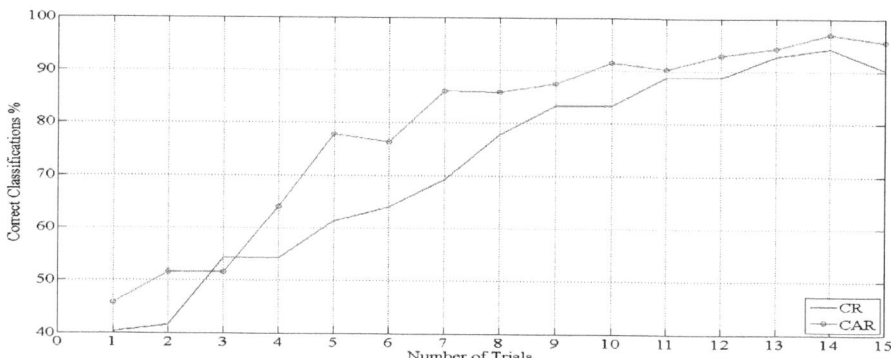

**Fig. 4.** CR & CAR comparison based on classification accuracy

## 4    Discussion

Based on the results we can see that CAR outperforms other re-reference methods. This can be explained with the concept of the ideal reference, where its activity is as neutral as possible. To measure the naturally we use the Power Spectral Density (PSD). As we can see in Fig.5 that the PSD for CAR is lower than other methods.

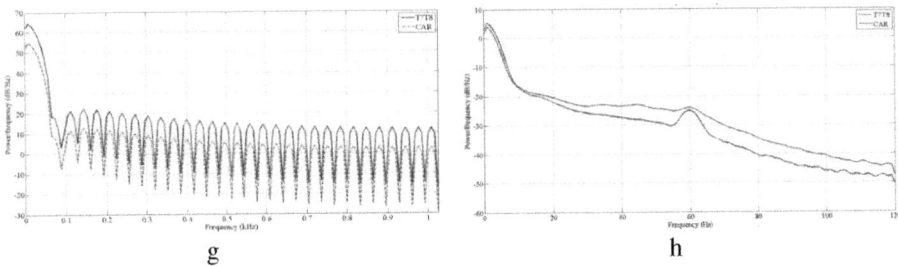

g                                            h

**Fig. 5.** Comparing PSD for CAR & T7T8 methods, for datasets 1 & 2

## 5    Conclusion

This paper illustrates that the common average reference (CAR) outperforms the other re-references techniques. The cost of the good performance of the CAR is the increasing of the number of electrodes, in such a way, to be uniformly distributed along the head. The experimental results comply with the mathematical justification of the best performance of the CAR as given in [11]. The experiments shown that CAR is the appropriate choice as it tends to be a neutral reference in the case that large number of electrodes has been used in the recording phase.

**Acknowledgement.** This work is part of "Design and Implementation of Brain Computer Interface (BCI) for Communication and Control" project funded by King Abdul-Aziz City for Science and Technology KACST.

## References

1. Hoffmann, U., Vesin, J.-M., Ebrahimi, T., Diserens, K.: An efficient P300-based brain–computer interface for disabled subjects. Journal of Neuroscience Methods 167, 115–125 (2008)
2. Ng, S.C., Raveendran, P.: Comparison of different Montages on to EEG classification. In: Biomed 2006. IFMBE Proceedings, vol. 15, pp. 365–368 (2007)
3. Dien, J.: Issues in the application of the average reference: review, critiques, and recommendations. Behavior Research Methods, Instruments, Computers 30(1), 34–43 (1998)
4. Teplan, M.: Fundamentals Of EEG Measurement. Measurement Science Review 2, sec. 2 (2002)

5. Salido-Ruiz, R., Ranta, R., Louis-Dorr, V.: EEG montage analysis in the blind source separation. Biomedical Signal Processing and Control (2010)
6. McFarland, D.J., McCane, L.M., David, S.V., Wolpaw, J.R.: Spatial filter selection for EEG-based communication. EEG Clin. Neurophysiol. 103(3), 386–394 (1997)
7. Krusienski, D.J., Sellers, E.W., McFarland, D.J., Vaughan, T.M., Wolpaw, J.R.: Toward enhanced P300 speller performance. Journal of Neuroscience Methods 167, 15–21 (2008)
8. Ranta, R., Salido-Ruiz, R., Louis-Dorr, V.: Reference estimation in EEG recordings. IEEE (2010)
9. Hu, S., Stead, M., Worrell, G.: Automatic identification and removal of scalp reference signal for intracranial EEGs based on Independent Component Analysis. IEEE Trans. Biomed. Eng. 54(9), 1560–1572 (2007)
10. Rrusienski, D., Schalk, G.: Documentation Wadsworth BCI Dataset (P300 Evoked Potentials) Data Acquired Using BCI2000's P3 Speller Paradigm (2004), http://www.bci2000.org
11. Bertrand, O., Perrin, F., Pernier, J.: A theoretical justification of the average-reference in topographic evoked potential studies. Electroencephalogr. Clin. Neurophysiol. 62, 678–695 (1985)

# Detecting Natural Gas Leaks
# Using Digital Images and Novelty Filters

Cícero Ferreira Fernandes Costa Filho[*], Roberlanio de Oliveira Melo,
and Marly Guimarães Fernandes Costa

Centro de Tecnologia Eletrônica e da Informacao,
Universidade Federal do Amazonas, Amazonas, Brasil
cffcfilho@gmail.com, roberlanio@yahoo.com.br,
marly.costa@uol.com.br

**Abstract.** This paper presents a new technique for detecting natural gas leaks in
the oil and gas industry. More precisely, the detection is done in wellheads of
industry installations. In the literature, other methods are already used, but with
some drawbacks. One technique detects gas leaks measuring the $CH_4$ concentration through the principle of catalytic combustion but suffers from reduced
life span and a narrow detection range of sensors. Another technique that measures infrared spectrum absorption suffers from high false negative values in the
presence of steam. The technique proposed in this study uses radiation in the
visible range that can be captured through CCD cameras already present in
*Closed-Circuit Television* systems used to monitor wells. The proposed method
uses the novelty filter concept to detect the leak and to identify the region where
it occurs. The proposed technique is a pioneering study of natural gas detection
with CCD in visible range. The results presented are promising, showing sensitivity and specificity equal to 100%.

**Keywords:** detection of natural gas leak, novelty filter, gas and oil industry.

## 1    Introduction

The oil and gas industry is one of the most complex and dangerous fields due to intrinsic characteristics of hydrocarbons, such as: toxicity, inflammability and explosion
velocity [1]. The occurrence of gas leaks in oil installations generates undesirable
financial and environmental consequences, and loss of human lives [2]. Constant
monitoring is necessary to avoid these undesirable consequences and there is a great
demand for the development of new systems for monitoring and controlling gas leaks.

Petroleum is a mixture of hydrocarbons in a solid, liquid or gaseous state, in standard temperature and pressure, according to molecule complexity and weight. Natural
gas is a byproduct of the petroleum found in a gas phase, being composed of a mixture of several hydrocarbons, whose molecules are in the form $C_nH2_{n+2}$, for n varying between one and four. Among the hydrocarbons present in natural gas, about 70%

---

[*] Corresponding author.

M. Kamel, F. Karray, and H. Hagras (Eds.): AIS 2012, LNCS 7326, pp. 242–249, 2012.

is methane $(CH_4)$, which presents a lower and upper limit of inflammability of 5% and 15%, respectively, and minimum ignition energy of $250\mu J$ [3]. Several methods used to detect natural gas are based on detecting the methane leak to the atmosphere. In the sequence, we give some examples of them.

The *Safety in Mines Research Establishment* (SMRS) [4] proposed the principle of catalytic combustion to measure the $CH_4$ concentrations present in the environment. This principle is based on temperature rises resulting from the heat generated from methane combustion in a catalytic surface employing the palladium as a sensor element. Due to ease in manufacturing and low cost, this device has been used for many years, until presently. These sensors, nevertheless, have a reduced life span and a narrow detection range [5].

The analysis of infrared (IR) spectrum absorption has been used more frequently in methane detection. The main reasons are: the IR detector has a life span of more than five years, stability and reliability. An IR detection system is comprised of an IR transmitter and receptors with electromagnetic spectrum $\lambda_{IR} = 2\sim5\mu m$. When IR radiation interacts with methane gas $(\lambda_{CH_4} \approx 3.5\mu m)$, a part of the energy is absorbed and the remaining energy is transmitted [6]. The energy absorbed increases the vibration of methane molecules and, consequently, increases the temperature of the gas. The gas concentration is obtained through the measure of the ratio between the incident and transmitted radiation [5]. The main drawbacks of this system include difficulties in installing and maintaining and the high false negative values in the presence of steam, because the IR radiation is also absorbed by this substance.

Another technique used increasingly in detecting natural gas leaks is digital image processing. In 1997, the *U.S. Department of Energy* (DOE) together with *Sandia of National Laboratories to National Security Missions* [7] proposed a system called *Backscatter Absorption Gas Imaging* (BAGI), whose basic principle was to illuminate a gas leak scenario, applying an IR laser, and then photograph this leakage using an IR camera. Systems employing this technology are very expensive [8-9], achieving values of U$ 80,000 when used for a single inspection, but not for continuous inspection.

Usually natural gas leaks appear to human beings as a white cloud or fog, because when the methane comes into contact with the atmosphere, its low temperature induces air condensation [10]. Considering this fact, this study proposes a natural gas detection method in *wellheads* of an *onshore* petroleum installation using a *Closed-Circuit Television* system with *Charge-Coupled Device* (CCD) cameras to monitor wells. As these systems are already available, no additional expense is necessary for hardware implementation.

The idea explored in this study was to use a technique known as a novelty filter [11] to investigate the presence of a natural gas leaks in digital images captured by CCD cameras.

This paper is organized according to the following sections: sections 2 and 3 describe the novelty filter concept and application for detecting natural gas leaks, respectively. Section 4 presents the results and discussion. Finally section 5 presents the conclusion.

## 2        Novelty Filter Concept

The Novelty filter concept was described by Kohonen [11]. In this sequence, the no-velty concept is described based on the classic orthogonalization method of Gram-Schmidt [12-13].

Let $\{x_1, x_2, \dots x_m\}$ be a set of $n$ dimensional Euclidian vectors which span a $m$ dimensional subspace $L \subset R^n$, with $m < n$. Considering the subspace $L$, an arbitrary vector $x \in R^n$ can be divided into two components, $\hat{x}$ and $\tilde{x}$, where: $\hat{x} \in L$ is the projection of $x$ on $L$ and $\tilde{x} \perp L$ is the projection of $x$ perpendicular to $L$. Vector $\hat{x} \in L$ represents the component of $x$ that is "known" by the subspace $L$ and can be represented as a linear combination of $\{x_1, x_2, \dots x_m\}$. Vector $\tilde{x} \perp L$ represents the new information, that is unknown by the subspace $L$ and cannot be represented as a linear combination of $\{x_1, x_2, \dots x_m\}$. Fig. 1 illustrates these two components in $R^3$ space.

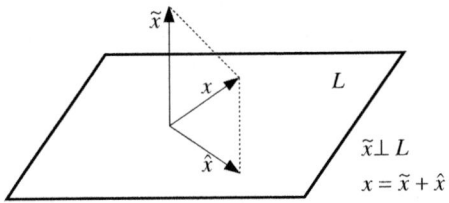

**Fig. 1.** Illustration of novelty filter concept in $R^3$ space

Thus, considering subspace L, $\tilde{x}$ is named novelty and the system that extracts this component from x and shows it as an output can be named the Novelty Filter.

To determine the components $\hat{x}$ and $\tilde{x}$ of a vector x, the Gram-Schmidt orthogonalization process was used, described as follows. Considering that we have a set of $m$ independent vectors $\{x_1, x_2, \dots x_m\}$, the L basis. From this set, we can obtain m orthogonal vectors $\{v_1, v_2, \dots v_m\}$, through the procedure described in (1) and (2).

$$v_1 = x_1 \tag{1}$$

$$v_k = x_k - \sum_{i=1}^{k-1} \frac{(v_i, v_k)}{(v_i, v_i)} \cdot v_i, k = 2,3, \dots m \tag{2}$$

where $\frac{(v_i, v_k)}{(v_i, v_i)} \cdot v_i$ is the projection of $x_k$ on $v_i$.

Given a sample vector x, its novelty component is calculated as the $(m + 1)$ step of the process described by (2), or $\tilde{x} = v_{k+1}$, as described in equation (3):

$$\tilde{x} = x - \sum_{i=1}^{m} \frac{(v_i, x)}{(v_i, v_i)} \cdot v_i \tag{3}$$

The magnitude of vector $\tilde{x}$, $\|\tilde{x}\|$, is used as a dissimilarity measure. The lower the magnitude of vector $\tilde{x}$, the closer it will be to the base $\{v_1, v_2, \dots v_m\} \in R^n$.

# 3    Novelty Filter Application for Detecting Natural Gas Leaks

The images of the wellhead area are obtained through the system CFTV previously described, placing the CCD camera in a fixed position. The spatial resolution of the CCD sensor used was 352x240 pixels. This image is shown in Figure 2a. Inside this image, a *Region Of Interest* (ROI) closer to the wellhead was selected, with sions 128x128. The objective was to reduce the noise from trees and the sky and to concentrate the novelty filter focus on the *Christmas tree* of the *wellhead*, where the leaks are more frequent. An ROI with no natural gas leak is shown in Figure 2b, while an ROI with a natural gas leak is shown in Figure 2c. A set of 60 images with no natural gas leaks was obtained and another set with 30 images with natural gas leaks was also obtained. These images were used as described: two training image groups were formed, one of them, called group $F_1$, with 20 images and the other, called group $F_2$, with 30 images, both of them with no natural gas leaks. A test image set was formed comprised of 30 images with no natural gas leaks and of 30 images with natural gas leaks.

(a)                                    (b)                    (c)

**Fig. 2.** Images used in the study:  (a) Original image; (b) ROI extracted close to the wellhead with no natural gas leak; (c) ROI extracted close to the wellhead with natural gas leak

Applying the Novelty Filter to detect natural gas leaks comprises the following steps:

1. Compose an original set of $m$ ROI images with no natural gas leaks as those shown in Figure 2b. These images are used to form $m$ vectors $\{x_1, x_2, \dots x_m\}$ of the novelty filter. To increase the robustness of the novelty filter, images with objects that eventually are present in the scene are included in the original set of images; For example, images with employees doing maintenance of the well were included.
2. The set of vectors $\{x_1, x_2, \dots x_m\}$ are obtained following the procedure described in [14]. All the images are converted to HSI color space. Then, for each one, an intensity matrix of 128x128 pixels is constructed containing the intensity component I of the HSI color space. In the sequence, each vector $x_i$ corresponding to each image is constructed concatenating the 128 columns of each intensity matrix, as shown in Figure 3. In this figure M=N=128, so the size of the vector is 16,384 components;

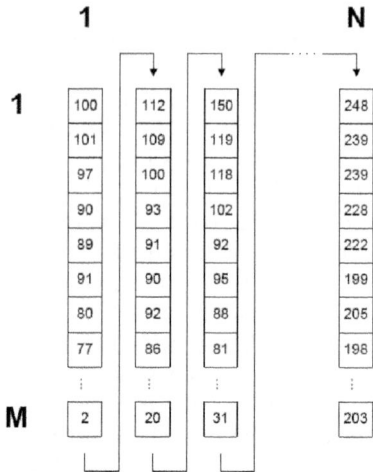

**Fig. 3.** Vectors of characteristics using intensity levels of the images

3. The set of orthogonal vectors $\{v_1, v_2, \dots v_m\}$ is obtained following the procedure previously described in section 2;
4. The novelty $\tilde{x}$ for each image of the test set is then calculated following the procedure previously described in section 2.

Once the novelty filter detects that a natural gas leak is present in one image it is possible to highlight the region of the image where the leak occurs. In this study this region is called the novelty region. The process of highlighting this region adopted is described in the following steps:

1. Convert the novelty vector $\tilde{x}$ into a novelty matrix $\tilde{X}$ with dimensions $128x128$;
2. Convert the novelty matrix $\tilde{X}$ into a novelty image with pixels in the range 0 to 255;
3. Apply a median filter to remove noise;
4. Binarize the image applying a threshold with a value equal to an average image value.

The application of this process is illustrated in the next section.

## 4    Results

As described in the previous section, two novelty filters were trained. The first one, novelty filter $F_1$, using a base with 20 images and another, novelty filter $F_2$, using a base of 30 images. The objective was to investigate the performance of the novelty filter when varying the number of images in the base.

After the training of the two novelty filters, the test images were evaluated. For each image test, the dissimilarity measure was calculated, the magnitude of the novelty, $\|\tilde{x}\|$. The range of values of the magnitude is defined as: $0 \leq \|\tilde{x}\| \leq 1$. As close as the magnitude $\|\tilde{x}\|$ is to 0, the likelihood of a natural gas leak is low. As close as the magnitude $\|\tilde{x}\|$ is to 1, the likelihood of a natural gas leak is high. In order to determine a threshold value to classify the images as having a leak or no leak, we used the Receiver Operating Characteristic (*ROC curve*) methodology, as proposed by Metz [15]. The *ROC curve* is traced with the horizontal axis representing the coordinate *1-specificity* and the vertical axis representing the coordinate *sensitivity*. The best operation point of an ROC *curve* is one situated at the upper and right side of the curve. This point corresponds to *sensitivity* = *specificity* = *1*. The points used for curve tracing (*1-specificity*, *sensitivity*) were obtained varying the threshold between 0 and 1, with steps of 0.02. One ROC curve was traced for each novelty filter trained (base with 20 and with 30 images). Figure 4 shows these curves.

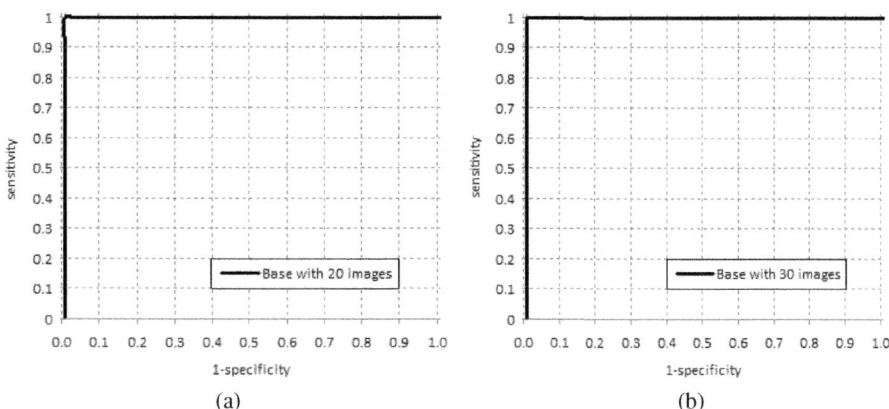

**Fig. 4.** ROC *curves*: (a) Novelty Filter $F_1$ (base with 20 images); (b) Novelty Filter $F_2$ (Base with 30 images)

Novelty filter $F_1$ correctly classified all 60 test images (30 with natural gas leak and 30 with no natural gas leak) with a threshold in the range: $0.3 < threshold < 0.38$. In the extremes, $threshold = 0.3$, the sensitivity value was equal to 0.9 and the specificity was equal to 1, and, $threshold = 0.38$, the sensitivity value was equal to 1.0 and the specificity was equal to 0.93.

Novelty filter $F_2$ correctly classified all 60 test images with a threshold in the range: $0.3 < threshold < 0.34$. In the extremes, $threshold = 0.3$, the sensitivity value was equal to 0.9 and the specificity was equal to 1, and, $threshold = 0.34$, the sensitivity value was equal to 1.0 and the specificity was equal to 0.97.

Both novelty filters, $F_1$ and $F_2$, reach an optimum performance, $sensitivity = specificity = 1$. The ranges that novelty filters present optimum performance (sensitivity=specificity=1), $F_1$ and $F_2$, are nearly equal.

Figure 5 illustrates the steps of the process described in the previous section to highlight the novelty region. Figure 5a shows an original image with a natural gas leak. Figure 5b shows the novelty image. Figure 5c shows the image resulting from

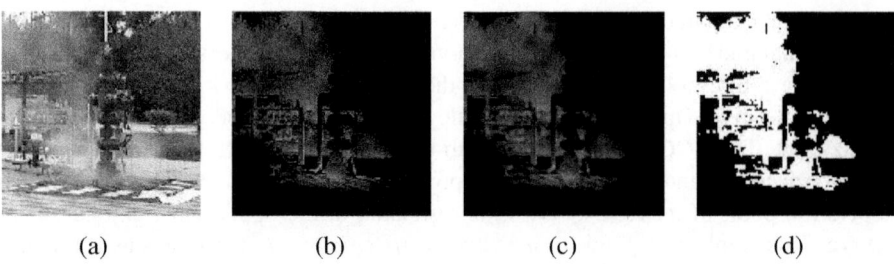

(a)                 (b)                 (c)                 (d)

**Fig. 5.** Detection of novelty region: (a) Original image; (b) Novelty image (c) Image resulting from median filter application; (d) Image showing novelty region

the application of a median filter. Figure 5d shows the binarized image with the novelty region (white pixels). The novelty region corresponds to the region where the natural gas leak occurs.

## 5    Conclusion

In this study a method was proposed for detecting natural gas leaks in a *wellhead* using images obtained with CCD cameras using the novelty filter concept. As cameras are already present in the *Closed-Circuit Television* system used to monitor the well, no additional expense is necessary for hardware implementation. The obtained results, sensitivity = 1 and specificity = 1 are promising and encourage new studies with the technique proposed herein. As this was a pioneering study of natural gas detection with CCD in the visible range, no other comparison can be made with the literature.

The following characteristics of the method are worth noting: the images must always be obtained in the same position and must be the same size.

Future studies will address the following questions: use a large base for the novelty filter, including other objects that may be present near the well; adapt the novelty filter to successful function with environmental changes, such as rain and lack of light (night). At night, a possibility is to employ IR illumination, which is also already available in *Closed-Circuit Television* systems used to monitor wells.

**Acknowledgments.** We would like to thank FAPEAM and FINEP (process 0329/08), for financial support.

## References

1. Souza, C.A.V., Freitas, C.M.: Perfil dos acidentes de trabalho em refinaria de petróleo. Rev. Saúde Pública 36(5), 576–583 (2002)
2. Liu, H., Zhong, S., Rui, W., Keqiang, L.: Remote helicopter-borne laser detector for searching of methane leak of gas line. In: Prognostics and System Health Management Conference, pp. 1–5. IEEE, Shenzhen (2011)

3. Zabetakis, M.G.: Flammability characteristics of combustible gases and vapors. Bulletin 627, Bureau of Mines, U.S. Government Printing Office, Washington (1965)
4. Firth, J.G., Jones, A., Jones, T.A.: The principles of the detection of flammable atmospheres by catalytic devices. Combustion and Flame 21(3), 295–414 (1973)
5. Fan, Z., Taishan, L., Liping, Z.: BP Neural Network Modeling of Infrared Methane Detector for Temperature Compensation. In: Proceedings of 8th International Conference on Electronic Measurement & Instruments, ICEMI 2007, Xi'An, China, pp. 4-123–4-126 (2007)
6. Krier, A., Sherstnev, V.V.: Powerful interface light emitting diodes for methane gas detection. Journal of Physics D: Applied Physics 33(2), 101–106 (2000)
7. McRae, T.G., Kulp, T.J.: Backscatter absorption gas imaging: a new technique for gas visualization. Journal of the Optical Society of America, Applied Optical 32, 4037–4050 (1993)
8. Kastek, M., Sosnowski, T., Piątkowski, T., Polakowski, H.: Methane detection in far infrared using multispectral IR camera. In: 9th International Conference on Quantitative Infrared Thermography, Krakow, Poland, pp. 1–4 (2008)
9. Kastek, M., Sosnowski, T., Piątkowski, T., Polakowski, H.: Methane detection in far infrared using multispectral IR camera. Institute of Optoelectronics, Military University of Technology, S.Kaliskiego 2, Warsaw, Poland (2008)
10. Ross, C.E.H., Solan, L.E.: Terra Incognita: A Navigation Aid for Energy Leaders. Penn-Well Corporation, Oklahoma (2007)
11. Kohonen, T., Oja, E.: Fast adaptive formation of orthogonalizing filters and associative memory in recurrent networks of neuron-like elements. Biological Cybernetics (BIOL CYBERN) 25(2), 85–95 (1976)
12. Kohonen, T.: Self-Organization and Associative Memory, 3rd edn. Springer, Heidelberg (1989)
13. Costa, M.G.F., Moura, L.: Automatic assessment of scintmammographic images using a novelty filter. In: Proceedings of the 19th Annual Symposium on Computer Applications in Medical Care, São Paulo, pp. 537–541 (1995)
14. Costa, C.F.F.F., Pinheiro, C.F.M., Costa, M.G.F., Pereira, W.C.A.: Applying a novelty filter as a matching criterion to iris recognition for binary and real-valued feature vectors. Journal Signal, Image and Video Processing 5, 1–10 (2011)
15. Metz, C.E.: Basic Principles of ROC Analysis. Seminars in Nuclear Medicine (Edition ENSP) 8(4), 283–298 (1978)

# Detection of Edges in Color Images: A Review and Evaluative Comparison of State-of-the-Art Techniques

Ajay Mittal[1], Sanjeev Sofat[1], and Edwin Hancock[2]

[1] Department of Comp. Science & Engg.,
PEC University of Technology,
Chandigarh, India
{ajaymittal,sanjeevsofat}@pec.ac.in
[2] Department of Computer Science,
University of York, York, UK
erh@cs.york.ac.uk

**Abstract.** We present an evaluative review of various edge detection techniques for color images that have been proposed in the last two decades. The statistics shows that color images contain 10% additional edge information as compared to their gray scale counterparts. This additional information is crucial for certain computer vision tasks. Although, several reviews of the work on gray scale edge detection are available, color edge detection has few. The latest review on color edge detection is presented by Koschan and Abidi in 2005. Much advancement in color edge detection has been made since then, and thus, a thorough review of state-of-art color edge techniques is much needed. The paper makes a review and evaluation of various color edge detection techniques to quantify their accuracy and robustness against noise. It is found that Minimum Vector Dispersion (MVD) edge detector has the best edge detection accuracy and Robust Color Morphological Gradient-Median-Mean (RCMG-MM) edge detector has highest robustness against the noise.

**Keywords.** Color edge detection, edge detection, synthetic methods, vector methods.

## 1    Introduction

Edge detection is a fundamental low level operation used in many computer vision and image processing applications. It simplifies image analysis by significantly reducing the amount of information, while at the same time preserving useful structural information about the objects in the scene [1]. A lot of research has been done in the area of gray scale edge detection. However, gray scale edge detection is not robust against metamerism and varying lighting conditions [2]. If metamerism is observed for adjacent objects, the edges of adjacence cannot be robustly extracted using gray scale edge detection. Color plays a significant role in the perception of object boundaries under such situations. It provides richer description of objects in the scene and greatly simplifies their detection, but its processing requires handling more data.

M. Kamel, F. Karray, and H. Hagras (Eds.): AIS 2012, LNCS 7326, pp. 250–259, 2012.

Although the present state of technology (faster processors and cheaper memories) facilitates processing of bulky data, efficient algorithms for color processing are required even then. More surprisingly, research in the problem of color edge detection seems to have been neglected as compared to the bulk of work done in gray scale edge detection so far.

Several reviews of the work on gray scale edge detection are available in the literature [3-5], but color edge detection has few. The latest review on detection of edges in color images is presented by Koschan and Abidi [6] in 2005. Much advancement in color edge detection has been made since then, and thus, a thorough review of state-of-art color edge techniques is much needed. This paper presents a review of techniques developed for detection of edges in color images. Their accuracy and robustness against noise is evaluated by testing them on a dataset of synthetic and natural color images.

## 2     Color Edge Detection

Edge detection in color images is different from that in gray scale images due to a fundamental difference in their nature. In contrast to a scalar value used to represent a pixel in a gray scale image, a pixel in a color image is represented by a color vector (which generally consists of three components, the tristimulus values). Thus, in color edge detection a vector-valued image function is processed instead of a scalar image function (as in gray scale edge detection). On the basis of the principle used for this processing, Koschan and Abidi [6] and Chen and Chen [7] classified the color edge detection methods into two categories:

1. *Synthetic methods or monochromatic-based methods:* These methods decompose the color vectors into different components (image decomposition), process each component separately and combine together the individually gained results (image recombination). According to the nature of recombination, Ruzon and Tomasi [8] classified the synthetic methods as: output fusion methods [7, 9-14] and multidimensional gradient methods [15-20].
2. *Vector methods:* These methods preserve the vector nature of color throughout the computation and use various features of three-dimensional vector space to detect edges in a color image.

### 2.1     Synthetic Methods or Monochromatic-Based Methods

Along with the model matching and edge decision steps of gray scale edge detection, these methods additionally use image decomposition and image recombination steps. The image recombination step can be inserted at different places in the edge detection pipeline, as shown in Fig. 1, and accordingly the synthetic methods are classified as: output fusion methods and multidimensional gradient methods.

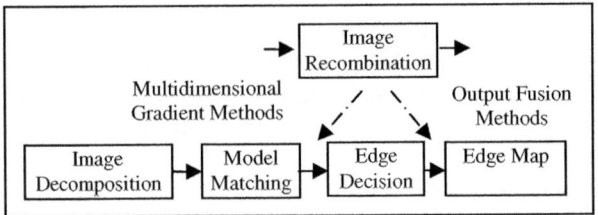

**Fig. 1.** Edge detection pipeline for synthetic methods

### 2.1.1  Output Fusion Methods

Output fusion methods are the most popular and the simplest ones. They perform gray scale edge detection independently in each color component and fuse the results to produce final edge map as shown in Fig. 2.

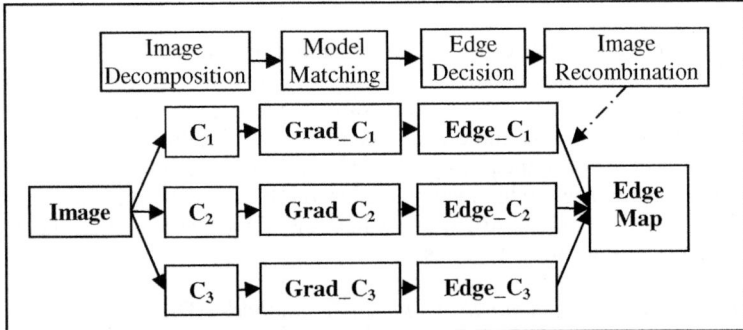

**Fig. 2.** Edge detection pipeline for output fusion methods

The first such method was proposed by Nevatia [9] in 1977. A number of output fusion methods for color edge detection have been proposed since then. These methods vary in terms of usage of different color spaces, edge detection operators and the fusion criteria. Table 1 enumerates some of the significant output fusion color edge detection methods presented till date.

**Table 1.** Output fusion methods for color edge detection

| Method | Year | Color Space | Edge detection operator | Fusion criterion |
|--------|------|-------------|-------------------------|------------------|
| Nevatia [9] | 1977 | YCrCb | Hueckel operator | Same Orientation |
| Shiozaki [10] | 1986 | RGB | Entropy operator | Weighted Summation |
| Hedley and Yan [11] | 1992 | RGB | Sobel operator | Summation |

**Table 1.** (*Continued*)

| Carron and Lambert [12] | 1994 | HIS | Sobel operator | Weighted Summation and Trade-off parameter between Hue and Intensity |
|---|---|---|---|---|
| Fan et. al [13] | 2001 | YUV | Second-order gradient, Entropy based threshold- | Logical OR |
| Cabani et. al [2] | 2006 | RGB | Basic declivity operator | Shortest of basic declivities in the corresponding R, G and B layers |
| Niu and Li [14] | 2006 | HSV | Direction Information measure | Weighted Summation |
| Chen and Chen [7] | 2010 | RGB | Improved Sobel operator | Weighted sum and adaptive thresholding using Otsu method |

### 2.1.2 Multidimensional Gradient Methods

These methods make a single estimate of the orientation and strength of an edge at a point, as shown in Fig. 3.

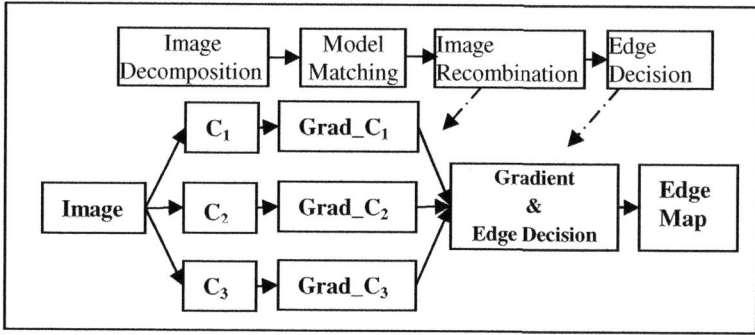

**Fig. 3.** Edge detection pipeline for multidimensional gradient methods

The first color edge detector belonging to this category was proposed by Robinson [15] in 1977. Various methods belonging to this category have evolved since then. These methods differ in terms of how gradients of the image components are combined into one. The methods along with the techniques used to find and combine gradients are listed in Table 2.

The synthetic methods for color edge detection are simple and produce better output than the traditional gray scale edge detection techniques. However, their main problem is how to combine the channels to give a final result. Moreover, their output is heavily dependent on the used color space. Several color spaces such as RGB, HSI, YUV, YCrCb have been used, but none of them dominates the other for all kind of color images.

**Table 2.** Multidimensional gradient methods for color edge detection

| Method | Year | Gradient operator | Combination method |
|---|---|---|---|
| Robinson [15] | 1977 | 24-directional derivatives | Maximum of all the gradients |
| Zenzo [16] | 1986 | Sobel gradient | Tensor of gradients: Sum of gradients, Root Mean Square (RMS) of gradients, Maximum of gradients |
| Cumani [17] | 1991 | Second order derivatives | Eigen vector approach |
| Drewniok [18] | 1994 | Canny edge operator | Integration of information from various channels in a well-founded way. Integration of Correlated information by averaging and integration of uncorrelated information by simple inclusion. |
| Chapron [19] | 1997 | Canny-Deriche gradient | Neyman-Pearson optimal decision rule |
| Soumya Dutta [20] | 2009 | Directional masks | Weighted Summation |

## 2.2    Vector Methods

In contrast to synthetic methods, vector methods preserve the vector nature of color and perform edge detection in vector space. Detecting edges in vector space offers a great potential and is used in state-of-the-art color edge detection methods.

Machuca and Phillips [21] proposed the first vector method for color edge detection that uses rotational and curvature properties of vector fields to identify edges. Vector methods proposed in [22-24] replace gray level differences of adjacent pixels in some way by vector differences and then perform edge detection. Color edge detection based on vector-order statistics (VOS) has been proposed by Trahanias and Venetsanopoulos [25]. Their approach is inspired by the morphological edge detectors [26, 27] proposed for gray scale images. These detectors operate by detecting the local maxima and minima in the image function and combining them in a suitable way in order to produce a positive response for an edge pixel. Since no exact equivalent of the scalar "min-max" operator exists for vector-valued variables, vector-valued ranking operator (R-ordering) is used. If R-ordering is specified for all the color vectors lying within a window and $X_i$ represents the $i^{th}$ vector in this ordering, they defined the Vector Range (VR) detector as:

$$VR = \|X_n - X_1\| \tag{1}$$

In a uniform area, where all vectors are close to each other, the output VR will be small. However, its response on an edge will be large. Thus, edges were obtained by thresholding the output VR. VR detector is very sensitive to noise and to increase robustness against noise several other VOS operators, i.e. Minimum Vector Range (MVR), Vector Dispersion (VD), and Minimum Vector Dispersion (MVD), are proposed [25]. MVR edge detection is similar to VR edge detection except that it employs the magnitudes of the differences of the k-highest vectors from $X_1$. MVR edge detector is robust against long tailed noise and is formulated as:

$$MVR = {}_{j}^{min}\{\||X_{n-j+1} - X_1\||\}, j = 1,2, \dots k, k < n \qquad (2)$$

VD edge detection is similar to VR edge detection except that instead of using first vector from the ordered list, a vector obtained from the linear combination (average) of the ordered vectors is used. VD edge detector is robust against short tailed noise and is formulated as:

$$VD = \left\||X_n - \sum_{i=1}^{l} \frac{X_i}{l}\right\||, l < n \qquad (3)$$

MVD edge detection is combination of MVR and VD edge detection. To detect edges it employs the magnitudes of the differences of the k-highest vectors from the vector obtained from the linear combination of the ordered vectors. It is defined as:

$$MVD = {}_{j}^{min}\left\{\left\||X_{n-j+1} - \sum_{i=1}^{l} \frac{X_i}{l}\right\||\right\}, j = 1,2, \dots k; \ k,l < n \qquad (4)$$

Evans and Liu [28] proposed a color edge detector based on morphological gradient operator. If $X = [X_1, X_2, \dots, X_n]$ be the set of $n$ vector contained within a structuring element $g$, the color morphological gradient (CMG) is defined as:

$$CMG = {}_{i,j\in X}^{max}\{\||X_i - X_j\||_p\} \qquad (5)$$

If the CMG uses $L_2$ norm i.e. p=2, its response is maximum of the distances between all pair of vectors in the set, i.e. the structuring element. An edge is found by appropriately thresholding the maximum response R of CMG. Since, CMG is very sensitive to noise robust color morphological gradient (RCMG) that makes use of pair wise pixel rejection scheme is defined as:

$$RCMG = {}_{i,j\in\{X-R_s\}}^{max}\{\||X_i - X_j\||_p\} \qquad (6)$$

where $R_s$ is the set of $s$ pairs of vectors removed. RCMG is shown to have better gradient estimation than CMG.

Similar to the sliding neighborhood vector methods proposed in [25, 28], Nezhadarya and Ward [29] recently proposed a gradient vector estimation operator for color images. A sliding neighborhood window of size $5 \times 5$, which provides a good compromise between the gradient localization and noise reduction, is used. The basic elements employed in each window are: highpass filter (used for gradient estimation), lowpass filter (used for noise reduction) and aggregation operator. To reduce the edge blurring effects of lowpass filter, the highpass filter and the lowpass filter are applied in perpendicular directions. The color gradient estimators, MVD [25] and RCMG [28], are used for highpass filtering and the median filter is used for lowpass filtering. Since the used filters are nonlinear in nature, the order in which they are applied change the result. The problem is solved by aggregating the results obtained by applying the lowpass and highpass filters in different orders (i.e. highpass filter first and then lowpass, and vice versa). The mean and max operators are used for aggregating the results. Using the combination of mentioned choices of highpass, lowpass and aggregation operators, four operators namely, MVD-Median-Mean, MVD-Median-Max, RCMG-Median-Mean (RCMG-MM), RCMG-Median-Max are proposed. It was experimentally shown that the RCMG-MM performs the best in estimating the gradient and detecting the edges in noisy color images.

## 3     Evaluation of Color Edge Detection Techniques

This section provides an evaluative comparison of the edge detection accuracy and noise robustness of various color edge detectors. Since, the output quality of synthetic methods is not comparable with that of vector methods; only the state-of-the-art vector methods are considered for the comparison. These methods are quantitatively and qualitatively compared on a set of synthetic[1] and natural color images. The accuracy of these detectors is evaluated on a synthetic image dataset using Pratt's Figure of Merit (FOM) [30]. The results of various edge detectors on one synthetic image are shown in Fig. 4. The FOM values of various edge detectors obtained for different synthetic images are given in Table 1. The noise robustness of various edge detectors

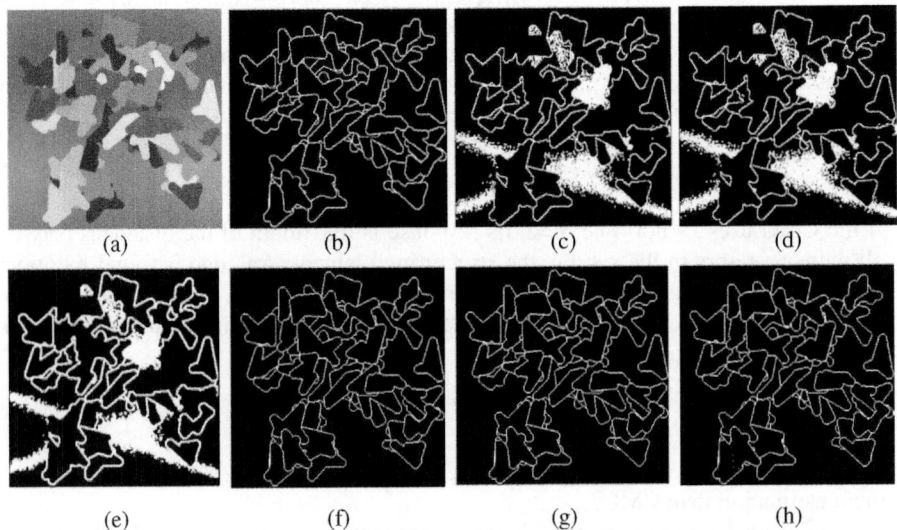

**Fig. 4.** Color edge detection, FOM values shown bracketed, threshold=2500. (a) Original synthetic image. (b) Its ground truth. (c) VR result (0.4503). (d) VD result (0.4503). (e) MVR result (0.4156). (f) MVD result (0.8153). (g) RCMG result (0.8126). (h) RCMG-MM result (0.8077).

**Table 3.** FOM values of various edge detectors on synthetic images

| Method | $FOM_1$ | $FOM_2$ | $FOM_3$ | $FOM_4$ | $FOM_5$ |
|---|---|---|---|---|---|
| VR | 0.7151 | 0.5427 | 0.8241 | 0.4503 | 0.5829 |
| VD | 0.7151 | 0.5427 | 0.8241 | 0.4503 | 0.5829 |
| MVR | 0.5825 | 0.4908 | 0.6489 | 0.4156 | 0.5144 |
| MVD | 0.8172 | 0.8109 | 0.7878 | 0.8153 | 0.8058 |
| RCMG | 0.8170 | 0.8154 | 0.7871 | 0.8126 | 0.8051 |
| RCMG-MM | 0.7928 | 0.7830 | 0.7312 | 0.8077 | 0.7874 |

[1] Synthetic images and their ground truth edge map is generated by method given in [31].

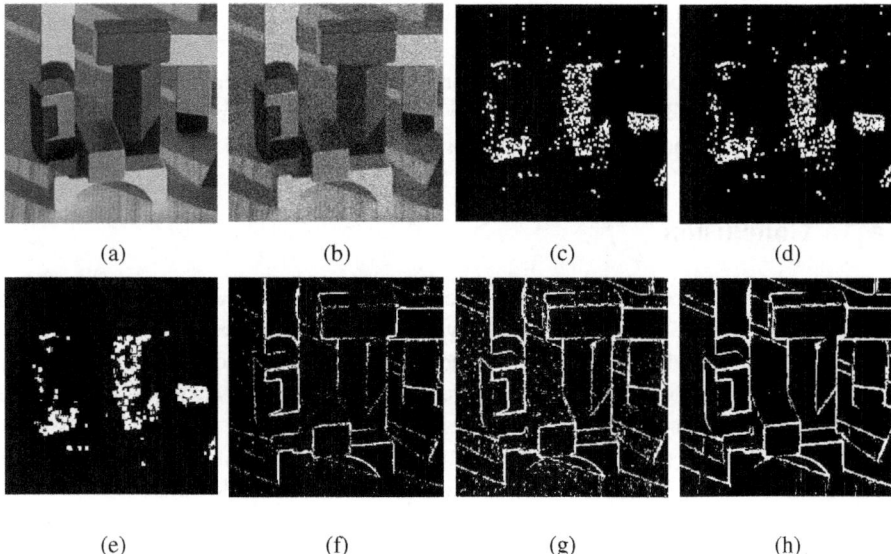

(a)          (b)          (c)          (d)

(e)          (f)          (g)          (h)

**Fig. 5.** Behavior of color edge detectors for a color 'blocks' image corrupted with uncorrelated Gaussian noise, SNR=20, FOM values shown bracketed. (a) Original image. (b) Noise corrupted image. (c) VR result (0.0439). (d) VD result (0.0439). (e) MVR result (0.0466). (f) MVD result (0.7809). (g) RCMG result (0.7783). (h) RCMG-MM result (0.8609).

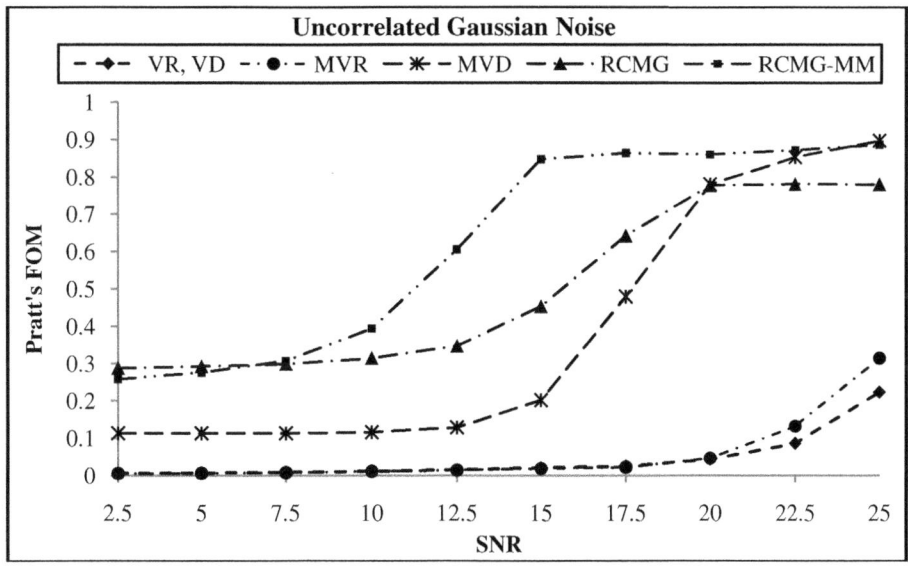

**Fig. 6.** Average FOM-SNR curves of color edge detectors

against uncorrelated Gaussian noise with varying SNRs is evaluated by comparing the edge maps, before and after adding the noise, using Pratt's FOM. The edge maps obtained from various edge detectors for one of the noise realizations is shown in Fig. 5. The average FOM-SNR curves of various edge detectors for ten different realizations of the uncorrelated Gaussian noise are shown in Fig. 6.

# 4    Conclusion

The synthetic methods for color edge detection are simple and easy to implement, but their output quality is not comparable with that of vector methods. Various vector methods for color edge detection have been evaluated. It has been found that VR and VD edge detectors give the same results, MVD has the best accuracy and RCMG-MM has highest robustness against the noise, amongst the methods in its class. However, all the vector methods have a shortcoming. The quality of their output is dependent upon the neighborhood window size and the threshold value used.

# References

1. Dutta, S., Chaudhari, B.B.: A statistics and local homogeneity based color edge detection algorithm. In: Proc. of 2009 International Conference on Advances in Recent Technologies in Communication and Computing, pp. 546–548 (2009)
2. Cabani, I., Toulminet, G., Bensrhair, A.: A fast and self-adaptive color setero vision matching: a first step for road obstacle detection. In: Proc. of Intelligent Vehicles Symposium, Tokyo, Japan, June 13-15, pp. 58–63 (2006)
3. Davis, L.S.: A survey of edge detection techniques. Comput. Graph Image Process. 4(3), 248–270 (1976)
4. Torrem, V., Poggio, T.: On edge detection. IEEE Trans. Pattern Analysis and Machine Intelligence, PAMI 8, 147–163 (1986)
5. Ziou, D., Tabbone, S.: Edge detection techniques-An overview. Dept. Math Informatique, Univ. Sherbrooke, Sherbrooke, QC, Canada, Tech. Rep. no. 1995 (1997)
6. Koschan, A., Abidi, M.: Detection and classification of edges in color images: A review of vector valued techniques. IEEE Signal Processing Magazine, 64–73 (January 2005)
7. Chen, X., Chen, H.: A novel color edge detection algorithm in RGB color space. In: Proc. of IEEE 10th International Conference on Signal Processing, pp. 793–796 (2010)
8. Ruzon, M.A., Tomasi, C.: Edge, Junction, and Corner Detection using Color distributions. IEEE Transactions on Pattern Analysis and Machine Intelligence 23(11) (November 2001)
9. Nevatia, R.: A Color edge detector and its use in scene segmentation. IEEE Transactions on Systems, Man and Cybernetics, SMC 7(11) (November 1977)
10. Shizoaki, A.: Edge extraction using entropy operator. Computer Vision, Graphics, and Image Processing 36(1), 1–9 (1986)
11. Hedley, M., Yan, H.: Segmentation of Color images using spatial and color space information. Journal of Electronic Imaging 1, 374–380 (1992)
12. Carron, T., Lambert, P.: Color edge detection using jointly hue, saturation and intensity. In: Proc. of IEEE International Conference on Image Processing, pp. 977–981 (1994)
13. Fan, J., Aref, W.G., Hacid, M.S., Elmagarmid, A.K.: An improved automatic isotropic color edge detection technique. Pattern Recognition Letters 22, 1419–1429 (2001)

14. Niu, L., Li, W.: Color edge detection based on direction information measure. In: 6th World Cong. on Intll. Cont. and Automation, pp. 9533–9536 (2006)
15. Robinson, G.: Color Edge detection. Optical Engg. 16(5), 479–484 (1977)
16. Zenzo, S.D.: A note on gradient of a multi-image. Computer Vision, Graphics, and Image Processing 33(1), 116–125 (1986)
17. Cumani, A.: Edge detection in Multispectral Images. CVGIP: Graphical Models and Image Processing 53(1), 40–51 (1991)
18. Drewniok, C.: Multispectral Edge Detection- Some Experiments on data from Landsat-TM. International J. Remote Sensing 15(18), 3743–3766 (1994)
19. Chapron, M.: A chromatic contour detector based on abrupt change techniques. In: Proc. of International Conference on Image Processing, vol. 3, pp. 18–21 (1997)
20. Dutta, S., Chaudhari, B.B.: A color edge detection algorithm in RGB color space. In: Proc. of International Conference on Advances in Recent technologies in Communication and Computing, pp. 337–340 (2009)
21. Machuca, R., Phillips, K.: Application of vector fields to image processing. IEEE Trans. Pattern Anal. Machine Intell., PAMI 5(3), 316–329 (1983)
22. Huntsberger, T.L., Descalzi, M.F.: Color edge detection. Pattern Recognition Letters 3(3), 205–209 (1985)
23. Pietikainen, M., Harwood, D.: Edge information in color images based on histogram of differences. In: Proc. of International Conf. on Pattern Recognition, Paris, France, pp. 594–596 (1986)
24. Solinsky, J.C.: The use of color in machine edge detection. In: Proc. VISION 1985, pp. 4.34–4.52 (1985)
25. Trahanias, P.E., Venetsanopoulous, A.N.: Color edge detection using vector order statistics. IEEE Trans. Image Processing 2(2), 259–264 (1993)
26. Rivest, J.F., Soille, P., Beucher, S.: Morphological Gradients. J. Electronic Imaging 2(4), 326–336 (1993)
27. Lee, J.S., Haralick, R.M., Shapiro, L.G.: Morphologic edge detection. IEEE Trans. Robot. Autom. 3(2), 142–156 (1987)
28. Evans, A.N., Liu, X.U.: A morphological gradient approach to color edge detection. IEEE Trans. Image Proc. 15(6), 1454–1462 (2006)
29. Nezhadarya, E., Ward, R.K.: A new scheme for robust gradient vector estimation in color images. IEEE Trans. Image Proc. 20(8), 2011–2220 (2011)
30. Pratt, W.K.: Digital Image Processing. Wiley, N.Y. (1991)
31. Daniel, M., Lord, S., Papon, J.: A survey of vector order statistical edge detectors and their ability to mimic the human visual system, technical report, Stanford Univ. (2008)

# An Efficient Scheme for Color Edge Detection in Uniform Color Space

Ajay Mittal[1], Sanjeev Sofat[1], and Edwin Hancock[2]

[1] Department of Comp. Science & Engg.,
PEC University of Technology, Chandigarh, India
{ajaymittal,sanjeevsofat}@pec.ac.in
[2] Department of Compter Science, University of York, York, UK
erh@cs.york.ac.uk

**Abstract.** An efficient method for color edge detection using a uniform color space and Euclidean distance between the color vectors is proposed. Two color vectors are perceptually distinguishable if Euclidean distance between them is greater than Just Noticeable Color Distance (JNCD) threshold. In the presence of noise, the threshold tends to be greater than JNCD and is adaptively determined from the statistical analysis of an image scan line. The edges are detected by determining and thresholding local extremities in each image scan line. The proposed operator is quantitatively evaluated on a set of test images using Pratt's figure of merit. As opposed to other color edge detectors, the proposed method is time efficient. It takes only 1 second to compute edge map of a 400×400 color image, which makes it suitable for real time problems like obstacle detection, color cheque processing, etc.

**Keywords:** Color image processing, Color edge detection, Vector methods, Color distance declivity, Obstacle detection.

## 1 Introduction

Obstacle detection is one of the main problems to be solved to ensure safe navigation for visually impaired people, robots, humanoids, intelligent vehicles, etc. Obstacle detection systems typically employ a vision based sensor consisting of a CCD camera that provides gray level images [1]. These systems are reliable under good lighting, but suffer from lack of visibility in poor lighting and weather conditions. To robustly detect obstacles in such situations, the obstacle detection systems use color cameras. Color cue is appropriate during poor visibility and environmental conditions [1]. It provides richer description of obstacles in the scene and greatly simplifies their detection, but its processing requires handling more data. Although, the present technology (faster processors and cheaper memories) facilitates processing of bulky data, an efficient color processing algorithm is still required so that it can be used under real time constraints. The current state-of-the-art color edge detection algorithms [2-4] are computationally expensive and are not suitable for real time problems like obstacle detection, etc.

M. Kamel, F. Karray, and H. Hagras (Eds.): AIS 2012, LNCS 7326, pp. 260–267, 2012.

This paper presents an efficient method for color edge detection and is structured into six sections. Section 2 presents an overview of various state-of-the-art color edge detection techniques. The proposed method is formulated and presented in Section 3. The experimental results of the proposed method are shown in Section 4. Finally, conclusions are drawn in Section 5.

## 2    Related Work

Edge detection in color images is different from that in gray scale images due to a fundamental difference in their nature. In contrast to a scalar value used to represent a pixel in a gray scale image, a pixel in a color image is represented by a color vector (which generally consists of three components, the tristimulus values). Thus, in color edge detection a vector-valued image function is processed instead of a scalar image function (as treated by the gray scale edge detectors). On the basis of the principle used for this processing, Koschan and Abidi [5] and Chen and Chen [6] classified the color edge detection methods as: Synthetic methods and Vector methods.

The synthetic methods decompose the color vectors into different components, process each component separately and combine the individually gained results. Their main problem is how to combine the results of the individually processed color components to give a final result. Moreover, their output is heavily dependent on the used color space. In contrast to the synthetic methods, vector methods preserve the vector nature of color and perform the color edge detection in vector space. Detecting edges in vector space offers a great potential and is used in the state-of-the art color edge detectors. The first such method was developed by Machuca and Phillips [7]. The rotational and curvature properties of vector fields were used to identify edges. Color edge detection based on vector-order statistics (VOS) has been proposed by Trahanias and Venetsanopoulos [2]. Their approach is inspired by the morphological edge detectors [8, 9] proposed for gray scale images. These detectors operate by detecting the local maxima and minima in the image function and combining them in a suitable way in order to produce a positive response for an edge pixel. Since no exact equivalent of the scalar "min-max" operator exists for vector-valued variables, Trahanias and Venetsanopoulos suggested the use of vector-valued ranking operator (R-ordering) for edge detection in color images. In R-ordering, each vector-valued observation is reduced as a function of a distance criterion to a scalar value $d_i$. An arrangement of $d_i$'s in ascending order, $d_1 \leq d_2 \leq \ldots \leq d_n$, associates the same ordering to the vector valued data $x_i$'s, $x_1 \leq x_2 \leq \ldots \leq x_n$. If R-ordering is specified for all the color vectors lying within a window and $x_i$ represents the $i^{th}$ vector in this ordering, they defined the color edge detector [Vector Range (VR) detector] as:

$$VR = \|x_n - x_1\| \tag{1}$$

In a uniform area, where all vectors are close to each other, the output VR will be small. However, its response on an edge will be large. Thus, edges were obtained by thresholding the output VR. Several other VOS operators were proposed to increase the robustness to noise, of which the Minimum Vector Dispersion (MVD) was shown to be most effective.

Evans and Liu [3] proposed a color edge detector based on morphological gradient operator. If $X=[X_1, X_2, ...., X_n]$ be the set of $n$ vector contained within a structuring element $g$, the color morphological gradient (CMG) is defined as:

$$CMG = \max_{i,j \in X}\{\|X_i - X_j\|_p\} \tag{2}$$

If CMG uses $L_2$ norm i.e. p=2, its response is maximum of the distances between all pair of vectors in the set, i.e. the structuring element. An edge is found by appropriately thresholding the response of CMG. Since, CMG is sensitive to noise, robust color morphological gradient (RCMG) that makes use of pair wise pixel rejection scheme is defined as:

$$RCMG = \max_{i,j \in \{X-R_s\}}\{\|X_i - X_j\|_p\} \tag{3}$$

where, $R_s$ is the set of $s$ pairs of vectors removed. RCMG is shown to have better gradient estimation than CMG.

Similar to the sliding neighborhood vector methods proposed in [2,3], Nezhadarya and Ward [4] have recently proposed a gradient vector estimation operator for color images. The neighborhood window of size $5 \times 5$, which provides a good compromise between the gradient localization and noise reduction, is used. The basic elements employed in each window are: highpass filter (used for gradient estimation), lowpass filter (used for noise reduction) and aggregation operator. The color gradient estimators, MVD [2] and RCMG [3], are used for highpass filtering and the median filter is used for lowpass filtering. Since the used filters are nonlinear in nature, the order in which they are applied change the result. The problem is solved by aggregating the results obtained by applying the lowpass and highpass filters in different orders (i.e. highpass filter first and then lowpass, and vice versa). The mean and max operators are used for aggregating the results. Using the combination of mentioned choices of highpass, lowpass and aggregation operators, four operators namely, MVD-Median-Mean, MVD-Median-Max, RCMG-Median-Mean (RCMG-MM), RCMG-Median-Max are proposed. It was experimentally shown that the RCMG-MM performs the best in estimating the gradient and detecting the edges in noisy color images.

The methods proposed in [2-5] use RGB color space and the Euclidean distance ($L_2$ norm) between the color vectors contained in a neighborhood window to detect edges. Liu and Chou [10] found that in non-uniform color spaces like RGB, XYZ, YUV, YCbCr, etc. equal Euclidean distance between pair of color vectors does not correspond to equal perceptual difference between the colors. Consequently, the methods using non-uniform color space and the Euclidean distance metric for color edge detection cannot rely on a single threshold. However, if the color space is uniform, the perceptual difference between any two colors can be ideally represented as the Euclidean distance between their color vectors. This uniformity can be used to quantify the perceptual redundancy of each color by a single threshold.

## 3    Proposed Color Edge Detection Method

The input RGB image is converted to CIE-Lab color space through standard nonlinear transformation. Although CIE-Lab is not perfectly uniform, it is nearly linear with the visual perception in a sense that the color distance between two colors is considerably correlated with their perceptual difference. The locus of colors which are not perceptually distinguishable from a given color form a sphere around the color coordinates in the space with radius equal to Just Noticeable Color Difference (JNCD), as shown in Fig. 1. The radius of the JNCD sphere for each color is almost the same because the Euclidean distance is closely correlated with the perceived color difference.

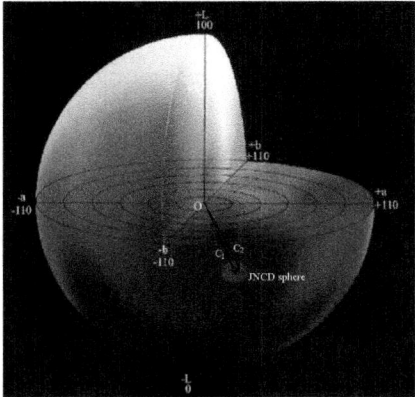

**Fig. 1.** CIE-Lab color space with JNCD sphere

Any two colors in CIE-Lab color space can be distinguished if their $\Delta L$, $\Delta a$ or $\Delta b$ values are out of tolerances. A single value $\Delta E$, defined as

$$\Delta E = \sqrt{\Delta L^2 + \Delta a^2 + \Delta b^2} \tag{4}$$

takes into account difference between the L, a and b values of two colors and provides Euclidean distance between the colors. The colors are perceptually distinguishable if

$$\Delta E > JNCD \tag{5}$$

Mahy et al. [11] evaluated the JNCD for CIE-Lab color space to be 2.3±1.3. Due to non-transitivity of the color indiscriminability (which states that if color $C_1$ is indiscriminable from color $C_2$, and the color $C_2$ from color $C_3$, the color $C_1$ need not necessarily be indiscriminable from the color $C_3$), the color distinction must be performed among a group of image pixels instead of image pixels pairs. The proposed method performs color distinction among the group of pixels that lie in an image scan line.

### 3.1     Implementation of the Proposed Method

Let $C$ be a $M \times N$ color image transformed from RGB to CIE-Lab color space and $C_k(L_k, a_k, b_k)$ represents the tristimulus values (components) of the color (vector) $k$ in CIE-Lab color space. The center (origin) of the color space (sphere) is (practically) represented by the color vector $C_O(50,0,0)$. The $L_2$ norm of $C$, defined as

$$E(i,j) = \|C\| = \|C_{ij} - C_O\|, i \in 1..M, j \in 1..N \tag{6}$$

represents the Euclidean distance of each pixel $ij$, represented by the color vector $C_{ij}$, from the origin $C_O$ of the color space.

To perform color distinction among the pixels of an image scan line $C_i$, the local extremities in $E(i,:)$ are determined. The set of contiguous pixels limited by two local extremities is termed as Color Distance Declivity ($CDD$). Thus, a color image scan line can be modeled as sequence color distance declivities, as shown in Fig. 2.

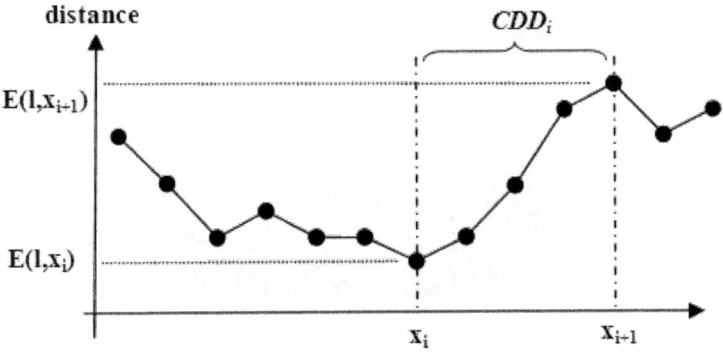

**Fig. 2.** Color image scan line modeled as sequence of color distance declivities

A color distance declivity $CDD_i$ in an image scan line $l$ is characterized by the following attributes:

i)   its starting position $x_i$ in the image scan line
ii)  its end position $x_{i+1}$ in the image scan line
iii) its amplitude, $d_i = E(l, x_{i+1}) - E(l, x_i)$
iv)  its position, $X_i$: The position of a declivity is determined by computing the mean position of the declivity points weighted by the gradients squared in the image scan line, as given by equation 7:

$$X_i = \frac{\sum_{x=x_i}^{x_{i+1}-1} [E(l,x+1)-E(l,x)]^2 (x+0.5)}{\sum_{x=x_i}^{x_{i+1}-1} [E(l,x+1)-E(l,x)]^2} \tag{7}$$

The declivities in an image scan line are classified as significant or non-significant by thresholding their amplitude. Non-significant declivities, i.e. low amplitude declivities, correspond to noise and non-significant elements in the image. Significant declivities, i.e. high-amplitude declivities, correspond to luminance and chroma edges.

## 3.2    Classification of Significant Color Distance Declivities

A color vector $C_2$ is perceptually indistinguishable from the color vector $C_1$ if it lies within $C_1$'s JNCD sphere. However, due to noise it is possible that the color vector $C_2$, which is actually indistinguishable from $C_1$, falls outside $C_1$'s JNCD sphere. In such scenario, classifying $C_2$, as distinguishable from $C_1$ leads to false edges. Thus, in the presence of noise, the actual perceptual color difference threshold, termed as Adaptive Just Noticeable Color Difference (AJNCD), tends to be larger than JNCD. It is adaptively determined from the statistical analysis of an image scan line.

To do this, we assume that the Euclidean distance $E(l, x_i)$ is composed of a deterministic signal $E'(l, x_i)$ and a random signal $\eta(l, x_i)$, which represents noise. The gradient $G(x)$ can be expressed by:

$$G(x) = E(l, x + 1) - E(l, x) = E'(l, x + 1) - E'(l, x) + \eta(l, x + 1) - \eta(l, x)$$

$$= G'(x) + \eta'(x) \tag{8}$$

where $G'(x) = E'(l, x + 1) - E'(l, x)$ and $\eta'(x) = \eta(l, x + 1) - \eta(l, x)$

In practice, most of the $G(x)$ entries are zero or have small absolute value, which corresponds to noise, and non-significant elements in the image. It has few high absolute values which correspond to significant elements, i.e. edges, in an image scan line. Thus, $G(x)$ is mostly due to noise i.e. $\eta'(x)$ and $G'(x)$ is negligible. Therefore, the distribution of $G(x)$ is similar to that of $\eta'(x)$, and vice-versa. To statistically determine the value of AJNCD, Chebyshev's theorem is used.

**Chebyshev's Theorem:** The probability that any random variable X will assume a value within k standard deviations of the mean is at least 1-1/k2 i.e. $P(\mu - k\sigma < X < \mu + k\sigma) \geq 1 - 1k^2$

To filter noise with at least 90% probability, the value of k is to be 3.2. Thus, a color distance declivity is marked as significant if

$$d_i < \min(\mu - 3.2\sigma, -JNCD) \text{ or } d_i > max(\mu + 3.2\sigma, JNCD) \tag{9}$$

where, $\mu$ and $\sigma$ are determined from the gradient vector of an image scan line.

# 4    Experimental Results

The proposed method is quantitatively evaluated on a set of synthetic and natural images using Pratt's figure of merit (FOM). A $200 \times 200$ synthetic image, as shown in Fig. 3 (a), is created to test and compare the proposed method with other operators. Color A and B differ only in intensity and have the same chromatic values. The same is true for the colors C and D. The thresholded binary results of RCMG and RCMG-MM operators, obtained by using default MATLAB threshold level of 0.5, are shown in Fig. 3(b) and (c), respectively. The result of the proposed method is shown in Fig. 3(d).

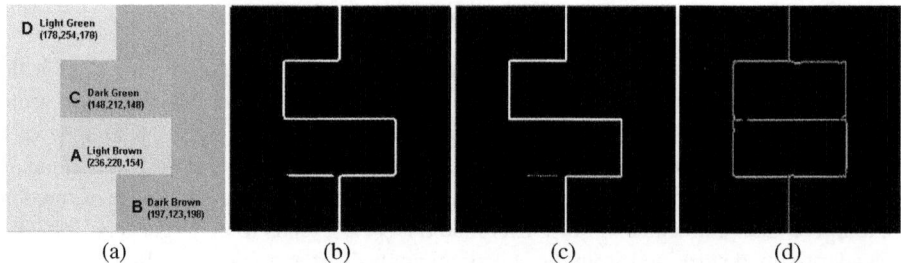

**Fig. 3.** Color edge detection result on synthetic image, FOM values are shown bracketed (a) Original image, with different color regions being labeled. (b) RCMG result (FOM=0.3942). (c) RCMG-MM result (FOM=0.3848). (d) Result of the proposed method (FOM=0.9468).

The outputs of the proposed method and other color edge detectors on a natural Kodak color image "House" (used in [5]) are shown in Fig. 4.

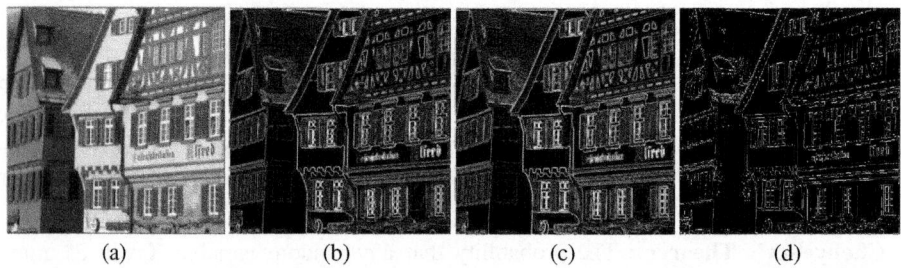

**Fig. 4.** Color edge detection results for the Kodak House image. (a) Original image. (b)RCMG result, s=8. (c) RCMG-MM result, s=2. (d) Result of the proposed method.

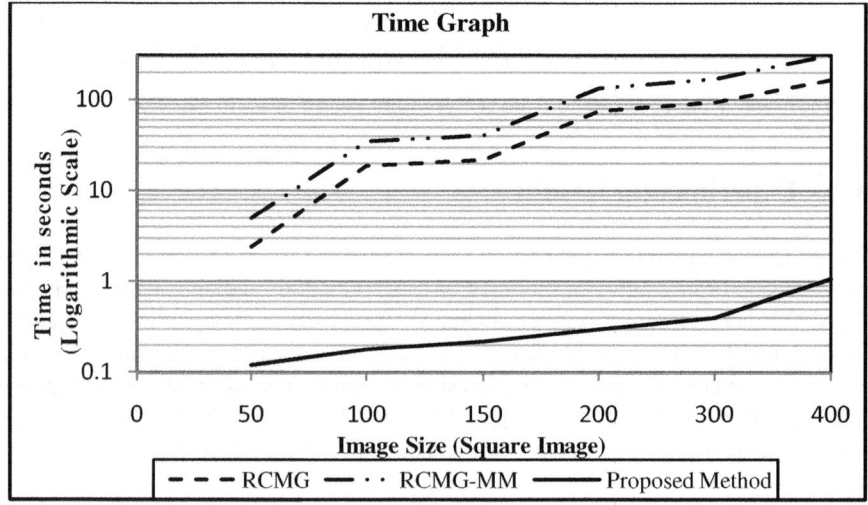

**Fig. 5.** Average time (in seconds) required by different operators to compute edge maps

The experimental results indicate that the proposed method produces thinner edge response and sharp corners as compared to other operators. The average time required by the proposed method and other color edge detectors to compute the edge maps of color images of different sizes is plotted in Fig. 5. The time axis in the figure is shown on logarithmic scale. It has been found that the proposed method takes on an average 1 second to compute the edge map of a 400×400 image, while the RCMG and RCMG-MM methods take on an average 166 seconds and 315 seconds to compute unthresholded edge maps of the images of same size.

## 5    Conclusion

A new efficient method for color edge detection is proposed. The proposed method belongs to the class of vector methods for color edge detection and uses uniform CIE-Lab color space and Euclidean distance between the color vectors to detect edges. As opposed to the recently proposed color edge detectors, which use sliding neighborhood window, the proposed method processes the color vectors present in an image scan line. The evaluations indicate that the proposed method results in thinner and better localized edges and sharp corners. The time measurements of the proposed method indicate that it is very efficient.

## References

1.  Cabani, I., Toulminet, G., Bensrhair, A.: A fast and self-adaptive color setero vision matching: a first step for road obstacle detection. In: Proc. of Int. Vehicle Symp., pp. 58–63 (2006)
2.  Trahanias, P.E., Venetsanopoulous, A.N.: Color edge detection using vector order statistics. IEEE Trans. Image Processing 2(2), 259–264 (1993)
3.  Evans, A.N., Liu, X.U.: A morphological gradient approach to color edge detection. IEEE Trans. Image Proc. 15(6), 1454–1462 (2006)
4.  Nezhadarya, E., Ward, R.K.: A new scheme for robust gradient vector estimation in color images. IEEE Trans. Image Proc. 20(8), 2011–2220 (2011)
5.  Koschan, A., Abidi, M.: Detection and classification of edges in color images: A review of vector valued techniques. IEEE Signal Processing Magazine, 64–73 (January 2005)
6.  Chen, X., Chen, H.: A novel color edge detection algorithm in RGB color space. In: Proc. of IEEE 10th International Conference on Signal Processing, pp. 793–796 (2010)
7.  Machuca, R., Phillips, K.: Application of vector fields to image processing. IEEE Trans. Pattern Anal. Machine Intell., PAMI 5(3), 316–329 (1983)
8.  Rivest, J.F., Soille, P., Beucher, S.: Morphological Gradients. J. Elec. Imag., 326–336 (1993)
9.  Lee, J.S., Haralick, R.M., Shapiro, L.G.: Morphologic edge detection. IEEE Trans. Robot. Autom. 3(2), 142–156 (1987)
10. Liu, K.C., Chou, C.H.: Perceptual Constrast Estimation for Color Edge Detection. In: Proc. of 6th EURASIP Conference on Speech and Image Processing, Multimedia Communications and Services, pp. 86–89 (June 2007)
11. Mahy, M., Eycken, L.V., Oosterlinck, A.: Evaluation of uniform color spaces developed after the adoption of CIELAB and CIELUV. Color Research and Application 19(2), 105–121 (1994)

# Author Index